What the experts say:

Roy Lindseth, founder of several geophysica[l ...] easy as a 'can't put it down' novel, yet is so e[...] surely will become the absolute reference for all [...]

Nigel Anstey, recipient of the Maurice Ewing [...]al and the Conrad Schlumberger award — *"I am very impressed. A fine piece of work and a useful contribution to the literature."*

Dave Kingston, an Exxon 'Rover Boy' — *"An excellent job. I believe this book will become a 'must have' book for present and future explorationists."*

Dietrich Welte, TU Aachen/Nuclear Research Centre Jülich/IES/Schlumberger — *"A very worthwhile and fascinating story."*

Sven Treitel, MIT/Amoco — *"Extremely well written and holds the reader's attention throughout."*

Sam Gray, CGG — *"Fascinating. Does a good job of compressing the history of our field into a few pages."*

Bert Bally, Shell/Rice University — *"A very good read."*

Leon Robinson, Humble Oil/Exxon — *"This is a very impressive tome. An amazing job to distil so much material."*

George Boyadjieff, National Oilwell Varco — *"A remarkable piece of work. I am sure it will fascinate for years to come."*

Jacques Bosio, Elf Aquitaine — *"Wonderful work that no one has undertaken before. It will be enjoyed by the whole petroleum industry."*

Walter Ziegler, Exxon/Petrofina — *"An eminently readable and informative book."*

Richard Bateman, Schlumberger/Amoco/Halliburton/Bridas/Gaffney Cline/Texas Tech University — *"A page turner, can't put it down. An incredible achievement and a reference for decades to come."*

Larry Bell, The Atlantic Refining Company/ARCO — *"The authors have added a lot of detail, particularly in identifying many pioneers often unrecognized who contributed to the evolution of the technology."*

Darwin Ellis, Schlumberger/Darwin Petrophysics — *"A big hit and an important addition to the oil-field literature — an essential reference for anyone working in the industry or contemplating it."*

Robert Freedman, Shell/Schlumberger — *"I very much enjoyed reading the book. The authors have done an outstanding job of writing a very readable history."*

George King, Amoco/BP/Rimrock Energy/Apache — *"Very good read — delightful commentary and history of an incredible bunch of determined characters. They changed the world."*

Donald Peaceman, Humble Oil/Exxon — *"I am truly amazed at how much territory you covered."*

EC Thomas, Shell/Bayou Petrophysics — *"I enjoyed reading your history and learned many facts I had no knowledge of. Overall it is a beautiful rendition. Bravo."*

Jan Meek, Shell/SBM Offshore/Heerema — *"A unique piece of work, a job well done!"*

Dr Mark Mau is a professional business historian and writer. Dr Mau previously conducted historical research for a European telecoms group and Deutsche Bank.

Henry Edmundson worked more than 45 years for Schlumberger, was founding editor of the Oilfield Review and now runs his own energy consulting business.

Abigail Whitehead graduated from the Bartlett School of Architecture, UCL, and the Royal College of Art, and is now an architect and freelance graphic artist based in London.

Groundbreakers

The story of oilfield technology and
the people who made it happen

MARK MAU AND HENRY EDMUNDSON

With illustrations by Abigail Whitehead

An environmentally friendly book printed and bound in England by
www.printondemand-worldwide.com

This book is made entirely of chain-of-custody materials

http://www.fast-print.net/bookshop

GROUNDBREAKERS:
THE STORY OF OILFIELD TECHNOLOGY AND THE
PEOPLE WHO MADE IT HAPPEN
Copyright © Mark Mau and Henry Edmundson 2015

All rights reserved

No part of this book may be reproduced in any form by photocopying
or any electronic or mechanical means, including information storage
or retrieval systems, without permission in writing from both the
copyright owner and the publisher of the book.

The right of Mark Mau and Henry Edmundson to be identified as the authors of
this work has been asserted by him in accordance with the Copyright, Designs
and Patents Act 1988 and any subsequent amendments thereto.

A catalogue record for this book is available from the British Library

ISBN 978-178456-187-1

First Published 2015 by
Fast-Print Publishing of Peterborough, England.

For

Lotte

Hannah and Lisa

Connie

Simon, Sigrid, Isabel

and all my colleagues
from a lifetime in the industry

Contents

Foreword	1
Preface	3
Chapter 1: Beginnings	7
The first wells	7
Early production	10
Chapter 2: The Birth of a Science	13
The roots of petroleum geology	13
Anticlines	15
Salt domes and Spindletop	17
Chapter 3: Drilling Gets Established	20
Rotary drilling takes over	20
Drilling mud makes its appearance	21
Derricks and drawworks	23
From fishtail to tricone bits	24
Chapter 4: Geophysics Enters the Fray	27
The first geophysical methods	27
The arrival of seismic exploration	32
Chapter 5: Discovering the Reservoir	40
Darcy's law of permeability	40
Mud logging	41
Taking core samples	43
The birth of well logging	46
Sidewall sampling	50
Chapter 6: First Steps Offshore	53
California and Venezuela	53
Out of sight of land	55
Seismic goes offshore	57
Chapter 7: The Service Industry Takes Hold	60
Invention of the casing shoe	60
Cementing	61
Controlling the well	65
Mud gets sophisticated	66
Chapter 8: Beginnings of Production Engineering	69
The first well tests	69
Measuring pressure and temperature	71
Early oil well perforating	73

Inventing oil-lift technologies	76
Electric submersible pumps	78
Acidizing	81
Separating oil, gas and water	83
Chapter 9: Exploring the World	85
Stratigraphy	85
The first discoveries in Saudi Arabia	87
Of doodlebuggers and roving geologists	90
From aerial photography to satellites	93
Sedimentology makes an impact	95
Understanding oil migration	98
The importance of plate tectonics	100
Chapter 10: The Proliferation of Well Logging	105
The beginning of log interpretation	105
The induction log	107
Enter nuclear physics	109
Acoustic logging	112
Chapter 11: Production Engineering Matures	117
Of World War II bazookas and shaped-charge perforating	117
The first well fraccing	120
Christmas trees	122
Chapter 12: Offshore Goes Deeper	122
Drillships and dynamic positioning	124
Semisubmersibles	125
Offshore blowout preventers and marine risers	128
Chapter 13: The Digital Revolution	131
Magnetic tape recording and noise reduction	131
The shift to digital recording and processing	133
Impact on seismic exploration	136
Migration of seismic data	137
Early detection of hydrocarbons with "Bright Spots"	140
The air gun replaces dynamite offshore	142
Vibroseis	144
Chapter 14: Petroleum Geology Changes the Game	146
Sequence stratigraphy	146
The concept of turbidite flow	147

Basin analysis comes into play	149
Petroleum systems	152
Soviet and Chinese contributions to geoscience	154
Chapter 15: Making the Reservoir Work	**157**
Waterflooding	157
Thermal recovery	158
Miscible displacement with gas	160
Chemical flooding	161
Chapter 16: Logging Breakthroughs	**165**
Shaly sands and other rock types	165
Imaging the borehole	167
Borehole geophysics	170
Nuclear spectroscopy	172
The miracle of nuclear magnetic resonance	174
From analog to digits	177
Chapter 17: Controlling the Well	**180**
Well kicks and high pressures	180
Oilwell firefighting	181
Well integrity for deep, hot wells	183
Mud cleans up	186
Chapter 18: Going Horizontal	**191**
Deviating from the vertical	191
Enter the positive displacement motor	194
Improving the roller cone bit	195
PDC bits	196
The arrival of topdrive	199
Horizontal and multilateral drilling	202
Chapter 19: Intelligent Drilling	**205**
The MWD struggle	205
The telemetry challenge	207
Logging while drilling	207
Geosteering	212
The emergence of geomechanics	216
Chapter 20: Modern Completion and Production	**219**
Sand control becomes important	219
Hydraulic fracturing meets sand control	222
Advancing hydraulic fracturing	223
Workover and coiled tubing	226

Pressure-controlled perforating	229
Modern pumping units	231
Analyzing production	233
Production logging and downhole tractors	236
Inflow control devices	239
Chapter 21: A Seismic Revolution	241
The decisive move to 3D seismics	241
The world of workstations	244
Amplitude variation with offset	246
Factoring in anisotropy	247
Better images with full waveform inversion	248
Marine seismics matures	249
Ocean-bottom seismics and shear waves	252
Chapter 22: Engineering the Oceans	255
Offshore production gets going	255
The growing importance of crane vessels	257
Massive concrete constructions	260
Drilling-inspired offshore production	262
Tension-Leg and Spar platforms	264
Enter FPSOs	268
The birth of subsea technology	269
Chapter 23: Reservoir Engineering Comes of Age	275
Research into porous media	275
Well testing widens its scope	278
Simulating the reservoir	281
Reservoir monitoring and 4D seismics	288
Crosswell monitoring	291
Juggling the data	293
Chapter 24: The Boom in Unconventionals	295
Steam-assisted gravity drainage	295
The shale gas bonanza	297
Shale oil in the Bakken	303
Chapter 25: Looking Ahead	306
The search continues	306
Latest and greatest in seismics	307
Electromagnetics as a complement to seismics	310
Expandable casing	312
From underbalanced to managed pressure drilling	313

Going superdeep offshore	315
Automating the drilling rig	316
Nanotechnology	319
An endless story	321
Epilogue	323
Timelines	327
Exploration	327
Drilling	335
Reservoir	341
Production	349
Acknowledgements	354
References	357
Interviews conducted for the book	357
Other sourced interviews	362
Bibliography	364
Internet sources	397
Glossary of acronyms used for organizations and companies	399
Glossaries of oilfield technical terms	400
Endnotes	401
Index	453

Foreword

ANYONE WORKING in the upstream oil and gas industry understands the importance of technology for finding new reserves and producing what's been discovered. Without continuous innovation, the world's thirst for oil and gas would have long outstripped its availability. Doomsday practitioners who predict the "end of oil" consistently fail to appreciate the ability of science and technology to extend the era of hydrocarbons.

It would be a mistake, however, to think that technology innovation is a recent phenomenon. From the earliest days, finding and producing oil and gas was fraught with obstacles and danger, and the pioneers were as challenged to innovate as we are today. They came from all walks of life, from academia, from their machine shops and sometimes from a life that had previously nothing to do with oil and gas. By ingenuity, perseverance and luck, these larger-than-life characters transformed wild and improbable ideas into viable businesses, many of which still bear their names. As the industry grew, research and development slowly became institutionalized with billions now spent every year. But every exciting idea still has to start from someone's inventive mind.

Mau and Edmundson have done an extraordinary job finding these people and telling their stories, and we are proud to have supported them. To our knowledge, this is the first comprehensive technology history of how the upstream got to where it is today. There have been plenty of political and economic histories of the oil patch, and books that record the history of particular institutions and companies, but none tell the full story of how the disciplines of geoscience, drilling and petroleum engineering developed, and how the industry reached its current level of scientific and technical sophistication.

In the four years it took to write this book, the authors have interviewed more than 120 distinguished scientists and engineers, many of them familiar names. Add to that an exhaustive search through relevant published material and the huge literature provided by our several professional societies. This book covers all bases, and with editorial independence from the beginning, shows bias to none. It will entertain and leave you excited about the industry's future.

Andrew Gould
Chairman, BG Group

Ashok Belani
Chief Technology Officer, Schlumberger

Preface

THERE IS no getting away from oil and gas. We need oil and gas to power our cars and fly our planes, to generate a huge chunk of the world's electricity for our homes, factories and hospitals, and to manufacture a vast array of indispensable materials that make modern life possible. Together, oil and gas provide more than half of the total energy it takes to power the world we live in.

Yet for such an important resource, its habitat remains tantalizingly out of sight. Nobody has ever gotten close to an oil or gas reservoir, let alone seen or touched one. Nobody has physically descended to the great depths of a typical oil or gas well. Therefore, every phase of exploitation has been by necessity at arm's length, crudely performed at first but over time increasingly sophisticated. This journey from blindman's bluff to today's smorgasbord of high-tech geoscience and petroleum engineering forms the backbone of this book.

Since the beginning, oilfield technology has addressed four main challenges. First is the geoscience challenge of deciding where to drill—deciding where oil and gas is most likely to be found. Then comes the challenge of drilling deep boreholes to see what's down there—a dangerous business because pressures in the subsurface can so easily get out of control. After that, the challenge is assessing the size of the oil or gas reservoir that the drill bit hopefully encounters. Finally, once a reservoir has been found, there remains the challenge of producing the maximum amount of oil and gas to the surface and into the waiting pipeline.

The adventure began in the mid-19th century with the first drilled oil wells in Azerbaijan and the United States. As with all adventures, the story is really about people—people who didn't think like others and who were willing to take risks, both intellectual and financial. Marcel Schlumberger, coinventor of well logging in the 1920s, believed the secret was "to have an independent mind, to think for oneself, not to follow fashions, not to seek honor or decorations, not to become part of the establishment." Petroleum geologist Marion King Hubbert was a maverick who rocked the oilfield establishment more than once, most famously with his prediction concerning the limits of US oil production. The story is also about people who were lucky enough to be in the right place at the right time. Napoleon Bonaparte famously asked for "lucky" generals because he believed luck attached to those who recognized opportunity and seized it.

The people you will meet were certainly lucky, but they were also larger than life, and some became exceedingly rich—people like drillbit innovator Howard Hughes Sr., cementing legend Erle Halliburton and petroleum geologist Everette de Golyer. Then there are the modern heroes of oilfield innovation such as Jon Claerbout, the major contributor to seismic imaging; Peter Vail, who bridged geophysics and geology with his seismic stratigraphy; topdrive inventor George Boyadjieff, who revolutionized the drilling process; and Jacques Bosio, who pushed horizontal drilling. Almost without exception, they had to fight against an industry both skeptical and reluctant to try new ideas. This book will introduce you to all of them.

Sometimes, seemingly random events triggered new thinking. The sinking of the Titanic directed several inquiring minds to invent ways to prevent iceberg collisions, and one of these developments led to today's technique of seismic exploration. Another impetus for seismic exploration came from scientists on both sides of the trenches in World War I trying to locate enemy artillery. The extraordinary use of flexible piping to transport fuel across the English Channel after D-Day to support the Allied advance eventually gave rise to the coiled tubing used today to service oil and gas wells. The refusal of local government to allow BP to create an artificial island in Poole Harbour, England, to tap the Wytch Farm oil field forced the company to become the first to embrace extended-reach drilling, in which offshore fields are exploited from land using wells that are drilled horizontally for several kilometers before hitting their target.

The story of oilfield, or upstream, technology is emphatically a high-tech one. The scientific and engineering sophistication is breathtaking and easily matches other comparable endeavors such as placing men on the moon. It is a story dominated by the three main players: the oil companies, from the small independents to the supermajors and national oil companies; the service industry now dominated by the big three—Schlumberger, Halliburton and Baker Hughes; and the world's academia. The preponderance of oilfield innovation has come from the West, but the Russians also played a huge part, especially in the early days, following parallel but not necessarily similar tracks.

From the mid-19th century to World War I, oilmen quickly developed the basics of drilling and invoked a primitive understanding of geology to pick their drilling locations. The oil seeps of Azerbaijan, Pennsylvania and Texas became natural targets for these early oil pioneers. Success was very much hit and miss, but given the fact that most of the world's oil fields were yet to be discovered, the hits were generally spectacular. Spindletop remains the epitome of this era, but soon to follow were successes in Venezuela, Indonesia, Mexico and Persia. The industry was spreading its wings. By World War I, the usefulness of oil was becoming clear: Henry Ford's Model T motor car rolled off the production line in 1908, and in the lead up to World War I, Winston Churchill, then First Lord of the Admiralty, made the momentous decision to convert the Royal Navy's ships from coal to oil. The oil era had begun.

The period from then until World War II saw multiple breakthroughs in geologic understanding and the birth of all the key geophysical techniques that progressively gave the industry actual images of the earth's subsurface, dramatically helping explorers pinpoint their targets. Seismic exploration was born, as were magnetic and gravity surveying techniques. Well logging made its entry, revolutionizing the correlation of geologic strata between wells, and then actually providing a direct indication whether or not the rock near the borehole contained oil and gas, or just water. The big discoveries during this time were in Bahrain, Kuwait and many areas of the US. The strategic importance of oil supply during World War II, as both sides struggled to secure supplies to their armament factories and troops in the field, has been noted in many histories of the period.

The post-World War II era was controlled by the big Western oil companies, the so-called Seven Sisters. To satisfy increasing postwar demand, an accelerated program of drilling led to the discovery of many huge fields, of which the largest of all time, the Ghawar field in Saudi Arabia, was found in 1948, although nobody at the time realized how big it was. Further discoveries followed in Algeria, Libya and Nigeria. Meanwhile, the search for oil headed into the ocean, the Gulf of Mexico becoming a hive of activity, followed by the North Sea in the 1960s. These prospects only became possible with the development of semisubmersible rigs, robust production platforms and eventually subsea technology. Gas also became a priority when a major discovery was made at Groningen in the Netherlands. The Soviet Union had not been idle either. By the 1970s, vast new oil and gas reserves were being discovered in Western Siberia.

Meanwhile, a series of technology leaps was transforming the industry. The invention of the transistor and the resulting miniaturization of electronics and computers led the industry smartly into the digital age, opening up endless possibilities for improving seismic data processing and ultimately the ability to image the subsurface. In parallel, seismic acquisition made the momentous shift from 2D to 3D. Two major breakthroughs in geology—plate tectonics and turbidites—were also transformational. Logging technology continued apace, with nuclear magnetic resonance logging and wireline testing making spectacular advances. With computer advances, reservoir simulators finally came of age. Lastly, drilling provided its own technology revolutions with the emergence of horizontal drilling, previously deemed out of the question by traditionalists.

Since the 1980s, oil and gas technology has continued its march forward. Every time further progress seemed impossible, someone has found a way. The most dramatic example today is the shale gas and oil revolution in the US, a country that had become reliant on others for energy, but is now becoming an energy exporter. Deploying a technique now called fraccing, first developed after the end of World War II to break open deep formations containing gas, George Mitchell and Harold Hamm found a way to produce gas and oil from shale, a rock previously thought impossible to produce. Their belief in the impossible has paid huge dividends to them and their country.

This book tells all these stories. But it is not a textbook—there are no equations or graphs. Nor is it a political or economic history. In short, this book limits itself to telling how inventive engineers and scientists pursued their dreams, and in the process made our modern, hydrocarbon-dependent world possible.

Chapter 1: Beginnings

> *On the confines toward Baku there is a fountain from which oil springs in great abundance in as much as one hundred ship-loads might be taken from it at one time. This oil is not good to use with food but it is good to burn, and is also used to anoint camels that have the mange. People come from vast distances to fetch it, for in all countries around there is no other oil.*[1]

MARCO POLO was clearly impressed by the petroleum seeps when his travels in the late 13th century led him to the Absheron Peninsula in what is now Azerbaijan. Oil had always oozed out of the ground in this part of the world, but it wasn't until the 18th century that systematic exploitation began. The method was to dig a shallow pit, wait for it to fill with oil and then bail the oil out manually. The method may have been primitive, but by the mid-19th century, annual production was recorded at 3,770 metric tons (28,000 barrels). In those days, oil was mainly used for lighting and water-proofing ships.

The first wells

On July 14, 1848, a far more efficient method of extraction was set in motion. That day, Count Mikhail Semyonovich Vorontsov, governor-general of the Caucasus, wrote a memo to his staff that would mark the birth of the modern oil and gas industry, "I hereby authorize oil exploration in the Bibi-Eybat sector, Baku district, Caspian Sea by means of earth drills and allocate 1,000 roubles for this purpose."[2] The same year, with funds in hand, a certain Major Alexeev succeeded in drilling a 21-meter well and struck a small quantity of crude. The first oil well had seen the light of day.[3]

Drilling the Bibi-Eybat well was exhausting work. The technique was to hit the earth with a sharp, heavy chisel suspended by rope from a hand-operated oscillating beam, then remove the cuttings with a kind of scoop and bucket arrangement. This rudimentary setup, called cable-tool drilling, had been practised since ancient times drilling for saltwater in China and Europe.[4]

A decade after Alexeev's success, oil seeps in North America began attracting attention. The first well drilled for oil was the work of a US prospector named "Colonel" Edwin Drake—the Colonel moniker was provided by an investor to impress the locals. Drake drilled his first well in Titusville, Pennsylvania, about 100 miles north of Pittsburgh, to a depth of 69 feet, by coincidence the same depth as Alexeev's well. Like Alexeev, Colonel Edwin Drake also used cable-tool drilling, this time driven by a five horse-power steam rig.

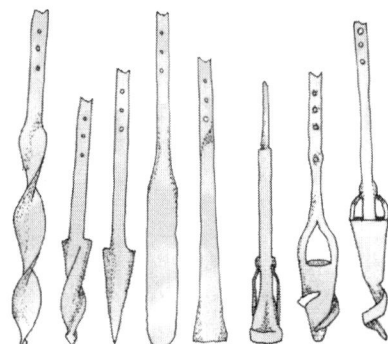

Early drilling tools, mid-19th century. From left: three boring tools; three spudding tool bits for shaping and straightening the borehole and breaking through hard rock, gravel and boulders; two boring tools that in theory could recover their own cuttings.

Production began on the August 27, 1859, when Drake's drilling foreman William A. Smith peered down the hole they had been drilling for several weeks and noticed oil standing near the top of the well. Smith grabbed a piece of tin gutter, rolled it into a tube, plugged one end, then lowered it on a rope into the well and retrieved the first oil. The next day Drake arrived, flushed with excitement at Smith's news, and improvised an iron water pump that had extra pipe sections to reach the bottom of the well. He then rigged the pump handle to the steam-driven oscillating arm that Smith had used to drive the cable-tool drilling rig, and thus began the first US oil production from a drilled well—into a metal washtub and later into empty whiskey barrels. Drake's production was about 25 barrels of oil per day. But like the Azeri oilmen before him, Drake had no idea if or when the flow of oil would stop. All he knew was how to drill, and the technique for doing this soon evolved.[5]

In the two decades following the Drake well, all the major elements of the cable-tool rig took form, enabling the US to lead the world in percussion drilling. Technical innovations included using water to circulate cuttings out of the hole and building rigs that could be easily moved from one location to the next. Drake was also the first to shore up his wells with a rudimentary casing made from wooden planks.

Samuel Smith, William A. Smith's son, recalled: "The casing was driven into the ground through the use of a battering ram lifted by an old-fashioned windlass. It was my job to operate the windlass and drive the casing in."[6]

The technical successes of the US oil industry soon spilled back to Azerbaijan when the oil industry there was opened to foreign investors in the early 1870s. Notable arrivals were the brothers Robert and Ludvig Nobel, two Swedish engineers who arrived in the Caucasus region in the mid-1870s. Over the next 25 years, the Nobel brothers ramped up the Azerbaijan oil industry, building refineries, commissioning oil tankers and constructing pipelines. But they also needed to drill and were quick to try out new methods from the West.[7]

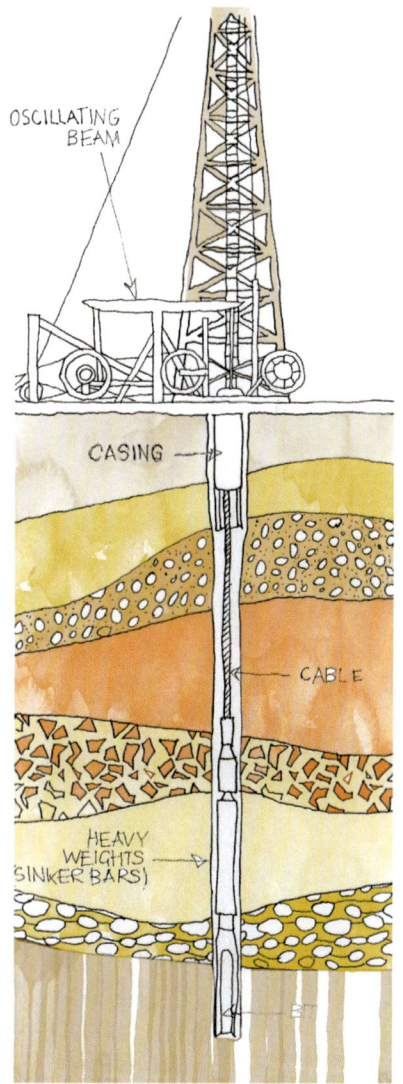

A typical cable-tool drilling rig and view of the borehole, late 19th century.

In 1878, they brought six drillers from the US to try the latest cable-tool drilling at Baku but soon deemed it unsuitable for the soft formations of the Azerbaijan oil fields. Instead, they developed the so-called Azerbaijan free-fall system.

In this variant of percussion drilling, the tools were allowed to fall freely to the bottom of the well and then get picked up. Nobel's engineers also introduced a system to clean out the borehole using hollow small-diameter piping. As drilling progressed, water was periodically forced down the pipe by a steam engine with sufficient pressure to push the debris up the sides of the casing. With these and other advances, no one could match the Nobel brothers for drilling prowess.[8]

Early production

Meanwhile in the US, Drake was battling the challenge of keeping his wells producing. He decided to borrow two inventions from the brothers David and Joseph Ruffner who in 1808 had drilled and completed the first brine wells in the US close to Charleston, West Virginia. The first was to use hollow tubing suspended in the well to transport the fluids to the surface. Drake used copper pipe two or three inches in diameter and in sections 12 to 14 feet long. He also added a gate-valve at the surface, the first infant step toward today's wellhead. The second Ruffner idea was to use a "seed bag," made of leather and containing seeds of various kinds, that was placed around the tubing and positioned near the bottom of the well. The idea was that as the seeds swelled, the bag would secure the tubing inside the borehole and provide a much-needed seal between production from the bottom of the well and all sorts of contaminating sediments and water flow near the surface.[9]

Seed bagging was critical for production, but the clumsy leather seed bags often failed to swell. In 1865, John Ross Cross from Chicago invented a system whereby a fibrous material was contained between two cast-iron hoops that could be placed anywhere in the well, then squeezed from the surface via a system of rods and wire ropes, to provide the necessary seal between the tubing and the borehole wall. But the process of lowering or raising the packer frequently compromised the fibrous packing material. A more reliable technique was needed, and it came from a young man named Solomon Robert Dresser.[10]

Dresser grew up on a farm in Michigan and would become a classic oilfield inventor-entrepreneur. In 1862, at the age of 20, he traveled to the oil fields of West Virginia and founded the Peninsular Petroleum

Company, redeveloping oil fields that had been abandoned during the American Civil War. He later moved to the booming Western Pennsylvania oil patch. Dresser spent hours hours trying to solve the packer problem, and by 1879, his early years tinkering with farm implements paid off. He took Cross's packer idea but substituted rubber for the fibrous material. Using a simpler mechanism than Cross had used, the rubber seal was squeezed and expanded toward the borehole wall by pulling on the tubing and contracted by lowering the tubing. In this manner, the packer could be moved up and down the well without destroying the rubber seal. A patent application was duly dispatched to Washington DC.

In anticipation of his patent being approved, Dresser rented space in the center of Bradford, Pennsylvania, and opened his doors for business on the very day the patent was granted, May 11, 1880. Dresser's business grew considerably in the following years. Soon his company portfolio included other oilfield hardware such as pipeline couplings and a new type of casing head. In 1903, he handed the business over to his son-in-law and entered politics as a Republican member in the US House of Representatives.[11] The Dresser company would continue to prosper.

Solomon Robert Dresser (1842-1911), oilfield inventor and self-made entrepreneur, later a Republican member of the US Congress. Depicted here on his electoral campaign button.

In most of Pennsylvania, where Dresser was active, there was insufficient pressure to propel the oil to the surface, so oil prospectors soon devised a pumping arrangement. With the drilling rig remaining at the wellsite, the steam-driven oscillating beam, initially used to lower and raise the cable tool for drilling, was then harnessed to a long wooden rod with a plunger attached to its bottom. As the rod was lowered and raised from the surface, the plunger pushed the oil up the tubing to the surface. An innovation around 1880 was to link several wells to one power source, using various mechanical contrivances to transmit the required oscillating movement to each rig. This eliminated the cost of maintaining steam engines at each well.[12]

In Azerbaijan, pumping was hardly needed. A gusher had been drilled in 1873, and scarcely a well could be bored afterward without finding oil in large quantities and generally rushing to the surface like a fountain. But even though the pressure was high, the Azeris had to deal with another problem. Almost without exception, Baku production was from unconsolidated sands, so the initial oil gush would carry with it vast amounts of loose sand, creating a mixture resembling to the locals something like their beloved caviar. In no time, the wells would fill with sand, production would drop and the only means of production was to bail out sand and oil together.[13]

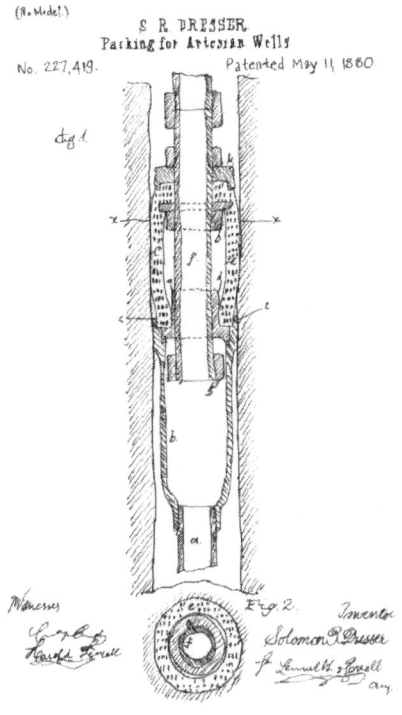

The fact that the early Azeri wells were bailed and not pumped meant that the early prospectors didn't use tubing or packers, nor did they have to construct wellhead termination points at the surface. Despite these hurdles, Azerbaijan quickly surpassed the US in oil production. By 1900 the Russian Empire became the world's leading oil producer with an output of 11 million metric tons per year (220,900 barrels per day).[14]

Solomon Dresser's oil well packer from 1880. This depiction from the original patent shows a vertical section of the packer and a sectional view from the top.

Chapter 2: The Birth of a Science

DURING THESE early days, there were still plenty of oil seepages to show the driller where to place his rig. But that's where the knowledge stopped. Azeri and US drillers alike had little awareness of the rock strata they were about to penetrate, let alone the formations that they hoped would contain oil. The science of rock or oil didn't exist then, although the foundations had been laid almost 100 years before.

The roots of petroleum geology

In 1795, James Hutton, an all-round talent from Edinburgh who had studied medicine and worked as a chemical manufacturer and farmer, published his "Theory of the Earth; or an Investigation of the Laws observable in the Composition, Dissolution, and Restoration of Land upon the Globe." Hutton had taken to roaming the Scottish landscape and found himself drawn to riverbeds, ditches, borrow pits, coastal outcrops and inland cliffs, wondering what great scheme lay behind them. It was punishing work collecting rocks as he once stated, "Lord pity the arse that's clagged to a head that will hunt stones."[1]

The origins of petroleum were not so easily resolved, although many at the time correctly suspected oil was contained in sedimentary rocks. Belsazar de la Motte Hacquet, a native of France and a physician, swept up by the Seven Years' War and deposited in the mining town of Idrija, Slovenia, became an authority on salt mines and the oil and gas commonly found in them. In 1793, he attributed the origin of petroleum to marine animal matter. This was substantiated five years later by the English chemist Charles Hatchett. Based on laboratory studies of bitumen extracted from seeps in Trinidad, Hatchett suggested in 1798 that these heavy oils were simply decomposed plant and animal matter. As for James Hutton, he postulated in his "Theory of the Earth" that petroleum was distilled from coal.[2]

In the bigger scheme of things, Hutton was the first to grasp the full significance and immensity of geologic time. He demonstrated that the hills and mountains of the present day are far from everlasting but have themselves been sculpted by slow processes of uplift and erosion—processes that are ongoing today. The self-made geologist observed that the sedimentary rocks on the earth's crust bore all the hallmarks of having accumulated exactly like those being deposited in the present. The vast thickness of sedimentary rocks, he surmised, spoke of cycles of erosion and sedimentation for periods of time that could only be described as inconceivably long. In other words, what one can see as rocks and structures on the surface today is in fact something in motion, originating not tens or hundreds of years ago but tens or hundreds of millions of years ago.[3]

Hutton's ideas were too profound for general comprehension by his contemporaries and gained few adherents until Charles Lyell arrived on the scene. In the late 1820s and early 1830s, the lawyer-turned-geologist wrote the seminal "Principles of Geology: Being an Attempt to Explain the Former Changes in the Earth's Surface, by References to Causes now in Operation." The three-volume work was essentially Hutton's theory, supported by the great wealth of observations that Lyell had made in his homeland Scotland as well as in England and Continental Europe. Charles Lyell crucially popularized James Hutton's geologic concepts and boiled them down to the memorable line: "The present is the key to the past."[4]

Charles Lyell (1797-1875), a father of modern geology who coined the phrase: "The present is the key to the past."

With these broad principles established, the next step was unraveling the sequence of geologic ages and dating them. The first step had been taken in 1669 when Niels Stensen, better known as Steno, a Danish anatomist and later a Catholic bishop, surmised the superposition principle that says that a rock layer overlying another is always younger than the layer below. The first to use this principle was William Smith, nicknamed the "Father of English Geology," who mapped stratigraphic layers observed in canal excavations across England, eventually making the first geologic map of England in 1815.[5]

The naming of rock strata came from everywhere. The Cambrian derived from a classical name for Wales. The Ordovician and Silurian were named after ancient Welsh tribes, because they were identified from stratigraphic sequences in Wales. The Devonian was named for the English county of Devon. The succeeding Carboniferous was named after the ubiquitous occurrence of coals within its sequences, while the Permian was named after the ancient kingdom of Permia by Sir Roderick Murchison during his extensive travels in Russia in the mid-19th century.

The Triassic was named in 1834 by the German geologist Friedrich August von Alberti from the three distinct layers of reddish sedimentary deposits found throughout northwest Europe. The Jurassic was named by French chemist, mineralogist and zoologist Alexandre Brongniart for the extensive marine limestone exposures of the Jura Mountains. The Cretaceous, from Latin *creta* meaning "chalk," was defined by Belgian geologist Jean Baptiste Julien d'Omalius d'Halloy in 1822 from observations of the extensive chalk beds in the Paris basin.

Dating rock strata, however, proved elusive. In 1841, William Smith's nephew, John Phillips, made a first attempt by combining Steno's depositional rule with observations of fossils found in various strata. Phillips's timescale provided a broad framework featuring epically long periods of geology, such as *Paleozoic* ("old life" from Cambrian to Devonian, 540 to 250 million years), *Mesozoic* ("middle life" from Triassic to Cretaceous, 250 to 66 million years) and *Cenozoic* ("recent life" from 66 million years to the present), formerly known as the Tertiary period.[6]

Anticlines

The new discipline of geology was soon embraced outside Europe. Canada's Geological Survey was founded in 1842 by Montreal-born William Edmond Logan, who came from the mining business with deep knowledge of coal deposition. He had also traveled to the oil springs in Gaspé, Quebec, "Here the connexion is evident between the oil springs and undulations of the strata which form the accumulation of the petroleum."[7]

Elsewhere, geologists were making similar observations. In 1855, an Anglo-Irish geologist, Thomas Oldham, working in Burma, pointed out that the oil from the Yenangyuang field, then being produced from wells dug by hand, was connected with the highest part of an upfold—or anticline—in the earth's strata. In the US, Ebenezer Baldwin Andrews, both priest and geologist, reported in 1861 that in western Virginia the productive wells were closely associated with the axial area of anticlines.

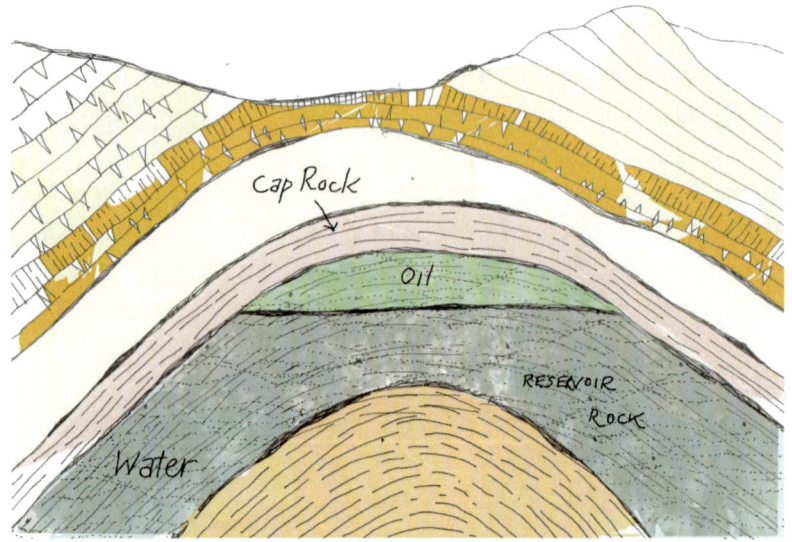

An idealized sketch of an anticline, a geologic phenomenon that traps oil. Oil, being lighter than water, accumulates in the reservoir rock under the crest of cap rock.

Both Oldham and Andrews had hit the anticline jackpot. Over millions of years, compression and tension of the earth's crust folds the rock layers, occasionally forming uplifts or anticlines that can often be recognized at the surface. Oil, being lighter than water, moves upward, and as long as the uplift is sealed with a layer of impermeable rock the oil gets trapped, resulting in a reservoir of oil-filled porous rock.

Although Andrews became known as the father of the anticlinal theory, another American, Thomas Sterry Hunt, a geologist and chemist who had been an assistant to William Logan since 1846, made similar observations in western Ontario. Just two months after the publication of Andrews's western Virginia report, Hunt reported that oil finds made at Enniskillen, western Ontario, were likewise associated with a broad, moderately folded anticline. Later, exploration around Petrolia, a stone's throw from Enniskillen, resulted in a gusher.[8]

The anticlinal theory thus became the backbone of oil geology, and it remained of crucial importance for many oil and gas discoveries in the 20th century. The first successful wildcat in the Middle East, at Masjid-i-Sulaiman in Persia in 1908, was located on an anticline. Thirty years later, in the classic textbook "Fundamentals of the Petroleum Industry," the US geologist Dorsey Hager stated unequivocally that "the anticlinal theory is as fundamental to the geologist as Newton's gravitational law is to the physicist."[9]

Salt domes and Spindletop

The other geologic idea of the era centered on underground salt domes, a concept that would precipitate an oil boom on the US Gulf coast at the beginning of the 20th century. This brought John Dustin Archbold, a senior executive of Standard Oil, to a tight spot when he promised to drink every gallon of oil produced west of the Mississippi River, the dividing line between known areas of oil production such as Pennsylvania, east of the great river, and the rest of the US, which had produced nothing. A key figure in this oil boom was Patillo Higgins, a self-taught geologist who lived near Beaumont in southeast Texas. Noting gas seepages that were emerging from a small hillock called Spindletop on the flat plain, Higgins began reading more about the infant science of geology. A paper by Israel White of the West Virginia Geological Survey that he read in the summer of 1892 convinced him that Spindletop was an anticline.[10]

During the next six years, Higgins drilled six wells, found nothing, and ended up heavily in debt. In a last-ditch attempt, he advertised for someone to help his drilling enterprise and recruited Captain Anthony Francis Lucas. Lucas had moved to the US from Austria, where he had studied engineering and then spent time in the navy. Lucas knew about drilling, but luckily for Higgins he also knew about salt.

Earlier in 1893, Lucas had taken a job with Myles and Company of New Orleans, superintending operations at a salt mine owned by the Avery family at Petite Anse, Louisiana, 30 miles south of Lafayette, now known as Avery Island. When exploring for salt on Avery Island, Lucas had found sulfur and traces of oil and gas. He had in fact stumbled upon a second type of geologic structure harboring oil and gas—the salt dome. The salt Lucas found was from the Jurassic age that in patches underlies the Gulf of Mexico and most of its coastal region. Salt is lighter than most rock, so over geologic time, these patches rise dome-like, distorting the overlying rock layers and creating traps that can accumulate hydrocarbons. In 1896, Lucas moved his salt explorations to Jefferson Island and the following year to Belle Isle, both in Louisiana, where he again found sulfur, oil and gas. Speaking later of his Belle Isle explorations, he wrote, "This led me to study the accumulation of oil around salt masses, and I formed additional plans for prospecting other localities. Thus I began my investigations into the occurrence of oil on the Coastal Plain."[11]

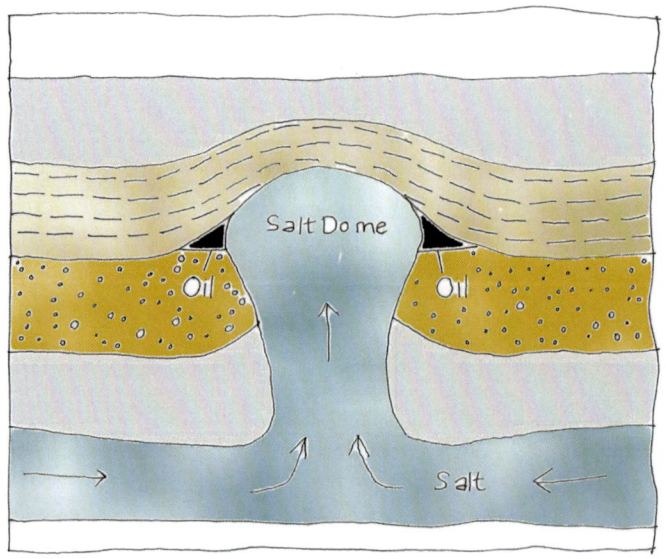

An accumulation of salt, a mineral lighter than most sedimentary rock, pushing up and forming a dome-like structure. Salt domes may be several kilometres high and on their flanks form traps where oil can accumulate.

By the time Lucas started working for Higgins at Spindletop, he had enough confidence to convince Higgins that it was salt domes and not anticlines that were worth investigating. His reasoning resulted in one of the most famous oil gushers in history, Spindletop, in 1901. Ten years later, the salt dome concept was well accepted in the Gulf region, counting 36 explored salt domes and 10 oil fields in southeast Texas and Louisiana and giving rise to a huge regional oil industry. Neither Higgins nor Lucas nor any other explorer of the time, however, had the technical means to find the remaining hundreds of salt domes hidden around the margins of the Gulf of Mexico. This would become possible only with the development of modern geophysics two decades later. During that time, drilling went through a minor revolution.[12]

Chapter 3: Drilling Gets Established

IN SPITE of early progress in both Azerbaijan and the US, drilling needed to be more reliable, faster and less prone to the vagaries of deep sediments that could stymie the most experienced practitioner. In particular, cable-tool drilling was not working in the soft formations of the southern US.

Rotary drilling takes over

A new approach was required, and it was provided by brothers M.C. and C.E. Baker of South Dakota. Their idea was to rotate the chisel or tool instead of using it as a percussion instrument. In the early 1880s, they developed the first rotary rig and used it to drill shallow water wells in the unconsolidated formations of the Great Plains. By 1895 the Baker brothers were using their rotary method for oilwell drilling, in the Corsicana field of Navarro County, Texas. An early convert to the Baker way of drilling was Captain Lucas at Spindletop. From the beginning, Lucas opted to drill with a steam-driven rotary rig and, in addition, adopted a double-pronged fishtail bit instead of the traditional single-pronged chiseling bit. The combination was a winner.[1]

Anthony Lucas (1855-1921), an Austrian-born engineer and former navy captain who not introduced the salt dome concept along the Gulf of Mexico coast in Texas and Louisiana but also pioneered the rotary drilling technique.

Spindletop helped promote rotary drilling as a viable alternative. In 1907, Shell began using rotary drilling on a modest scale in Romania, and in the US, rotary drilling gained a reputation for making hole fast in the Gulf of Mexico.[2] Standard Oil of California was impressed by results in Texas and Louisiana, and in 1908, the company hired six drillers from the Gulf Coast and bought three complete rotary outfits to drill the hard formations of California.

Nevertheless, rotary drilling remained a niche activity compared with cable-tool drilling, and Standard Oil of California had to overcome some human as well as technical issues, as the local newspaper *Bakersfield Californian* noted on May 13, 1909: "The main difficulty with the companies which previously tried the revolving drill, and abandoned it, seems to have been with the drillers. There is a deep-seated jealousy and half-hatred which exists between rotary and standard [cable-tool] driller, much like the feeling between cattlemen and sheepmen, and the standard operators who were put on the rotaries generally neglected the work, while genuine rotary men were hard to find. The Standard [Oil Company of California] seems to like the rotaries, however, and more will probably be added soon."[3]

Another milestone in rotary drilling technology was the invention of the rotary table and kelly, first used in 1915. The origins of the powered turntable in the middle of the drill floor went back to Spindletop. The primary function of the rotary table was to transmit torque to the drillstring via the kelly, a section of pipe with a square cross-section that slotted through a similar shape on the rotating table.[4]

At first, rotary tables were driven by chain from a sprocket on the hoist, or drawworks. But in 1918, Victor York and Walter G. Black of Standard Oil Company of California were granted a patent for driving the rotary table with a shaft. This innovation guaranteed the ongoing success of the rotary drilling method. By 1930, rotary rigs had replaced cable-tool rigs in most places, except for drilling very shallow wells.[5]

Drilling mud makes its appearance

Another challenge at Spindletop was keeping the hole open. Drilling the unconsolidated mix of sand and clay locally known as gumbo, the sides of the borehole would repeatedly collapse and fill the well with debris and cuttings. Two brothers, Curt and Al Hamill, both drillers at Spindletop in 1900, circulated the gumbo out of the hole with fresh water and shored up the well with wooden casing. But it was a losing battle.

Lucas' Spindletop well flowing out of control, January 1901.

One day, when Al was absent getting wood for shoring up the wells, Curt and workman Peck Byrd made a discovery. Both men noticed that while they were drilling through the gumbo, the fresh water used to clean the hole would become muddy, and this in itself helped stabilize the formation. They therefore decided to muddy up the fresh water before circulating it and see whether this would work even better. To try the idea, Curt Hamill recruited a local farmer, the Reverend John C. Chaney, who was already engaged to periodically clean out the slush pit containing the drilling water.

With the pit empty, Chaney plowed the clay at the bottom of the pit to a depth of about a foot, then filled the pit with water. After that, he drove some of his cattle into the water, walking them back and forth for a few hours until the liquid became thick with mud. The drilling crew then added some water to this viscous mixture. The result not only stabilized the well but also provided enough viscosity to circulate out the cuttings. Thus, with the help of a few cows, the Hamill brothers launched the era of drilling mud.[6]

In Russia, a rotary rig equipped with a drilling mud circulating system—similar to the Spindletop setup—drilled its first well near Grozny in Chechnya in 1902. But rotary drilling was slow to take off in Russia, and indeed, any type of innovation more or less ceased thereafter because of numerous political upheavals such as World War I and the small matter of a revolution.

Derricks and drawworks

A key breakthrough that cemented the popularity of rotary drilling was the introduction of the standard derrick in 1908 by Lee C. Moore. He had acquired a patent—originally granted to T.A. Neill, a field superintendent for the South Penn Oil Company—on a method of joining steel, tubular sections for constructing derricks. At the time, derricks were built using wood, each time from scratch. Moore's easy-to-construct tubular sections could be reused, creating a huge efficiency gain. A second major development from Lee C. Moore was his 1937 jackknife cantilever-type drilling mast, which could be erected as a single piece.[7]

However, the Achilles heel of the new rotary drilling rigs was the drawworks, the central winch system that lowered and lifted the drillstring. By 1930, the drawworks had evolved from single-speed (1910) to two-speed (1920) and then to four-speed. It had grown to a considerable size and typically consisted of 70 or more distinct parts that had to be carefully dismantled and rebuilt each time the rig was moved.

Drilling floor on an early rotary rig in Texas, around 1915, showing the driller manning the brake of the drawworks. The chain on the right turned the rotary table, bottom left.

Hu Harris, head of the drill-tools division of the Texas-based Humble Oil and Refining Company, recalled the problems: "In those days it took three to five men two days or more to set up the drawworks and about the same time to tear it down. Another objectionable feature to erecting the drawworks on each well was that the parts, especially the brakes, were not always assembled properly. Drawworks brakes were the source of considerable complaint and if they were not installed correctly, there was always the danger of injuring the driller. Also, in dismantling the drawworks and moving to another location many of the small parts were lost or misplaced."[8]

Harris decided to design a drawworks with fewer pieces. By assembling the shafts, chains and guards as a unit, he created a drawworks with just a few large parts saving days between rig moves. Harris's idea was immediately copied by others, and by 1934 rig manufacturers were systematically building unitized drawworks.[9]

From fishtail to tricone bits

As rotary drilling took off, attention turned to the drill bit. The traditional fishtail bit that Lucas used was proving unreliable when drilling hard formations. The solution was found by one of the many engineer-entrepreneurs who journeyed south to ride the oil boom in southeast Texas. This young man in his 20s had let go a career in the lead and zinc mining industry in Joplin, Missouri, to try his luck in oil. His name was Howard Hughes Sr.[10]

Howard Hughes Sr. (1869-1924), inventor of the roller-cone drill bit.

All things mechanical fascinated Hughes. In his youth he spent hours tinkering with watches, clocks and engines of every sort "to see what made them go," remembered his brother Rupert. Hughes recognized the need for a completely new drill bit for the rotary rig. Visiting a machine shop in 1907, in Sour Lake, Texas, where some conventional drilling tools were being repaired, he noticed a grinder with two outer wheels moving in one direction and an inner wheel moving in the opposite direction. Hughes figured this type of contrary motion might work for drilling, and in subsequent experiments eventually came up with a two-cone rock cutting tool that, rather than scraping against the formation like a fishtail bit, actually crushed and ground the rock. Early experiments enabled Hughes to successfully file for a US patent. In 1909, Hughes, together with his longtime business associate Walter Sharp, formed the Sharp-Hughes Tool Company, renamed Hughes Tool Company in 1912, to market and develop the new bits.[11]

Meanwhile, wells were being drilled deeper, and it became increasingly difficult to gain smooth passage up and down the hole. In 1917, Hughes and his engineers came up with the idea of adding a reamer, a large-diameter sub placed some way above the bit, with cutters to keep the hole in gauge. By the time Hughes died prematurely of a heart attack in January 1924, there were 73 patents to his name, all connected with rotary drilling. Hughes's innovative rock bits had a near monopoly in the early days and made his only son, the eccentric business magnate and aviator Howard Hughes Jr., one of the wealthiest men on earth.[12]

During the Depression in the early 1930s, Hughes Tool continued to stay ahead of the competition. In 1932, Hughes Tool commercialized staggered teeth and cantilevered bearing shafts to make the hole faster and cheaper. These minor improvements lead to Hughes Tool's most famous innovation, the tricone bit. Bit producers had long been striving to find an alternative to the standard two-cone model. The two-cone bit had problems in gumbo or shale, which caused the cones to ball up or lock. Many bit manufacturers were experimenting with four cones, except Hughes Tool. Their new bit came with three cones mounted centrally with teeth that intermeshed and self-cleaned. Substantially outdrilling the two-cone bit, without balling up, the tricone bit soon won acceptance in West Texas, and in 1933 Hughes Tool was granted a US patent. The Tricone bit set the standard

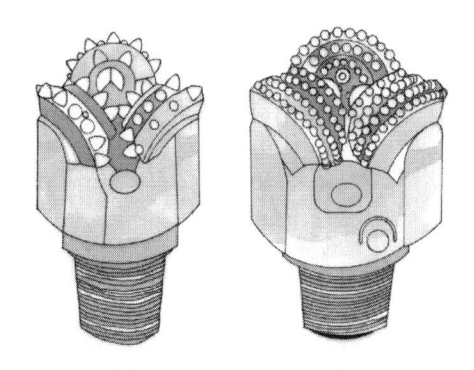

Modern roller-cone bits: a milled tooth bit and an insert bit.

for developing even faster drill bits through the 1930s and 1940s.[13]

By this time, oilfield engineers had a forum for exchanging ideas and technical news. In 1871, the American Institute of Mining Engineers (AIME) was founded in Pennsylvania, and by 1913 a standing committee on oil and gas was created to provide a professional home for them. Later, this evolved in a semi-autonomous Petroleum Division that was renamed the Society of Petroleum Engineers (SPE) in 1957. From a handful of members, the society has mushroomed to several thousand members.[14]

Chapter 4: Geophysics Enters the Fray

UNTIL THE 1920s, drilling for oil and gas was still very much hit and miss. Geology provided some clues in terms of anticlines and salt domes, but otherwise it was follow your nose or see what your successful neighbor was doing. It took more science to provide the first crude pictures of the subsurface and increase the odds of an oil strike. Four techniques in particular revolutionized exploration for locating the next big oilfields. Together, they signaled the beginnings of exploration geophysics.

The first geophysical methods

Magnetics, the first technique, was based on the fact that some rocks, generally those containing iron, have a relatively high natural magnetism. This causes minute local effects on the intensity and direction of the earth's magnetic field at the surface. These effects can be measured, and the resulting map of magnetic anomalies can be related to structure below the surface. In the early 1920s, prospectors began using a simple magnetometer invented by Swedish engineers Thalén and Tiberg in 1870 to locate iron ore deposits. In its simplest form, a magnetometer measures azimuth, the horizontal direction, and dip, the vertical inclination, of the earth's magnetic field. For ten years or so, the magnetometer became a favorite for mapping potential oil structures. Especially in South Texas, oil operators had some success using the magnetometer because it reacted to basaltic plugs often associated with oil. However, the magnetometers produced only rather crude maps of the subsurface and were mainly used as a reconnaissance tool. Their use declined beginning in the early 1930s when other, more precise geophysical methods came into play.[1]

Detecting variations in gravity on the surface was the second technique. Since the density of rocks varies quite widely, gravity surveys reacting to these density differences promised an alternative view for the explorationist. The earliest gravity measurement, actually of the gravitational constant, was made by Henry Cavendish in 1797 using a variant of the so-called torsion balance invented 20 years earlier by French scientist Charles-Augustin de Coulomb, the celebrated French 18th century polymath. The torsion balance comprises balanced masses suspended from a thin fiber. The fiber acts as a very weak torsion spring and twists when for any reason the masses are disturbed. As the fiber twists, an optical system indicates the angle of deflection, which translates to the force of gravity acting on the masses.[2]

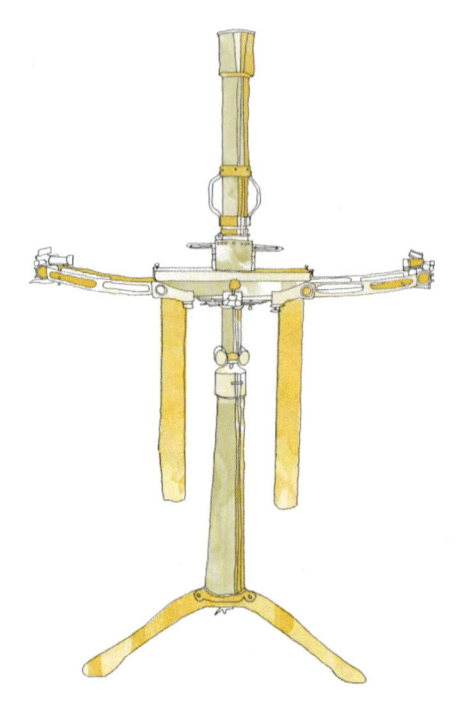

A double-beam torsion balance from 1902. This type of instrument would become a standard tool in the oil industry after World War I.

Baron Roland von Eötvös, a Hungarian professor of experimental physics at the University of Budapest, was the first to use the torsion balance to interpret subsurface geologic structure. Eötvös studied law, but in a letter to his father in 1867, he announced, "I was born with ambition and a sense of duty not only to one nation but towards the whole of humanity. In order to satisfy these urges and to retain my own individual independence, my aim in life will be best achieved, as far as I can see at present, if I follow a career in science."[3] In 1901, Baron Eötvös took his torsion balance on the frozen Lake Balaton in Hungary, and was able to map the irregular surface of the lake bottom. He also mapped the subsurface extension of the Jura Mountains in France. By this time, the Baron was becoming known internationally. He even had a mountain named after him in the Italian Dolomites. But it was the director of the Hungarian Geological Survey, Hugo de Böckh, an Eötvös contemporary, who was the first to use gravity to look for anticlines and salt domes, in the early 1910s in Transylvania, Romania.

Baron Roland von Eötvös (1848-1919), a professor of experimental physics at the University of Budapest, who was the first to use the torsion balance to interpret subsurface geological structure.

About this time in the US, a young man named Everette Lee DeGolyer began his own journey in oil exploration. DeGolyer liked to learn everything from scratch. While a mining student at the University of Oklahoma, where he began his studies in 1905, he earned money during the summer break working as a cook at a local US Geological Survey (USGS) camp. Mixing with the geologists, he picked up enough geology to delay his graduation for a couple of years and instead join the Mexican Eagle Oil Company as a field geologist. For some years the company had been exploring up and down Mexico without success. DeGolyer was asked to pick the next drilling location, and in 1910 Mexican Eagle struck oil at what became the famous Potrero del Llano no. 4 well near Tampico on the Mexican east coast. The well came in a raging monster and would later record 100,000 barrels per day. A year later, a revolution in Mexico would result in the nationalization of foreign-owned companies and drive out their foreign personnel, including DeGolyer.

DeGolyer moved back to the US, set up a geologic consultancy business in New York City and became interested in the new geophysics. In early 1914 he learned of Eötvös's torsion balance and immediately ordered one from Hungary. With the outbreak of World War I, however, the instrument never made it, and the commercial application of gravity surveying by the US oil industry was delayed by almost a decade. Meanwhile, petroleum companies in Europe continued to use the tool successfully. Deutsche Erdöl AG surveyed prospective salt domes in the North German Plain from 1916 to 1918. Some of their structural interpretations were corroborated by drilling after the war. Shell also carried out successful gravity surveys in Egypt, Borneo and Mexico between 1919 and 1922.[4]

Everette Lee DeGolyer (1886-1956), legendary oil-finder from Oklahoma. Trained as a geologist, he quickly learned the exploration geophysics of the day and made a fortune, first with his Amerada Petroleum Corporation, then with Geophysical Research Corporation (GRC) in the late 1920s, and finally with Geophysical Service Incorporated (GSI) from 1930.

In 1919, DeGolyer formed his own oil company, Amerada Petroleum Corporation, and in November 1922 he was finally delivered two sets of torsion balances and immediately tested them at Spindletop. More tests followed, and then in 1924 he surveyed a prospect at Nash in Brazoria county, Texas, and found a salt dome. Drilling started in December 1924, and on January 3, 1926, Amerada struck oil.

Some of the secrets of DeGolyer's success may be found in his love for books. He was a prolific and meticulous collector of rare books and in the course of his lifetime would buy nearly 90,000 volumes, not only in science and technology but also business, history and literary classics. Once, while in San Francisco, DeGolyer purchased a small but valuable pamphlet from book dealer Lew Lengfield and told him to mail it to his home in Dallas. "It's small," said Lew, "Why don't you stick it in your pocket and take it with you?" "Oh, no," replied DeGolyer, "I don't want anything to happen to it as I'm going by plane you know."[5]

One of DeGolyer's purchases was Conrad Schlumberger's "Etude sur la Prospection Electrique du Sous-Sol," describing a third geophysical technique. This publication explained how to map subsurface structure by making electrical resistivity measurements on the earth's surface. Conrad Schlumberger, a physicist and professor at the Ecole des Mines in Paris, had conceived this idea in 1912 and conducted early experiments in the grounds of the family estate in Val Richer, Normandy, France. His brother Marcel proved a perfect collaborator, being a gifted engineer who was fixing his father's car at the age of 14 and who knew how to make Conrad's ideas work in the field.

Conrad Schlumberger (1878-1936), performing surface electrical experiments to map the subsurface at his parents' summer residence Val-Richer in Normandy in 1911. One hundred years later, an archaeologist used the same method to map the walls of a buried monastery in the same property.

In 1919, their father, a wealthy businessman and owner of a textile factory, realized his sons's joint potential and made a covenant with them. He agreed to fund the brothers 500,000 francs, almost a million US dollars in today's money, to develop a business for measuring underground rocks, but demanded at the same time, "The scientific interest in research must take precedence over financial interest." At first, measuring the electrical resistivity of rock was assumed to have application only for finding ore deposits. But recognizing that salt was highly resistive, Conrad and his brother Marcel started looking for salt domes. In 1923, they successfully tested their new method at the prolific Ariceştii field near Ploesti, Romania.

Three years later, the Schlumberger brothers carried out a large electrical survey in Meyenheim, Alsace, outlining the crest of an elongated underground arch nearly seven kilometers long where the salt layer bowed up. These experiences encouraged Conrad and Marcel to found a company in Paris in July 1926 under the name Société de Prospection Electrique—the genesis of the Schlumberger company. Even though their electrical method was successful to a degree in the mid-1920s, its moment of glory was temporary and in particular, along with magnetics and gravity, failed to compete with the up-and-coming fourth geophysical technique, seismic exploration.[6]

The arrival of seismic exploration

Seismic exploration owes its origins to seismology, the recording of earthquake tremors. Through the ages, mankind has devised a variety of instruments for detecting earthquakes. The first recorded seismograph was constructed in the year 132 AD by the Chinese astronomer Zhang Heng. His so-called "frog seismograph" was a large bronze urn with eight dragon heads gazing outward in different directions. Each dragon held a ball in its mouth. A bronze frog, with mouth open, was located under each dragon around the base of the urn, and a delicate inverted pendulum was hidden inside the urn. When a seismic event occurred, the pendulum swung a little and tapped a mechanism that dislodged one of the balls. The ball fell from the mouth of the dragon into the mouth of the frog below, landing with a great clang that announced the earthquake. Knowing which frog had received the ball would indicate the direction of the earthquake.[7]

The first modern device to detect earth movement was constructed in 1885 by John Milne, a British mining geologist and advisor to the government of earthquake-plagued Japan. Milne's seismograph consisted of a heavy mass suspended like a pendulum from a frame firmly fixed in the ground. During an earthquake the framework moved with the earth, while the suspended mass remained relatively stationary. A pen fixed to the mass then drew the earthquake's characteristic oscillating signature on a turning roll of paper that was attached to the frame.[8]

A 2,100 year-old Chinese "frog seismograph". Hidden inside the urn was a heavy pendulum. Depending on the direction of the earthquake, the pendulum would displace one of the balls placed in the dragon mouths, dropping it into the respective frog's open jaws. This would indicate the direction the tremors came from.

During World War I, German mine surveyor Ludger Mintrop used a portable seismograph of his own design to locate Allied artillery firing positions. Mintrop detected earth movement using a highly sensitive carbon-grain microphone, a precursor of the modern geophone. Curiously, on the other side of the trenches, Captain of the French artillery Conrad Schlumberger and others were trying the same technique with some success and succeeded in locating "Big Bertha," the 100-kilometer range gun that regularly fired on Paris. Their technique used three seismographs, spaced some distance apart from each other, facing the artillery. Triangulating the times that it took for the sound waves to reach the seismographs established the position of the enemy's artillery.[9]

The sound waves in this instance were refracted waves. This type of wave enters the earth at an angle but then changes direction at shallow strata and proceeds parallel to the earth's surface to get picked up by a seismograph. After the war, Mintrop began applying the refraction method to petroleum exploration by using a dynamite explosion to create an artificial sound wave. In 1919, he applied for a patent called "Method for the Determination of Rock Structures" and in 1921 founded a company he named Seismos and was soon engaged by Gulf Oil to conduct seismic refraction surveys along the Texas coast line.

Key to success in the Gulf of Mexico was knowing the acoustic properties of salt. Seismic waves travel faster through salt than most sediments, so to locate the salt domes the Seismos crews started analyzing the recorded squiggles of their seismographs for particularly fast travel times. Mintrop's fame peaked in June 1924 when a Seismos crew discovered the Orchard salt dome in Fort Bend County, Texas. This find was the first success using the new seismic method in the US and was spectacular enough to begin displacing magnetics and gravity.[10]

Ludger Mintrop (1880-1956), a German mine surveyor who in 1921 founded Seismos, the first company to market the refraction seismic method for oil exploration.

Naturally, this led to an extensive campaign of refraction shooting, with competitors such as Petty Geophysical Engineering from San

Antonio, Texas, joining the fray. By 1930, most of the shallow domes on the Gulf Coast had been discovered. The following year, Seismos ended its refraction operations, although the technique continued to be offered by other companies and in 1956 was instrumental in the discovery of the Hassi Messaoud field, Algeria's largest. In 1934, Seismos reorganized itself and began offering the services of another even more promising seismic technique called reflection seismics.[11]

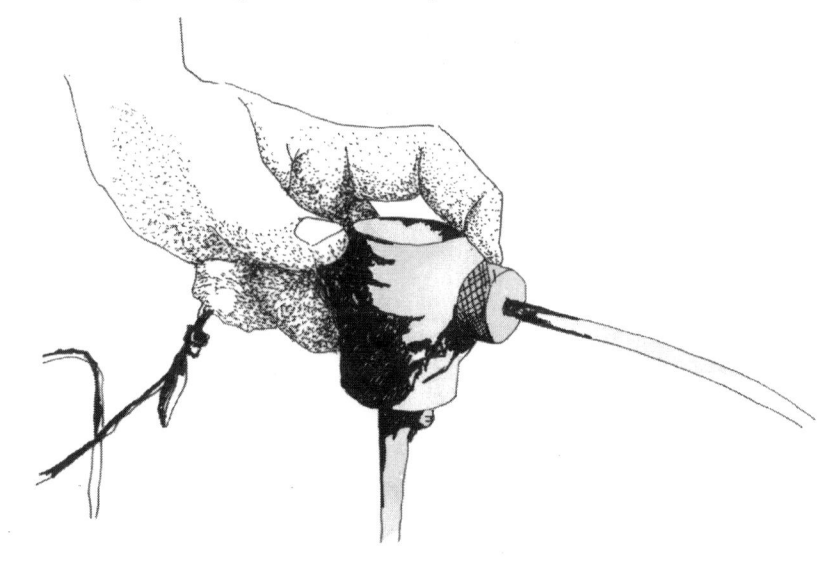

A geophone from the 1980s. Geophones translate ground motion, detected through a spike placed firmly in the ground, into an electrical voltage and send its signal via a cable, right, to the recording truck.

This technique owes its origins to the sinking of *RMS Titanic* when it collided with an iceberg in the North Atlantic, resulting in the loss of more than 1,500 lives. The famous disaster inspired Reginald Aubrey Fessenden, a Canadian inventor who had been working for Thomas Edison, to construct a device that would detect icebergs by emitting a sound wave and timing the return of the reflected echo. In January 1913, Fessenden filed a patent and on April 27, 1914, aboard the US Coast Guard cutter, *Miami*, in the North Atlantic, was able to demonstrate that his device could detect icebergs up to 12 miles away.[12]

But his inventive mind didn't stop there. In the spring of 1913, Fessenden and his assistants began experimenting near Framingham, Massachusetts, and succeeded in detecting both refracted and reflected waves from the subsurface. These tests resulted in another Fessenden patent in September 1917 entitled "Method and Apparatus for Locating Ore Bodies." The news of Fessenden's work quickly spread through the growing US geoscience community.[13]

Meanwhile, the US government had also been sponsoring research to use sound waves in artillery detection and sent physicist John Clarence Karcher from the US Bureau of Standards's Sound Section in Gaithersburg, Maryland, to the front lines in France to investigate. Returning to the US, Karcher had another look at Fessenden's patents and started discussing with his colleague William Peter Haseman, a physics professor on leave from the University of Oklahoma, the idea that sound waves generated by explosions and reflected by the subsurface might be used for identifying petroleum-bearing structures.

Over the next two years, the Sound Section of the Bureau of Standards prepared for tests and on April 12, 1919, obtained the first seismic reflections from strata beneath a Maryland rock quarry. Encouraged by these results, Karcher and others from the University of Oklahoma formed the Geological Engineering Company in April 1920. A year later, in June 1921, Karcher and his Geological Engineering colleagues conducted a reflection experiment at Belle Isle in Oklahoma City and obtained a clear reflection from the interface between two known strata, the Sylvan Shale and the Viola Limestone, a hard limestone cap rock under which producers later discovered several major oil reservoirs.[14]

But Geologic Engineering soon folded due to a collapse in the price of oil, and engineers still had work to do to resolve problems detecting the weak reflected signals. However, a number of innovations in the mid-1920s improved the reflection method considerably. Mechanical seismographs were superseded by a variety of electric geophones, the most popular being the moving-coil electrodynamic type. The development of vacuum-tube amplifiers made it possible to strengthen the reflected signals, and electronic filters were developed to eliminate extraneous vibrations. It also became possible to record on the same strip of photographic paper the vibrations from a number of seismographs set up at different points on the surface.[15]

In 1925, oil prices had rebounded, and DeGolyer, still in charge of Amerada, decided to create a research arm called the Geophysical Research Corporation (GRC) and asked Karcher to lead it. Karcher's first move was to acquire Fessenden's patent and services as a consultant, but his main task was to improve the reliability of the reflection seismic method. In 1930, GRC used the reflection technique at Seminole, Oklahoma, to discover three reservoirs, securing its place as the most efficient and practical method for hydrocarbon exploration. Although the older refraction method remained well suited for finding salt domes, the reflection seismic technique generated more precise observations over complex geologic structures and could pinpoint the location—or at least provide clues—of a potential oil pool.[16]

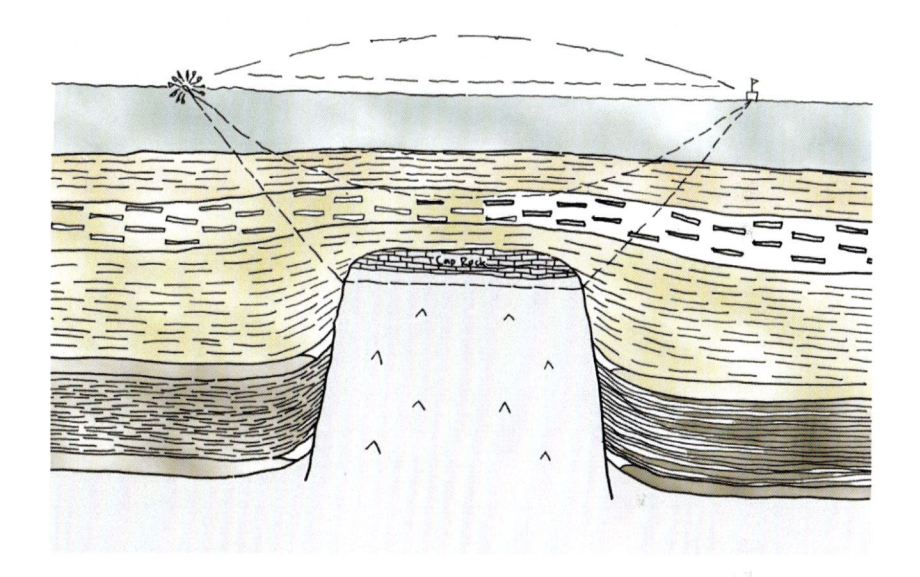

Refraction seismics. The lower of the four dashed lines than run from the point of explosion at the surface, left, to the receiver, right, shows the path of a refracted seismic wave through a salt-dome structure.

By the 1930s, four companies were offering reflection seismology: Geophysical Service Incorporated (GSI), formed in 1930 by spinning off GRC from Amerada, run by DeGolyer and Karcher; Compagnie Générale de Géophysique (CGG), set up in 1931 by Conrad and Marcel Schlumberger jointly with the French government and another French geophysical company, Société Géophysique de Recherches Minières; Seismograph Service Corporation, also set up in 1931, by electrical engineer William G. Green who had previously worked in both GRC and GSI with Karcher; and finally Western Geophysical, created in 1933 by Henry Salvatori who worked for Karcher in GSI but left to establish his own company. All these founding fathers would acquire great wealth over the next few decades, some became philanthropists, and some went into politics such as Salvatori.[17]

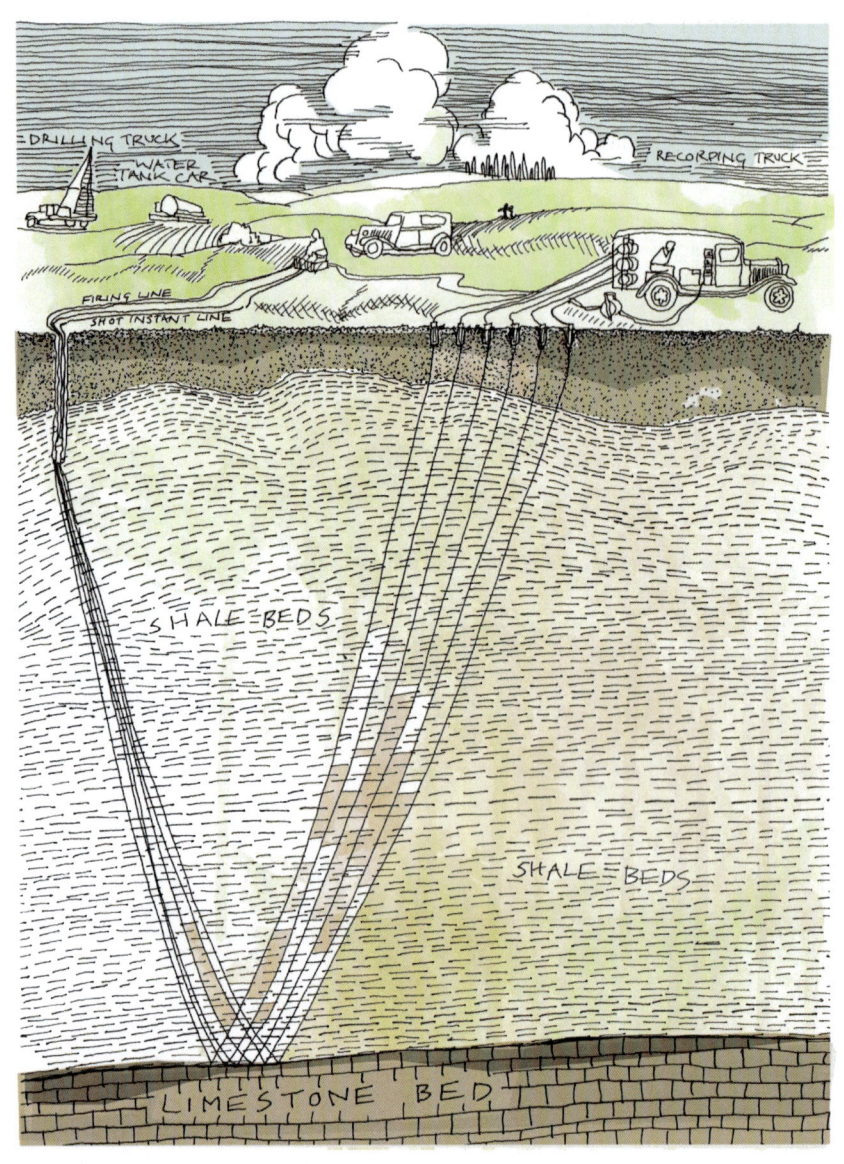

Reflection seismics. From a firing point, left, detonations create seismic waves that travel down through the subsurface until they hit a boundary, here between shale and limestone beds, from where they reflect back up again and are detected by geophones at the surface.

Reflection seismology convinced the industry that geophysics had arrived, and on March 11, 1930, thirty geophysicists met at the University Club in Houston to cement this coming of age and to found what would be later become the Society of Exploration Geophysicists (SEG). At first, their technical meetings were held alongside those of the American Association of Petroleum Geologists (AAPG) that DeGolyer and friends from Tulsa, Oklahoma, had launched in 1917. Both professional societies would grow from a handful of members at their founding to many thousands today.[18]

Chapter 5: Discovering the Reservoir

THE EARLY pioneers didn't devote much thought to the nature and behavior of oil reservoirs. Their idea of developing a field was to drill as many wells as possible and then produce them at maximum capacity, placing the wells on pump when the natural flow dried up. However, as more oil was discovered and production took off, a few curious souls started to wonder how the oil seeped through the rock to the well, indeed how large was the reservoir so recently discovered. The age of reservoir engineering was dawning.

Darcy's law of permeability

Fortuitously, the key scientific foundations had been established by French engineer Henry Darcy in the 1850s. Darcy spent his entire career at the Corps des Ponts et Chaussées (Office of Bridges and Roads), a government agency in his hometown Dijon, France. He suffered ill health and died prematurely at the age of 54 from pneumonia, but his ailments did not blunt his sense of humor. Corresponding to a young colleague while preparing a publication, he wrote, "It was written to me that you had toasted recently to my health. I fear that this obligation will be imposed on you for a long time, but the wine of Burgundy is good and I feel a little less sorry for this devotion than the one you will need to read these proofs."[1]

Darcy's decisive innovation was published in an appendix to his work "Les Fontaines Publiques de la Ville de Dijon" in 1856. In this work, which summarized the result of experiments he had made by flowing water through a sand-filled cylindrical tube, he postulated that flow through the sand was linearly proportional to the pressure drop across the sand. Darcy called the constant of proportionality "permeability."[2]

The oil field began to wake up to Darcy's ideas when the US Bureau of Mines was established in 1910, and four years later added a Petroleum Division to study and understand the physical processes involved in oil production. The key breakthrough, however, would come from the USGS, where Perley Gilman Nutting, who previously worked for the US Bureau of Standards and Eastman Kodak and had joined the Survey in 1924, started looking afresh at Darcy's permeability definition. Crucially, Nutting worked out how to introduce viscosity, a measure of the resistance of a fluid to being moved. The viscosity idea had come from French physicist Jean Léonard Marie Poiseuille in 1842, and the unit of viscosity "poise" was named after him. In his 1930 AAPG paper "Physical Analysis of Oil Sands," Nutting introduced a viscosity term in Darcy's equation and in the process suggested a simple method of measuring the permeability of oil sands using small rock samples.[3]

Nutting's work paved the way for the first standard method for measuring permeability. In their seminal 1933 paper "The Measurement of the Permeability of Porous Media for Homogeneous Fluids," Ralph Wyckoff, Morris Muskat, Holbrook Botset and Donald Reed of Gulf Research and Development Company in Pittsburgh, Pennsylvania, described a detailed technique for the measurement of the permeability of porous media. They named the permeability measurement unit a "darcy."[4]

Morris Muskat (1906-89), a pioneer who helped lay the theoretical foundations of reservoir engineering.

Mud logging

As the industry started coming to grips with picturing what was going on in the reservoir, their attention focused on the rock being drilled through. The first clues were gained by inspecting the cuttings extracted from the well during drilling. With the advent of rotary drilling, the cuttings were collected from the circulating mud exiting the well and then analyzed by a geologist on the wellsite, a practice called mud logging. Early mud logging focused on looking for telltale signs of oil and gas, including watching for oil sheen in the mud returns and looking for gas coming out of the mud as it depressured. But attention was also paid to the rock type. At first, mud loggers relied on the naked eye or, at best, a hand lens. The cuttings were identified by main lithological type—sandstone, limestone and shale, for example—and described in terms of color and texture.[5]

The next step for mud loggers was harnessing paleontology. The importance of examining fossils and organic remains had been recognized as early as 1790 by Polish geologists carrying out surface investigations of petroliferous areas in the Carpathian mountains. Throughout the 19th century, Polish geology continued to be at the forefront of paleontological research, culminating in the efforts of Józef Grzybowski, a professor of geology at the University of Krakow. In the 1890s, Grzybowski was investigating mud returns from oil wells in the Carpathian mountains and reported that one group of marine microfossils called foraminifera was especially suited to providing a record of the sediments age.[6]

Two companies, Humble Oil and Rio Bravo Oil Company, a subsidiary of the Southern Pacific Railroad, were quick to embrace paleontology; Humble Oil's first paleontologist was Alva Ellisor, based at its Fort Worth headquarters. Humble's embrace of paleontology was not without its detractors. The traditionalists believed that the chronologic range of microfossils was too wide to make them useful in age determinations, but a new generation of micropaleontologists, many women among them, fought to overturn this belief and, for that matter, the concept that the oil field was exclusively a man's world.

Alva Ellisor (1892-1964), a micropaleontologist who worked for Humble Oil in Fort Worth during the 1920s.

In December 1921, a 26-year-old Esther Applin, who had joined Rio Bravo Oil Company that year as a micropaleontologist, presented ongoing studies at a Paleontological Society meeting in Amherst, Massachusetts, suggesting that microfossils could be used to date oil-bearing Gulf Coast formations. Professor J.J. Galloway of The University of Texas at Austin stood up and objected: "Gentlemen, here is this chit of a girl, right out of college, telling us that we can use foraminifera to determine the age of a formation." Four years later Applin was vindicated when she coauthored a paper with Alva Ellisor and Hedwig Kniker of The University of Texas's Bureau of Economic Geology, demonstrating conclusively that the chronological sequence of oil-bearing zones in the Gulf Coast could be established using microfossils. Applin, Ellisor and Kniker's work helped advance the scientific study of micropaleontology and provided a badly needed index of fossils for oil drilling operations.[7]

Taking core samples

Paleontologists working in the 1920s not only had drilling cuttings at their disposal but also drilling cores, large-size samples of the reservoir rock obtained using a special drilling tool. The first coring tool was invented by French engineer and tunnel designer Rodolphe Leschot, around 1863. He conceived the idea of a hollow tubular tool set with diamonds cutters at one end and with circulating fluid passing through. Leschot used it for drilling faster blast holes while tunneling through Mount Cenis on the France-Italy border for a railway. In the 1870s, this simple device was replaced by a double-barreled tool with inner and outer barrels separated by ball bearings. This allowed the inner barrel to remain stationary to receive a core while the outer barrel was rotated by the drillstring to cut the core. These double-barreled tools were not equipped with diamond cutters until Milan Bullock from Chicago added them in 1892.

The Bullock coring tool worked best in hard formations, so in soft-rock regions, such as the US Gulf Coast and California where many companies were active, the search for an effective coring tool was still on. In 1919, Shell trialed a double-barreled core drill developed expressly for loosely consolidated formations; this was an invention of Jan Koster of the Holland Geological Survey. But the core drill wasn't strong enough and soft formations easily balled up inside it. However, one of Shell's Californian geologists, a mining engineer called John "Brick" Elliott—nicknamed for his shock of red hair—did not give up.

He quit Shell in 1920, set up his own consulting company in Los Angeles and within a relatively short period developed an improved double-barreled coring tool that featured reaming teeth that prevented the tool from balling up. Elliott's core drill was capable of recovering core two inches in diameter and several feet long. At first, oil company executives were skeptical, but in August 1921 Elliott successfully recovered cores from a well at the Huntington Beach oil field. His new coring tool became the basis for all rotary core barrels used today.[8]

Learning to extract information from cores paralleled these advances. At first, inspection by the field geologist was qualitative. But soon, the physical properties of the cores became of interest. As early as 1880, pioneering petroleum geologist and engineer John Franklin Carll took cores from the Pennsylvania Venango Sands—the formation where Colonel Drake had struck oil at a very shallow depth in 1859—and began visually estimating their porosity. In 1885, Frederick Haynes Newell, a mining engineering student at the Massachusetts Institute of Technology (MIT), forced water, kerosene and crude oil through small discs cut from cores, in an effort to understand the fluid-flow properties of the rock, but failed to link this to permeability.[9]

By the early 1920s, determining the amount of pore space in a rock, or porosity, had turned quantitative in the hands of Arles Melcher, a geologist at the USGS's Physical Laboratory. In 1924, Melcher measured porosity from cores across a complete section of the Bradford Sand, near Custer City, Pennsylvania. And in 1925, Melcher measured the flowrate of crude oil as a function of pressure in cores from a variety of Oklahoma oil sands. A year later, Melcher's work was further refined by Charles Fettke of the Pennsylvania Geological Survey, who developed many of the classic laboratory methods for core analysis. The same year, William Russell of the South Dakota Geological Survey introduced a new method for the determination of porosity that became standard. The method compares the weight of a core sample filled with air with the weight when filled with a saturating liquid.[10]

Commercial core laboratories were now starting to emerge. The first was founded in 1928 in Bradford, Pennsylvania, by petroleum engineer-geologist Paul Torrey. Torrey's lab determined the porosity of the reservoir cores, and also oil and water saturation. However, these early saturation measurements were palpably unreliable because they failed to take into account changes caused by the lifting of the core out of the well.[11]

During that same year core analysis spread to academic research. In 1928, the Pennsylvania State College, today's Pennsylvania State University, established training programs in oil production with help from local oil producers. George Fancher, James Lewis and Kenneth Barnes ran the first courses and embarked on a unique three-year core collection and analysis project. Fancher, Lewis and Barnes collected and analyzed more than 400 sandstone cores, storing them in a college building that became known as the "Pennsylvania Core Depository." This project was the first large-scale, systematic determination of the properties of porous media from cores.[12]

Geologists examining core samples at a drilling site in Montana in the early 1980s. The geologist to the right is using a small hand lens to examine the testure of the sample. Hands-on physical analysis of cores remains a crucial element of formation evaluation.

In 1933, the research trio published their results in the now classic article "Some Physical Characteristics of Oil Sands" in the college's *Mineral Industries Experiment Station* bulletin, providing extensive references to the various core analysis methods and techniques evolved up to that time. With relation to permeability, they demonstrated a difference between permeability to air and to water, suggesting that discrepancies were largely due to hydration of clays present in the cores. And they presented many original porosity determinations, including the determination of so-called effective porosity, which is the volume of pore space that is interconnected and contributes to flow, as opposed to total porosity that also includes deadends and isolated noncontributing porosity. Core analysis now had a firm scientific basis and was key for understanding the reservoir.[13]

The birth of well logging

But coring and the subsequent analysis of the cores remained a time-consuming and cumbersome business. What was needed was a shortcut, a quick alternative that could yield at least some of the results coring provided. In the late 1920s, just such a miracle occurred, and given its provenance in a picturesque wine region of Alsace, France, this miracle became known as "carottage électrique", or electrical coring.

Near the small village of Pechelbronn in Alsace, oil had been excavated by hand since the 1740s. By the 1920s, the Pechelbronn oil fields counted more than 3,000 wells, and the local oil company was drilling more every day. Following the slump after World War I, demand was on the rise again, and in June 1926, the company opened a new refinery that could handle 80,000 metric tons per year. The question facing the company directors was whether all these new wells would produce enough to feed the refinery. In early 1927, the Pechelbronn company, which had already contracted Schlumberger for surface electrical surveys, discussed with Conrad Schlumberger the idea of making resistivity measurements in the borehole to see if this could help the company geologists obtain a better understanding of the oil-bearing formations.[14]

The opportunity came at the right time for Conrad and his brother Marcel. They had just lost a lucrative contract with Roxana Petroleum, a Shell subsidiary, after delivering disappointing results along the US Gulf Coast using surface electrical prospecting. At the same time, they were facing competition from other geophysical methods such as magnetics, gravity and especially seismics.[15]

In fact, Marcel Schlumberger had already tried resistivity measurements in the borehole in 1921. In March that year, he made resistivity measurements over a few feet at the bottom of a 760-meter hole in Molières-sur-Cèze in southern France. The results were inconclusive, but the feasibility of a downhole resistivity measurement had been proved. The geophysical community was skeptical. The

German geophysicist Richard Ambronn had been doing similar experiments in Germany and maintained that below a certain depth, all geologic formations were so compact as to become infinitely resistive. Nothing of the kind had been observed with Schlumberger's surface measurements, but then they did not penetrate very deeply.[16]

On September 5, 1927, Henri-Georges Doll, Conrad's son-in-law and two colleagues, Roger Jost and Charles Scheibli, proceeded to a Pechelbronn well called Diefenbach 2905 and conducted the first electrical logging operation in an oil well. The well was 500 meters deep, and they logged an interval of 140 meters, starting from a depth of 279 meters. They rigged up a hand-operated winch that lowered into the hole three insulated wires—cables of the type used for lighting fixtures—tied together here and there by friction tape. The longest of the wires was used to inject current into the well and formation, with a return at the surface. The other two wires, shorter and of slightly different lengths, measured the resulting potential field and provided the resistivity measurement. Measurements were made point by point at intervals of one meter; the entire operation took five hours. The result was a resistivity log that distinguished between the many layers of sand and shale pierced by the borehole. Doll, Jost and Scheibli repaired to the local tavern for a celebratory dinner.[17]

Henri-Georges Doll (1902-91), Conrad Schlumberger's son-in-law, one of the pillars of the company's early years. Here he is seen in uniform in paris in 1923 at the École Polytechnique, the elite institute of technology and science.

Continuing through 1928, resistivity logging was conducted throughout the Pechelbronn oil field, and the resulting correlations of resistivity from one well to the next revolutionized the stratigraphic understanding of the field. Soon the Pechelbronn Oil Company was raising the capacity of its new refinery to 100,000 metric tons per year. In 1929, the new logging technique went global. Schlumberger logging crews were engaged by Shell for their explorations in Venezuela, the US and the Dutch East Indies, and by the Soviet Union for the oilfields of Grozny, Chechnya and Baku, Azerbaijan.

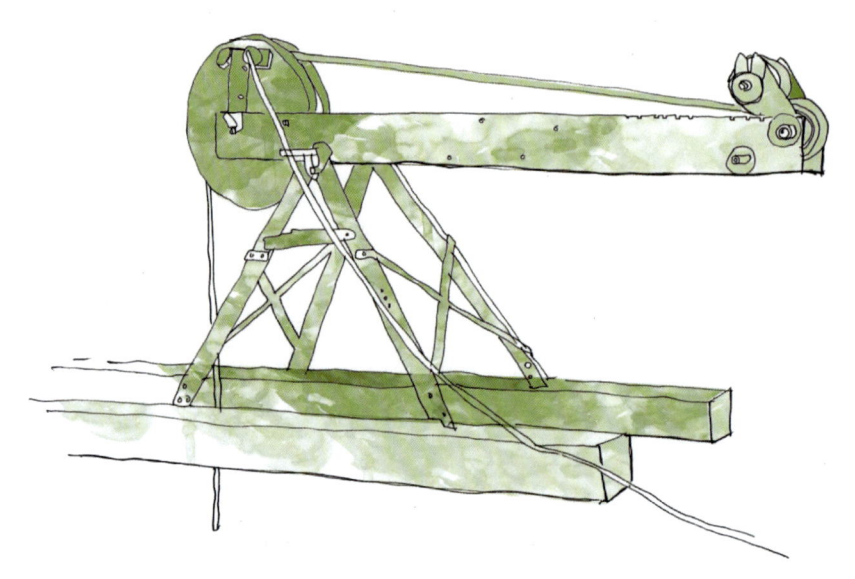

The cable pulley used in the first oil well logging operation in 1927, near Pechelbronn in Alsace, France. The pulley was placed on top of the hole and conveyed the cable up and down the well. The cable connected to a winch powered by a truck engine.

Meanwhile, Conrad was playing with new ideas for his well logging. Realizing that oil was infinitely resistive to electricity, he postulated that if a zone was oil-bearing and reasonably thick, logging with a longer spacing between the wires measuring potential would see deeper into the formation and record a higher resistivity; the phenomenon couldn't be observed in Pechelbronn because the oil zones were too thin. So he asked his handful of field engineers to try the idea and report back. Marcel Jabiol was the solitary Schlumberger engineer in northern Sumatra at the time and within a month had tried Conrad's proposal and observed exactly what had been predicted. His telegram in June 1930 with the good news triggered celebrations in the small Paris office. Electrical logging could now locate oil from the borehole.[18]

The well where the first oil well logging operation took place on September 5, 1927. The derrick was protected by a wooden shell to give the workers protection from the weather. The winch for the logging cable can be seen between the truck and the two workers.

The company's fortunes, like everyone else's, were rocked by the world depression that had hit in November 1929. But by the end of 1932, drilling revived and with it the acceptance of Schlumberger's electrical logging. Also helping was the introduction of a new logging measurement called the spontaneous potential (SP). Natural electrical potentials in the subsurface had been discovered by the British geologist, natural philosopher and inventor Robert Were Fox in 1830 in ore deposits in Cornwall, England. In the borehole environment, naturally occurring potentials are caused by electrochemical interactions between the borehole fluid and adjacent sand and shale formations.[19]

Conrad Schlumberger had received a French patent on the SP in 1929, claiming it could be used to locate permeable strata, but found no practical application. A year later and quite by chance, Doll observed natural potentials while logging in the Seminole oil field in Oklahoma. With the battery disconnected, he noticed the needle of the potentiometer vibrating back and forth as the electrodes were being lowered into the well. Experiments on the SP phenomenon followed at Pechelbronn, and by 1930 it was concluded that the SP could differentiate permeable beds, such as sand and limestone, from impermeable formations, such as shale. The combination of SP and resistivity curves turned out to be of much greater value than the resistivity log alone in locating and scoping out production possibilities.[20]

SP measurements proved to have other benefits as well. In 1935, Doll had the idea of adding SP electrodes to the arms of a three-arm caliper tool and attaching it to the so-called teleinclinometer tool that Conrad and Marcel Schlumberger had developed in 1932 for measuring borehole deviation and direction. The three SP measurements distributed around the borehole combined with teleinclinometer data promised the first downhole measurement of the rock strata dip and direction. The combination was tested in Long Beach, California, and successfully commercialized in Louisiana in 1941.[21]

Sidewall sampling

Even though electrical well logging showed promise for evaluating downhole formations, the technique did not eclipse coring. Logging may have facilitated stratigraphic correlation and given early pioneers a way to distinguish between shale and porous rock and between hydrocarbon- and water-bearing rock, but it wasn't quite the same as having a piece of rock in front of you. Coring was the answer, but it was still expensive.

An alternative and cheaper solution to coring was sidewall coring, in which a device is lowered into the hole that can sample the formation sideways from the borehole at any desired depth. The earliest attempts in the mid-1920s used the drillstring to convey such a coring tool. At the desired depth, a boring device or knife arrangement would be pushed obliquely into the borehole wall to cut a sample out of the formation. In the early 1930s, the first sidewall-coring tools that could be lowered on a cable came on the market. An early version, made by the Sperry Sun Well Surveying Company, based out of Philadelphia, was a miniature drilling tool that could be projected laterally, penetrate a few inches into the formation to take a sample. However, these devices were mechanically complex and unreliable. Then in the early 1930s, Marcel Schlumberger had the idea of using an explosive charge to shoot a cup into the formation, the cup being attached with strong cables to a logging tool and retrieved by simply pulling on the tool.

Marcel Schlumberger (1884-1953), adjusting a prototype of his sidewall coring tool in the basement workshop of the company's headquarters in Paris in 1935.

After trying numerous configurations of cup, explosive charge and cable, Marcel came up with a workable solution that was tested in Pechelbronn in 1935 and patented in 1936. Most of the early experiments were conducted in a test well accessible within the basement of the fledgling Schlumberger headquarters in the seventh arrondissement of Paris. His tool could fire any number of cups, or bullets, into the formation at different depths and, after each firing, pull the sample back into the borehole. The samples were then pulled to the surface and presented for inspection and analysis. Marcel Schlumberger took his prototype sample-taker to the US Gulf Coast in April 1936 for trials and performed the first commercial job in southwest Texas in September 1936. Since then, sidewall coring has become standard in the oil field, and the original design has hardly been improved on.[22]

Chapter 6: First Steps Offshore

WHILE EXPLORATION matured on land and men's understanding of the reservoir took its first steps, the industry had been quietly moving into the sea. Geologists had no doubt that the same conditions that produced gushers on dry land would be equally prevalent offshore. It was simply a matter of figuring out how to transport the entire industry into the water.

California and Venezuela

The first well drilled offshore was at Summerland Beach, north of Los Angeles. There, in the late 1890s, members of a Spiritualist community led by Henry Lafayette Williams began drilling onshore to capture oil and gas seepages. They noticed that the nearer the well was to the ocean, the more it produced. In the tidal area, gas even bubbled to the surface, indicating that underground reservoirs stretched beyond the shoreline. Reaching them, however, seemed an insurmountable challenge.

Williams then had the idea of building a wooden staging dock, perpendicular to the beach, to drill from. Williams built the first such pier and placed a cable-tool drilling rig at the end of it in 1897, thus claiming the earliest offshore drilling and production platform. The power generators and other supporting equipment remained along the beachfront. His first three piers extended some 1,350 feet from the shoreline, with water depths reaching 35 feet. Williams's crew pounded their way 455 feet down to two oil sands.[1]

Henry Lafayette Williams (1841-1899), A California spiritualist who pioneed offshore drilling in 1897.

Williams's days were numbered, however. In 1898, during an inspection of his four active onshore wells, Williams stumbled and fell into an abandoned well. Weakened by his injuries, he fell ill with pneumonia, was taken to San Francisco for medical treatment and passed away on January 13, 1899, at the age of 58. His family grieved over his death but others didn't. He was venerated for his oil drilling, but many in his community not persuaded by his spiritualist leanings were glad to see the back of him.[2]

Meanwhile, other Summerland entrepreneurs started to build piers, jutting seaward beyond the crashing surf. By 1900, there were 12 piers protruding into the ocean, and 22 operating companies drilling in search of oil. By 1903, 198 wells were in operation. Summerland wells continued to produce until the late 1930s.[3]

These offshore wells had a spin-off effect for drilling in shallow inland waters. Most important was Caddo Lake in Louisiana. The discovery of the Caddo field goes back to 1870 when a water well near the lake's south shore accidentally hit a gas reservoir. In 1902, another water well hit natural gas in significant quantities. Soon bona-fide oil prospectors came to the region hoping for another Spindletop. Among the first was the Gulf Refining Company of Louisiana, whose drilling superintendent, Henry A. Merlat, began constructing a wooden drilling platform in the lake. In the spring of 1911, Gulf started drilling operations on its first oil well in inland waters.[4]

The next milestone was the development of movable offshore drilling units in the mid-1920s. The Texas Company, founded just after the Spindletop discovery and later to become Texaco, first proposed augmenting wooden-piled drilling platforms with mobile steel barges that could hold some of the drilling machinery. The reason was simple: wooden drilling platforms were expensive and easily damaged, and if a well was dry, crews had to remove the entire rig from the platform, return it to dry land or transport it piecemeal to drill the next well. Mounting the rig on a barge reduced mobilization costs as well as overall platform deck size. When the Texas Company tried to file a patent, the company was shocked to learn that a certain Captain Louis Giliasso, US citizen and merchant mariner turned driller working in Lake Maracaibo in Venezuela, had already claimed the idea.

The Lake Maracaibo fields were discovered in 1917 by Shell with moderately successful wells drilled on the lakeshore. Then, in 1923,

Shell drilled a 100,000 barrel-a-day gusher and operators everywhere scrambled to get the best shoreline blocks. A drilling frenzy ensued. Shell continued working the shoreline, while Gulf and Lago Petroleum concentrated on the lake itself. Between 1924 and 1928, Lago Petroleum began drilling wells in the lake using wood piles and later concrete piles to support the rig.[5]

During this first Venezuelan oil boom, Giliasso used his shipping know-how to design a submersible barge that could rapidly move a drilling unit between locations, and he was awarded a patent for the idea in 1928. Giliasso tried to sell the barge idea to several Lake Maracaibo operators, but all doubted that a barge resting on a mud bottom could ever be refloated. The suction, they said, would hold the barge fast to the bottom. Giliasso argued his case but could never persuade anyone to try it.

The Texas Company, on the other hand, was convinced it would work but needed Giliasso's patent. Following an exhaustive search, the company located the former mariner in Panama running a bar. The Texas Company bought the rights to the patent and in 1934 completed the construction of the first submersible steel drilling barge, naming it *Giliasso* in the captain's honor. It was towed to a shallow, open-water section of Lake Pelto on the Louisiana coast, about 40 miles southwest of Houma, and the Texas Company started drilling. The *Giliasso* was a triumph of innovation and efficiency, cutting the move from one well to the next to just two days. Other oil companies were soon copying the design.[6]

Out of sight of land

The *Giliasso*, however, was just the beginning. Growing energy demand, particularly following World War II, moved oil producers to drill offshore and do so more efficiently. Most urgent was a complete offshore drilling rig that was maneuverable in the open ocean. Mobile barges could transport equipment from one platform to another, but the drilling platforms themselves were permanent, and installing permanent piled platforms for every exploration well was getting far too expensive and time-consuming.[7]

In 1946, Oklahoma-based Kerr-McGee Oil Industries, named for its two principals Robert Kerr and Dean McGee, developed a mobile drilling platform and drilled an exploration well 12 miles off the coast of

Louisiana. It was the first oil well ever drilled out of sight of land, albeit in water only 18 feet deep. The Kerr-McGee drilling platform consisted of a small platform with derrick and drawworks, and a separate floating tender to store the drilling equipment. Both could be moved separately and then rejoined to drill another well.[8]

Then along came British marine engineer John T. Hayward, chief engineer at Barnsdall Oil Company in Tulsa, Oklahoma. In 1949, Hayward designed and built the first integral, movable offshore drilling rig, the submersible *Breton Rig 20,* that could work in up to 20-feet water depth and be moved from location to location with minimal delay and setup cost. *The Breton Rig 20* drilled its first well in late 1949 to nearly 11,000 feet and continued drilling in the Gulf of Mexico until it was retired in 1968.

But the industry did not rush to embrace the Hayward platform. One enthusiast, however, was Alden J. "Doc" Laborde, formerly a Kerr-McGee employee. Laborde tried to interest Kerr-McGee in a new Hayward-type barge capable of drilling in waters up to 40 feet and withstanding winds of 70 miles per hour but met with rejection. Undeterred, Laborde eventually persuaded a group of investors led by Charles Murphy Jr. of Murphy Oil to form the Ocean Drilling and Exploration Company (ODECO), based in New Orleans. In late 1953, ODECO built its first offshore drilling rig—a floating, submersible drilling vessel—at a cost of US$ 2.5 million, named it *Mr. Charlie* after Charles Murphy Sr., and started drilling for Shell off the coast of Louisiana.[9]

Meanwhile, the concept of the jackup began to emerge. Curiously, the idea can be traced back 100 years to an 1869 *Scientific American* article describing a certain Samuel Lewis's invention of a "submarine drilling machine," intended for drilling shot holes in the treacherous rocks of Hell Gate in New York Harbor to prepare for rock-removal by explosions, "It consists of a steamboat, and a device whereby the vessel, when the drills are at work, may be raised entirely above the waves, at which time its weight is supported by six adjustable pillars."[10]

The notion of a self-elevating platform for oil drilling was first realized in 1954 by Colonel Leon B. Delong of the US Army Corps of Engineers, who had designed temporary marine platforms during World War II. In 1952, he formed the Delong Corporation to produce the first jackup barges and drilling rigs for the offshore industry. Jackups elevate their platforms out of the water by extending, or jacking downwards, long cylindrical or triangular legs to the seafloor to create a temporary platform. Raising the legs then allows the jackup to be quickly floated to the next drilling location.[11]

A modern jackup offshore drilling rig. The derrick is on the left with the marine riser leading to the sea-floor and the wellbore.

Seismic goes offshore

As drilling offshore matured, geologists tasked with picking well locations faced a dilemma. On land they could walk new prospects and study outcrops in minute detail. Offshore they were stymied, that is until seismic exploration learned to function in the marine environment.

The early offshore seismic surveys were made in shallow water using makeshift adaptations of land equipment. The first trial took place in 1938 when a Shell crew commandeered three 35-foot fishing boats, one for shooting and two for recording, and did a test survey four miles off the Louisiana coast in 65 feet of water. With no accurate means of navigation, the crew had to rely on spotters on the beach who triangulated the boats's positions and communicated via two-way radio. A simple stick of dynamite lowered into the water provided a sound source, and the reflected signal was picked up by water-tight geophones attached to a steel plate heavy enough to remain seated on the seafloor. Superior Oil and Mobil used a similar setup in 1944 exploring for salt domes offshore Louisiana, and in March 1946, Shell dispatched its first, fully fledged offshore seismic party, Party 88, from the docks at Grand Isle, Louisiana.

It was Kerr-McGee that made the first offshore seismic survey out of sight of land, in 1946. This time the geophones were mounted on a cable, also designed to sit on the ocean floor. With the recording boat anchored, the geophone cable was lowered overboard. Meanwhile, the shooting boats fanned out and circled the recording boat at a radius of up to six miles, detonating dynamite charges every mile. If the water was shallow enough, they dropped the dynamite into pipes drilled into the seafloor. In deeper water, crews bundled up 50 pounds of dynamite, inserted a cap, and tossed the charge overboard.[12]

A recurrent problem for Kerr-McGee was ocean currents or snagged debris breaking the cable. In 1947, German scientist Eugen Merten, director of the new Shell Geophysical Research Laboratory in Houston, solved the cable problem by floating it rather than weighting it to the ocean floor. But the real breakthrough came from Roy Paslay, who had been engaged in anti-torpedo research for the US Navy during World War II and afterwards joined the National Geophysical Company founded by the brother of Henry Salvatori. Paslay and colleagues patented an oil-filled float or streamer containing hydrophones—devices that are pressure sensitive as opposed to geophones that are motion sensitive—to pick up the incoming seismic energy, and in-built vacuum-tube amplifiers to maintain signal strength along the streamer. Their initial streamer consisted of eight 300-foot sections, each section containing three hydrophones.[13]

A seismic vessel setting off a dynamite charge in the Gulf of Mexico in the 1950s.

Meanwhile fishing boats were replaced by modified Word War II mine sweepers and other war-surplus boats. The first single-boat acquisition system—in other words shooting and recording equipment on one vessel—was launched by Marine Instruments and deployed off the coast of Galveston, Texas, in 1947. Because the idea of a marine seismic service was so new, their boat was not always in demand. So in quiet periods, Marine Instruments became the first company to gather speculative or "spec" data, shooting seismic surveys in interesting areas and hoping to sell the data to prospective clients sometime in the future.[14]

Chapter 7: The Service Industry Takes Hold

AS OIL and gas exploitation took shape, an assortment of tinkerers, entrepreneurs and self-taught engineers began inventing and supplying the industry with badly needed technologies. In many cases, the fledgling enterprises created by the early pioneers transformed over the decades into huge corporations whose reach extends into every oil and gas field on the planet. Dresser, Hughes Tool, Schlumberger and GSI are some early examples. But many more were to follow, and collectively they formed the beginnings of the oilfield service sector.

Invention of the casing shoe

One such player was Reuben "Carl" Baker of Coalinga from San Joaquin Valley, California. Born in 1867, Carl Baker began his career as a cable-tool driller and in 1898 started his own drilling business. Baker was a capable machinist and resourceful designer. He was especially interested in services associated with casing. Since Drake's well, installing casing had become standard, but Drake's wooden shoring was now replaced by wrought-iron or cast-iron pipe. Carl Baker's breakthrough invention was a new type of casing shoe, patented on July 16, 1907.

An early casing shoe advertisement of Baker Oil Tools.

A casing shoe is a short assembly, typically a heavy steel collar with a cement interior that is screwed to the bottom of a casing string. Casing shoes had long been used on the bottom of casing strings to strengthen the end of the pipe and drive the casing string through tight hole. Carl Baker had been making casing shoes since 1890. Whereas early shoes were thick and heavy, his 1907 casing shoe had a serrated beveled edge that could enlarge the borehole wall as it was lowered in the hole.

To begin with, Baker's new casing shoe was built for cable-tool rigs since rotary drilling hadn't yet arrived from Texas. When the rotary rig was introduced into the San Joaquin Valley around 1908, Baker quickly adapted his casing shoe for rotary use. Then in 1913, he created the Baker Casing Shoe Company, which launched the development of a long line of casing accessories and would become Baker Oil Tools in 1928, a future pillar of the service sector. The design of the Baker casing shoe remained virtually unchanged for 40 years.[1]

Carl Baker lived until September 29, 1957. Speaking at the 50th anniversary celebration of the Baker casing shoe patent just two months before he passed away, he remarked, "I conceive practically all of my inventions while I lie in bed at night. I do not get up to make notes or sketches at the time. Instead, I work out all the details of the inventions mentally and it may be a day or even several weeks before I make a sketch of the device."[2]

Cementing

A serious problem right from the beginning was water seeping from behind casing and entering a well via the casing shoe. The casing shoe had some ability to shut off water, but drillers were nevertheless forced to be creative. A common solution was to wrap various seeds in heavy canvas or leather around the bottom joint of the casing and wait for them to swell, and in the best outcome they provided a barrier.[3]

The idea of using cement to create a seal between casing and formation was first tried in Russia by a certain Romanovsky in 1859 in a water well. In the US, the idea can be attributed to John R. Hill with his 1871 patent "Improved Mode of Closing the Water Courses Encountered in Drilling Oil Wells." The patent describes putting cement into a borehole, setting a casing, then waiting for the cement to set, and eventually, as the patent explains, "The drill cuts out the cement from the bore of the well but leaves the water courses closed with said cement." Experimenting in this fashion, Wallace Hardison and Lyman Stewart of Hardison & Stewart Oil Company were the first drillers to put cement in a well in Pico, California, in 1883. The quality of the cement obviously wasn't that good because water soon began entering the well.[4]

By the late 19th century, a new type of cement was coming on the market. Portland cement was invented and named by John Aspdin, a bricklayer and inventor from Leeds, England. For some years he had been experimenting with various cement formulations, and in 1824 his efforts were crowned by a British patent entitled "An Improvement in the Mode of Producing an Artificial Stone," in which he coined the term portland cement. He named it thus because the produced solid resembled a limestone quarried on the Isle of Portland on the south coast of England. Unlike earlier cements, portland cement was made by burning a blend of limestone and clay, and crucially, it could harden in an underwater environment.[5]

By 1890, Hardison and Stewart cofounded an oil company called the Union Oil Company of California, later to be renamed Unocal, and in 1903 decided to try the new portland cement. Frank F. Hill, a director of production for Union Oil, was the first to use the new cement. Frustrated with leakage from unconsolidated sands in a well in the Lompoc region of California, Hill dumped 20 sacks of portland cement mixed with water into the hole. He then raised the casing 30 feet, capped the top and lowered the string back to the bottom. Air pressure forced most of the cement up the outside of the casing into the annulus. Hill still had to drill out the cement inside the casing, but the ruse worked. Later, he tried pumping cement down some tubing with a packer near the bottom. That eliminated most of the redrilling of cement set inside the casing, and thus began the era of modern cement jobs.[6]

Nevertheless, cementing was still seen as a costly procedure. In 1910, Almond A. Perkins, who owned The Perkins Oil Well Cementing Company, made the key breakthrough. In the Perkins method, portland cement was mixed with water to form a slurry. A plug was then inserted into the casing and pushed downhole in front of the slurry. Behind the slurry came another plug, this time pushed down by water. The first plug had the job of expelling the mud up the annulus between the casing and the formation, while the second did exactly the same with the cement. The first plugs used by Perkins were cast-iron with belting discs, with the addition of a leather cup on top of the second plug. Perkins's two-plug cementing method sped up operations no end, eliminating the need for redrilling cement that had set in the borehole. The Perkins's cementing business continued to thrive throughout the 1910s as its cementing technique was adopted throughout the US, and Perkins was therefore always in need of recruits. One young man he hired in 1918 was Erle Palmer Halliburton, who joined as a truck driver.[7]

Erle Halliburton was a diminutive man, but his energy and self-confidence made him seem larger than life. Erle was a quick student and a hard worker, and he was soon promoted to cementer. Halliburton had plenty of ideas of his own, and these brought him in constant conflict with his boss. Perkins grew so irritated at Erle's interference that in 1919 he fired him. Years later, Erle would say, "The two best things that ever happened to me were being hired, and then fired, by the Perkins Oil Well Cementing Company."[8]

Cementing legend Erle Halliburton (1892-1957) at the age of 28.

Freed from Perkins, Erle Halliburton immediately established his own cementing company. After borrowing a wagon, a team of mules and a pump, he built a wooden mixing box and started cementing oil wells around Duncan, Oklahoma. His company was rather inconveniently called the New Method Oil Well Cementing Company, later to be renamed the Halliburton Oil Well Cementing Company in 1924.

For Erle Halliburton, increasing the efficiency of the well-cementing process was paramount. Mixing cement with water could be done only in small batches, with workers stirring each batch with hoes and shovels. Mixing enough cement and pouring it down the pipe before the cement began to harden was as difficult as it was critical. To speed up the mixing and pouring process, Erle invented what he called the Jet Mixer. Using this device, workers had only to empty bags of cement into a large tub. The Jet Mixer would automatically add water and stir. From this device, the cement was pumped directly into the casing.

The Jet Mixer did the job so well that it created another problem. Cement was available only in sacks each weighing 94 pounds, the limit of what a strong man could handle, but cementing could consume a thousand sacks in only a few minutes. No man, regardless of his ability, could open sacks fast enough to keep up. Erle Halliburton solved the problem by inventing the Sack Cutter, which quickly and conveniently opened the sacks and dumped the cement into the mixer.[9]

In January 1930, Halliburton established a chemical laboratory at Duncan, Oklahoma, and appointed Count Hayden Roberts as head of research. Roberts started with nine researchers: five chemists, a physicist and three engineers. It was a modest beginning for what would become the industry's premier laboratory for cementing research and development. The laboratory was used primarily to test the properties of various cement mixes, and they had plenty to work on. Throughout the 1930s, wells were being drilled deeper and into hotter zones, and chemical additives had to be developed so the slurries pumped into the wells could flow and set in increasingly harsh conditions.

With cement still delivered in sacks, achieving a precise mix of cement and additives proved almost impossible. The solution introduced by Halliburton was to store the cement and additives at central plants and distribute in bulk by truck, eliminating the tedious handling of individual sacks and sack cutting. More important, bulk storage also offered the advantage of providing moisture-proof storage, and at the plant the cement and additives could be measured and proportioned for individual jobs. Halliburton opened its first bulk cementing plant in 1940 in Salem, Illinois.[10]

Another key development of the 1930s was offshore cementing. In 1938, Halliburton floated a barge from Louisiana into the Gulf of Mexico to a rig in the Creole field, and performed its first offshore cementing job. Though it was a new procedure, the crew was in familiar territory. For more than a decade, they had been cementing wells in the swamps and marshes of Louisiana. The crew drove their trucks onto shallow-draft barges, loaded the barges with bags of cement, and then floated the barges to the site of the well.[11]

Controlling the well

"It puffed and it blowed and it roared, and the earth about it fairly trembled with agitation. No one dared to approach it even within the circuit of the falling spray of oil and water." In the early days, oilwell blowouts like this one, as reported by local oilman Orange Noble in Pioneer, Pennsylvania, in 1863, were quite a spectacle and costly. Noble had to offer US$ 50 to anyone willing to enter the derrick and attach the discharge pipe to divert the flow of oil into tanks.[12]

It wasn't for the environment but for the economic loss sustained by blowouts that engineers started thinking about blowout preventers (BOPs). The earliest BOP was patented in 1882 by Mike A. Lanagan from New York. The device Lanagan christened "Safety Attachment for Oil Wells and Tanks" was designed to shear the drilling cable and then seal the wellbore with a gate-valve. "It is a well-known fact," Lanagan wrote in his patent letter, "That if the flow of oil and gas from the well can be quickly stopped or diverted so that the flames cannot reach the same, it is a comparatively easy matter to arrest their progress. In consequence of the intense heat surrounding the burning well, it is also necessary to provide an apparatus which can be easily and expeditiously handled at a safe distance from the well."[13]

Mike Lanagan's device was built for cable-tool rigs and it wasn't much in demand. One reason was the absence of any environmental and safety culture; another was the price. A third reason related to the concept itself. Lanagan's blowout preventer was designed for oil wells already on fire, not for preventing fires.

In 1903, Harry R. Decker, a well-known figure in the heady days following Spindletop, was granted a patent for the first ram-type blowout preventer. His BOP was similar in operation to Lanagan's gate-valve BOP but used a pair of opposing steel plungers moving toward the center of the wellbore to close and seal the well. Two men, Harry S. Cameron and James Abercrombie, made the first commercial application of Decker's patent. They would have a lasting impact on the drilling industry.

Cameron, a drilling tools manufacturer, and Abercrombie, an oilman and wildcatter who was once nearly killed by a blowout, created Cameron Iron Works in 1920. They were perfect partners: Cameron was a machinist who could work miracles with metal, and Abercrombie was a man with big ideas who could motivate others to make things happen. Their greatest innovation came in late 1921 when the Monarch Oil and Refining Company gave Cameron Iron Works a contract to find a way to control the increasing gas pressure in deep wells. Repeated attempts to solve this problem encountered in many wells around the world had failed. But through Abercrombie's persistence, he and Cameron built the first ram-type BOP and brought it to market in 1924.[14]

Soon, Cameron discovered he had a competitor. Housing developers in Santa Fe, California, had accidentally struck the nation's second-largest oil vein, triggering a California oil boom in 1923, and William D. Shaffer saw an opportunity to manufacture and sell BOPs. As N.H. LeRoy of Shaffer Tool Works recalled in 1954, "It was during the second Santa Fe boom that Mr. Shaffer conceived the idea of a ram-type blowout preventer to seal around the drillpipe. Mr. Cameron in Texas was also developing a ram-type preventer and both units were developed about the same time. The name SHAFFER or CAMERON immediately became associated with blowout preventers. A French oil operator recently told us that he thought the word "SHAFFER" was an English word meaning preventer which is typical of how expressions originate in the oil field."[15]

Mud gets sophisticated

Two decades after Spindletop, it became apparent that drilling mud had more roles to play than just removing drill cuttings and maintaining the borehole wall in good shape. It had to counteract the high pressures of the fluids found deep in the subsurface, and for that, it often had to be heavier than the conventional muds the Hamill brothers had envisaged.[16]

In 1922, B.K. Stroud, supervisor of the Mineral Division of the Louisiana Department of Conservation, strongly recommended that "drillers should frequently weigh samples of mud." He warned that success or failure in drilling a well in the Monroe, Louisiana, gas field depended on controlling gas pressure with weighted mud. Stroud recommended adding a heavy mineral such as iron oxide or pigment-grade barite, technically barium sulphate. At the time, barite was sourced from National Lead, a company that used the compound to make ink.[17]

Barite solved the weighting problem but created a new challenge that was apparent to Phillip Harth, a sales manager at National Lead. He noticed that when barite was added to mud, it would settle rather than remain in suspension. What was needed was a viscosifying agent to keep the mixture uniform, and after some trials Harth settled on bentonite, a clay consisting predominantly of montmorillonite, which swells when exposed to water. The clay additive worked, Harth obtained a patent and bentonite became a standard mud ingredient forever after.[18]

The invention of oil-base mud in the late 1930s and early 1940s proved to be an even bigger breakthrough. Oil-base drilling fluids, which use crude oil or refined products such as diesel or stove oil as the circulating medium, were developed to overcome some major disadvantages of water-base muds, in particular the destabilization of shales and the dissolution of salt formations. The first trial of an oil-base mud dates to 1935. Humble Oil prepared an oil mud and used it with mixed success to drill a troublesome shale interval in the Goose Creek Field in Texas. During the late 1930s, Standard Oil and Shell separately pursued oil-base muds, but it was finally George L. Miller who made available the first commercial oil-base mud when he formed the Oil Base Drilling Company in Los Angeles in 1942.[19]

In addition to all the advances in mud chemistry, there were also mechanical issues. During the drilling process, drilling fluid has to be separated from cuttings so the fluid can be reused to drill. Prior to 1930, the reclaiming of mud fluid was accomplished by fluming the mud and cuttings through ditches into the mud pit. Most of the cuttings settled as the fluid traveled through the ditches, and relatively clean mud eventually ended up in the pit. From this pit, the drilling fluid was sucked up and reused.

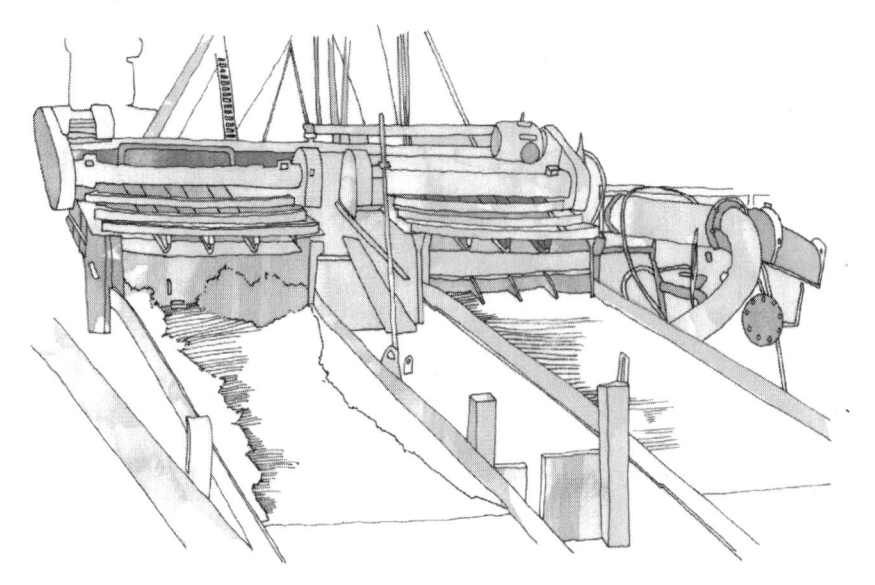

Shale shakers, for removing cuttings carried out of the hole by the mud.

In 1929, the first mechanical devices for cleaning drilling mud were introduced in the Kettleman Hills oil field of California by the Link-Belt Company, a manufacturer of mining and ore-dressing equipment. A wire-cloth screen vibrated while the drilling fluid flowed across the top of it. The liquid phase of the mud and solids smaller than the wire mesh passed through the screen and were reused, while the large cuttings fell off the back of the device and were discarded. The shale-shaker soon became an integral part of every rig.[20]

Chapter 8: Beginnings of Production Engineering

AS IMPORTANT as knowing about the rock they drilled through, oil companies also needed to ascertain whether their well would produce, and, if so, what it would produce and how much. In the early days, there was little need for such understanding. The oil simply spewed out of the hole, and collecting it without waste was the priority. The advent of rotary drilling and circulating mud changed this picture, and prospectors soon demanded a controlled way of testing production.

The first well tests

At first, there seemed to be an insuperable problem. During drilling, pressure in the borehole was maintained by a head of drilling fluid precisely to prevent premature production, so how to test a well's potential to produce oil? Several solutions were found in the 1920s, all relying on an arrangement of packers and valves on the end of the drillpipe to create a temporary completion of the well. The first method was invented by Mordica Johnston and his brother Edgar from Weatherford, Texas. In 1926, after a number of improvisations, the Johnston brothers fitted a conical packer near the bottom of the drillpipe and mounted a spring-controlled retaining valve above it. The drillpipe was run into the hole with the valve closed, preventing any drilling fluid entering the pipe. Then the packer was set above the formation to be tested, and weight applied to the drillpipe. This sheared some retaining straps and opened the valve, allowing the well fluids below the packer to produce to the surface. When the tester came out of the hole, the valve closed under the action of the spring, and the fluid that had entered the pipe remained there. Mordica Johnston received a patent on his drillstem tester in January 1932.[1]

About the same time, John Simmons of El Dorado, Arkansas, patented a different type of a drillstem tester and began experimenting with a prototype. Erle Halliburton, who was always on the lookout for new ideas, heard that it was on on display in the lobby of the Garrett Hotel in El Dorado and went to see it. He was impressed and sought out the inventor. Contrary to the Johnston well tester, in which the valve was opened by applying weight, Simmons's tester opened the valve by rotating the drillpipe. Halliburton and Simmons sat down for a drink, and an hour later Simmons left with a check for US$ 15,000 as payment for transferring the rights of his invention. The Simmons patent, issued in October 1933, was assigned to Halliburton.[2]

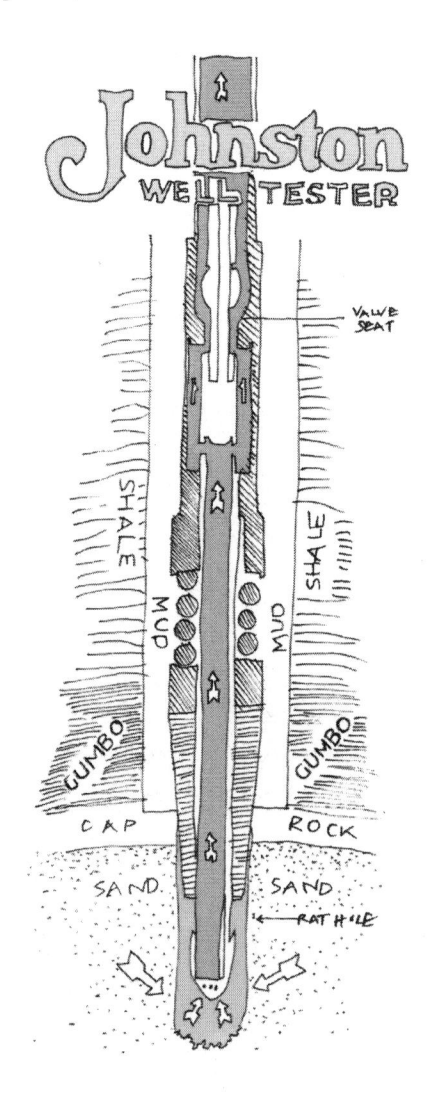

It didn't take long before Erle Halliburton and the Johnston brothers were embroiled in a legal battle over their patents. Eventually, the US Patent Office felt compelled to arbitrate. Halliburton needed John Simmons to testify, but Simmons was nowhere to be found. Then, out of the blue, Erle Halliburton received a letter from Australia. Simmons had run out of money and wanted Halliburton to pay for his trip home. Erle paid up, Simmons returned and testified, and the Patent Office recognized the validity of the Simmons patent. Simmons later got hired by Halliburton.

An advertisement for the Johnston Well Tester, from 1927. Copy accompanying the ad epitomized the spirit of a service company: "And what does it cost? The Johnston well Tester is not for sale. But we have a man in every field who personally conducts tests for $150 each."

Meanwhile, the Johnstons continued to test wells, so Halliburton promptly sued them for patent infringement. After lengthy court proceedings in both Texas and California, the judgment eventually escalated to the Supreme Court, and it finally decided against

Halliburton. Infuriated by the decision, Erle Halliburton declared, "If the courts will not sustain my patents, I am not going to respect anybody else's."

Halliburton had already set a precedent in electrical well logging. In 1934, his company had started logging operations in the US, in some areas taking up to a quarter of Schlumberger's business. Schlumberger filed suit against the Halliburton company and at a joint meeting in Houston in 1938 tried to reach a settlement. But Erle Halliburton, in a voice that thundered through the boardroom, said, "I'll tell you how to settle this lawsuit. You Frenchmen go back where you belong, and let Americans run American business!" That particular meeting came to an abrupt end, but the lawsuit continued. When it finally ended in September 1942 at the US Court of Appeals for the Fifth Circuit in New Orleans, the judge Samuel Sibley ruled against Schlumberger, writing, "The question as to each patent is: Does it merit the monopolization of the present art of electric logging?" Sibley's answer was an emphatic "No!" Halliburton was now recognized legally as a competitor in the rapidly expanding well logging business. Marcel Schlumberger was mortified, but Doll saw it as inevitable and for the good: there was nothing like competition to spur innovation.[3]

Measuring pressure and temperature

As the gushers faded into the past, and well production typically became more modest, a key challenge was sustaining well production during the life of the well. For that, the production engineer, as this role was becoming known, needed measurements. A good starting point was downhole pressure and temperature, two parameters that affected the efficiency of a well's production.

At first, downhole pressure was estimated by measuring fluid level in the well and calculating the resulting pressure head at the bottom of the well. An early attempt to determine fluid level was made by C.E. Beecher and Ivan Parkhurst, of the Cities Service Oil Company, in a Kansas well in 1925. They lowered a float device on an electrical wire into the well; when the top of the fluid was reached, the float rose and made contact with a terminal causing a doorbell to ring at the surface.[4]

Temperature measurements started earlier, in 1869, when Lord Kelvin measured the temperature at a depth of 347 feet in a water well in Blythswood near Glasgow, Scotland. The first temperature measurement in an oil well was performed in November 1912 by John Johnston and Leason Heberling Adams, physical chemists working at the Geophysical Laboratory of the Carnegie Institution, Ohio. They used a maximum-reading mercury thermometer, which they lowered in the well to a depth of 3,000 feet, seeing for the first time a systematic increase of temperature with depth, later called the geothermal gradient. In 1916, they conducted a similar experiment in a well near Mannington, West Virginia, this time with a nickel-wire thermometer that had an accuracy of better than 0.01°C. The geothermal gradient turned out to be about 5°C per 1,000 feet.[5]

Shortly thereafter, C.E. van Orstrand, a physical geologist working for the USGS, measured temperature in eight wells in West Virginia and in another well near Bessie, Oklahoma. Van Orstrand used the same equipment as Johnston and Adams, taking readings to 3,000 feet, but was interested in finding out what affected the geothermal gradient, for example the effect of fluid entry into the borehole, particularly gas. Another area of research was how subsurface structure, for example a salt dome, affected temperature. Orstrand's first experiments in this direction were in Oklahoma in 1919, working with George Matson, chief geologist of the Gypsy Oil Company, based in Tulsa.[6]

Oil companies were rather slow to pick up on pressure and temperature, with Gulf Oil being one of the first to make a bottomhole temperature survey in 1927, followed by Amerada Petroleum Corporation in 1928. However, these were still crude one-off affairs, but things were about to change. In 1929, Charles Millikan, Amerada's chief petroleum engineer, developed a gauge that could be lowered into the well to directly measure and record downhole pressure. Pressure was measured by a plunger set against a spring. The spring then moved an arm that traced a curve on a rotating drum. This design and its derivatives served the industry for more than 50 years.[7]

From 1930 onward, many oil companies, such as Shell, Standard Oil of California, Humble Oil and Sinclair Oil and Gas Company, adopted Millikan's gauge, particularly in the East Texas oil fields during the days of proration. Periodic pressure surveys made in key wells allowed engineers to draw up pressure-contour maps that were used to determine allowable production from each well. However, oil companies soon realized the wider benefits of pressure measurements, such as identifying reservoir fluids and determining open-flow well potential. In 1934, Amerada and Shell started packaging pressure gauges in their drillstem tests.[8]

Throughout this period, temperature measurements improved significantly. Again, Amerada took the first step. In 1931, the company initiated the first subsurface measurement of both pressure and temperature, recording both on a rotating drum. Schlumberger was next. By 1932, Marcel Schlumberger and Henri-Georges Doll had developed a new logging tool that could record temperature continuously versus depth and in addition laid the foundations for interpreting geothermal gradients. There were a host of possible applications, for example identifying the top of cement behind casing, and detecting fluid entries and leaks. By the mid-1930s, both pressure and temperature measurements were becoming standard.[9]

Early oil well perforating

Although cementing effectively created a seal between the casing and formation, the practice created an obvious dilemma: How to reach the hydrocarbons on the outside of the pipe. Somehow a way had to be found to pierce holes in the casing to allow the hydrocarbon to flow.[10]

The first method, patented by John Swan of Marietta, Ohio, in 1910, consisted of a cutting tool, called a "perforator," that was lowered into the hole on a string of tubing. The tool had a star-shaped toothed wheel with cutting points—a sort of rolling knife known as a "ripper" or "splitter." The weight of the tubing forced the knife out so that it cut a slit in the casing at any desired length. But the openings were irregular in size and their location in the well subject to error. Sometimes they were made at a casing joint, which weakened the whole casing string.[11]

In 1926, Sidney Mims, an oilman from Los Angeles, came up with the idea of shooting through casing with steel bullets. Mims's patent shows a steel cylinder with chambers, each one containing a powder charge and bullet, that could be lowered on a cable into the hole. His design made sense but never worked in practice. Meanwhile, outside the US, others had similar ideas. In Romania, Colonel Delamare Maze carried out experiments with a gun perforator for the Astra Română and Steaua Română companies in 1928. Although results were poor, Astra Română continued to experiment in the Moreni oil field near Ploesti. These early gun perforators were plagued with problems, of which premature ignition of the gun powder and the risk of splitting the casing were just the tip of the iceberg.

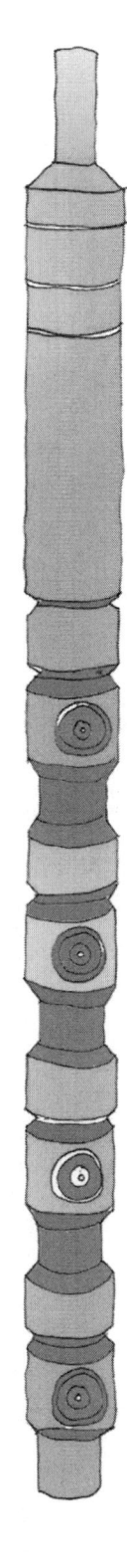

The breakthrough would come in California. Toward the end of the 1920s, two enterprising oilfield tool salesmen, Bill Lane and Walt Wells, were sufficiently impressed by Maze's efforts and Mims's patent to try their own luck. Lane and Wells traced Mims's address to an Elks Club in Los Angeles, a fraternal order in those days restricted to men only, and met him there late one night. Mims agreed to sell his patent to Lane and Wells.[12] The two salesmen then set to work.

Producing a reliable gun proved difficult. To operate in a high-pressure well environment, each gun had to be sealed to keep the powder charge dry. Also, a special powder mixture was needed to combat high well temperatures. The electric cable suspending the perforator into the well required special insulation to withstand the well conditions; otherwise, the electric firing impulse would not travel to the perforator. The bullets had to fire one-at-a-time, in sequence. If, by accident, they all fired at the same time, the casing would be damaged. Controlling the firing required a device situated in the gun and operated electrically from the surface. Finding a steel alloy that could withstand the pressures built up by the detonations was another challenge. Not least, the size and shape of the bullets had to be redesigned.[13]

The first perforating gun, built by Lane-Wells in 1932. The gun had four cartridges and fired bullets individually by electrical detonation of the powder charges. The powder employed was the same used in rifle cartridges and would explode at relatively low temperatures, limiting the application of the gun to shallow well depths.

By 1932 Lane and Wells had a system ready to go and formed the Lane-Wells Company in Vernon, Los Angeles. They had been lucky winning Union Oil for their first trial; no other company would take the chance in a live well. In December of that year, they tested their new perforating gun in Union Oil's 2,500-foot La Merced no. 14 well in the Montebello oil field near Los Angeles. The well had gone dry from the original openhole section and was ready for abandonment, so Union Oil decided it was worth trying the Lane-Wells perforator in a zone higher up in the cased section of the well where oil was expected. The ungainly-looking contraption dangling above the wellhead was described by contemporaries as a "string of coconuts." Bill Lane lowered the device into the well, and eight days later, after 87 shots had been fired in 11 runs, the well started flowing again at the rate of 40 barrels a day. News of this test spread rapidly, and by 1934 operators along the US Gulf Coast were trying it out.[14]

The practice of completing wells with gun perforations spread rapidly. For several years, Lane-Wells could not keep up with the demand for its services so established service companies started jumping in, including Schlumberger. This brought the well-logging company into a double dispute with Lane-Wells who were simultaneously trying to invade Schlumberger's logging market. Since both well logging and perforating depended on accurate depth measurement, Lane-Wells argued that whoever perforated should also log. Otherwise the differing equipment and cable specifications of the various contractors would lead to perforating off depth. Lane-Wells was therefore determined to enter the logging business and to that end acquired a patent in 1937 that supposedly presented electrical logging in a subtly different way than the Schlumberger patents.

Marcel Schlumberger had a horror of litigation, but the de-facto head of Schlumberger US operations, Eugène Léonardon, insisted on taking Lane-Wells to court in Los Angeles, where Lane-Wells was based. As both parties hunkered down for a good fight, Marcel dispatched Henri-Georges Doll to California to keep an eye on things. But more than just patent infringement was going on. Schlumberger also had ambitions. The company was keen to invade the Lane-Wells perforating territory. Outside the US this was no problem, but inside the US the company was blocked by the Lane-Wells perforating patents. While Léonardon rolled up his sleeves in court to debate logging, Doll secretly negotiated with Lane-Wells for a cross-licensing deal that allowed both companies to both log and perforate. The protagonists celebrated at the Coconut Grove nightclub in Hollywood.[15]

Inventing oil-lift technologies

Oil producers dealt with a variety of problems, but perhaps the most serious was the decline in reservoir pressure after the initial strike. The first solution for maintaining production was a sucker-pump driven mechanically from the surface. Another idea was to inject gas into the well, lightening the oil so it could rise more easily.

A modern sucker-rod pumping unit, also known as a beam pump or nodding donkey. It is the oldest and still most widely used type of artificial lift in onshore operations. The beam and crank creates a reciprocating motion on the sucker-rod string that connects to the downhole pump assembly containing a plunger and valve assembly.

The first attempt at lifting oil with a gas used air. This was in 1864, in five wells near Titusville, Pennsylvania, the same area where Colonel Drake first struck oil five years earlier. The lifting installations were called blowers, and the first air-lift patent was granted on November 22, 1864, to Thomas Gunning of New York. Air was compressed at the surface and pumped to the bottom of the well through a string of tubing. But the mixing of oil and air was hard to control. In 1892, US engineer Julius Pohlé of New Jersey patented a method of introducing the air in controlled stages. Pohlé's innovation was soon adopted around the world. In 1899, British geologist and engineer Arthur Beeby-Thompson, working for the European Petroleum Company, introduced air lift to the Baku oil fields. The new technology made quite an impact. In 1902, a Baku oil company well at Bibi-Eybat that had previously produced 3 to 5 metric tons a day by bailing shot up to 327 metric tons a day once air lifting was installed.[16]

About this time, air lifting was introduced along the US Gulf Coast. Then in 1911, Union Oil had the idea of using natural gas produced from their wells rather than air to lift production. The first Union Oil installation of natural gas lifting was in the Cat Canyon oil field in Santa Barbara County. Gas lifting had several advantages: first, gas being naturally at a high pressure was cheaper to compress for lifting than air was; second, it eliminated a safety concern in that air could absorb the lighter components of oil and become a safety hazard at surface; and third, it eliminated casing and tubing corrosion that oxygen would otherwise cause, particularly in the presence of salt water.[17]

Over the next 15 years, acceptance of gas lift leveled off, but with increasing oil demand during World War I and the 1920s, the technique soon found favor again. Some crucial improvements helped. The first innovation was due to Andrew Lockett of New Orleans and Joseph McEvoy of Houston, the latter working for Walter Sharp, the business partner of Howard Hughes Sr. In 1907 and the following year, they were granted patents for a spring-loaded kick-off valve that remained closed until a certain differential pressure was achieved, at which point the valve opened and released the pumped gas. This guaranteed sufficient pressure to kick-start production.[18]

Gas lift soon spread beyond the US. In the newly established Soviet Union, gas lift was introduced in 1923 and by 1932 accounted for 92% of all Baku oil production. In the US, Shell started using gas lift in 1924 in Oklahoma, and in 1926 The Atlantic Refining Company was so committed to gas lift that it built a manufacturing plant developing and building gas-lift valves. Atlantic Refining subsequently proved one of the key players in improving gas-lift technology.

In 1932, Jordan & Taylor Inc. of Los Angeles had the idea of installing many gas-lift valves in the same tubing string at intervals of a few hundred feet. This made it easier to prime the well. The top valve opened first. Once the pressure in the oil column above the valve decreased sufficiently, the valve closed, triggering the next valve to open, and so on until the entire production string was producing. Around 1935, Jeddy Nixon of the Wilson Supply Company in Houston figured that these gas-lift valves could be actuated by wireline, rather than with a prefixed differential pressure. He did a test run with a Halliburton wireline unit and succeeded in actuating several valves in succession as his actuating tool was lowered into the tubing. The wireline-operated gas lift turned out to be a great success.[19]

Electric submersible pumps

Gas lift, however, was not the universal solution for lifting oil. In many places, the traditional sucker-rod pump still reigned supreme, but as wells got deeper, sucker-rod pumps became unwieldy and even unworkable. An alternative that emerged during the 1920s was the electric submersible pump (ESP). This daring innovation dated back to Czarist Russia in Dnipropetrovsk, today part of Ukraine. In 1911, at the age of just eighteen, Armais Arutunoff founded the Russia Electrical Dynamo of Arutunoff Company. Two years before, Arutunoff had demonstrated a prodigious technical aptitude, using his mother's scissors to cut laminations from sheet metal to make an electric motor. At his new company, Arutunoff's management style was eccentric to say the least. He treated everyone in the company as family but was a stickler for punctuality and enforced it by furnishing his office with only three chairs, divided up for the entire day on a first-come first-served basis.[20]

Arutunoff's company produced electric pumps and sold them successfully for dewatering mines and pumping out ships. His pumps were the first that could be submerged in water. When the Russian Revolution caught up with Arutunoff, he decamped to Germany where in 1919 he established a company called Reda Motor-enverwertungsgesellschaft m.b.h., REDA being an acronym for his previous company in Russia. But the German hyperinflation in 1923 forced him to move once more, this time to Los Angeles. Again he had to start over, and wanting to develop and improve his pump motor, he approached The Westinghouse Electric Corporation for funding but was turned down. They proclaimed that his submerged pump motor was "impossible under the laws of electronics."[21]

Armais Arutunoff (1893-1978) and an early version of his electric submersible pump (ESP) motor in the machine shop in Bartlesville, Oklahoma, 1926. Arutunoff looks small, but he wasn't as Shell's Roger Hoestenbach who worked with him in the 1950s recalled: "He was a big Russian, about 6 foot 6 inches. He was very proud of what he had done."

Two years later, Arutunoff met sucker-rod salesman Samuel van Wert, and together they adapted Arutunoff's pump for the oil field. They approached an oil operator in Baldwin Hills, California, and procured a well to test a first prototype. After several tries and numerous modifications to the motor they made a successful test. But work on the prototype and field test had put both partners in debt. Needing funding and exposure to more oil producers, van Wert attended the 1926 American Petroleum Institute (API) meeting in California—the API had been founded in 1919 to promote and protect the US oil industry, both technically and commercially. As luck would have it, Clyde Alexander, who worked for Frank Phillips of Phillips Petroleum, was attending the meeting and happened to be looking for a high-volume lift method for the company's wells in Kansas and Oklahoma. The two parties met, Alexander witnessed the field test, and on June 15, 1926, a contract was signed for the supply of an ESP system.[22]

In March 1928, the first two ESP systems were ready, and one of them was tested in a Phillips well in the El Dorado oil field near Burns, Kansas. Arutunoff's early devices consisted of a 3-phase, 2-pole induction motor with an outside diameter of either 5 3/8 or 7 1/4 inches. Maximum power was 105 horsepower, and the length of the motor was about 20 feet. Attached to the motor and directly above it was a seal unit to prevent the leakage of fluids into the motor housing. Above the seal was a multistage centrifugal pump that lifted the oil to the surface. The complete ESP unit—motor, seal and pump—was run into the well on the bottom of the tubing string, and electricity was supplied from the surface to the motor by a special three-conductor cable. To this day, these are the main components of the ubiquitous ESP.[23]

The test unit in the Phillips well ran 24 hours a day for 16 days and was deemed a success. As part of the contract, Phillips Petroleum exercised an option in the contract and in 1928 created the Bart Manufacturing Company with 51% of the shares owned by Phillips and 49% by Arutunoff. Due to the debt incurred establishing the manufacturing plant, however, Phillips Petroleum divested itself of Bart soon after. Once again in need of money to keep the Bart Company going, Arutunoff contacted one of Frank Phillips's good friends, Charley Brown, a Bart stockholder and executive of Marland Oil

Company, and got a loan. On March 15, 1930, they dissolved Bart Manufacturing and created a new company, the REDA Pump Company.[24]

It was a tough time to start a new company. First, there was competition from other downhole pumps, such as a piston-operated pump driven by hydraulic pressure from the surface; Arthur Gage of Alta Vista Hydraulic Company, Los Angeles, had built a prototype and field tested it for General Petroleum Corporation at Santa Fe Springs, California, in June 1924. Then there was the harsh economic climate of the Great Depression in the early 1930s. To survive, REDA had the idea of renting as well as selling their ESP units. In 1932, still on the brink of collapse, REDA was saved by Phillips Petroleum buying 50 ESP units for their Oklahoma City field where high-volume lift was required to make the field profitable.

Five years later, REDA had done better than survive. Their ESPs were credited with lifting 2% of US oil production, and Arutunoff was getting some press attention. In 1936, the *Tulsa World* newspaper described an ESP rather quaintly as "an electric motor with the proportions of a slim fencepost which stands on its head at the bottom of a well and kicks oil to the surface with its feet."[25]

Acidizing

Hermann Frasch was a German chemist who emigrated to the US with his parents at age 16 in 1868. He studied pharmaceutical chemistry at the Philadelphia College of Pharmacy and soon became known for his daring and original experiments. His interests turned gradually to chemical engineering which was then coming into prominence, and in 1874 he established his own research laboratory. He was especially good at making use of things normally viewed as waste byproducts, such as paraffin wax produced in oil refining, which he purified to manufacture candles. Frasch's fame rests to a large degree on his process for extracting sulfur from mineral deposits.

Frasch's expertise in petroleum chemistry eventually led to the first chemical stimulation of an oil well. By 1894, Frasch was chief chemist at the Solar Oil refinery in Lima, Ohio. In his spare moments, he pondered why there were such large variations in production from the Lima field wells, which produced from limestone rock. He was sure that these differences must result from variations in pore size in the rock and the connectivity between pores. He then began thinking about dissolving parts of the formation to enlarge the pores and develop better connected channels.

In neighboring Pennsylvania, one method to get more oil out of the rock was to drop a stick of nitroglycerine down the well and hope the explosion would loosen things up a little. Hermann Frasch sought a slightly more controlled technique and suggested to J.W. van Dyke, general manager of Solar Oil, to try acid. Van Dyke agreed and in 1895, Frasch treated an oil well with hydrochloric acid for the first time. But the difficulties of pumping the acid without corroding oil well hardware discouraged further progress. Solar Oil did not pursue further development, nor did other oil companies operating in the Lima oil field. Nevertheless, in March 1896, Frasch was granted a patent for the acid idea, entitled "Increasing the Flow of Oil-Wells."[26]

Acid was tried again in the late 1920s by various companies—Gulf Oil in Kentucky, The Gypsy Oil Company in Oklahoma and Ohio Oil Company—but with scant success. The breakthrough for acid came in 1930 when Pure Oil Company, based out of Pittsburgh, Pennsylvania, initiated a collaboration with Dow Chemical, the chemical engineering giant in Midland, Michigan. Dow Chemical had a huge surplus of hydrochloric acid and, unable to find new commercial outlets, started dumping it in abandoned oil wells. Dow Chemical engineers noticed that some of these wells were reviving, and Pure Oil soon heard about it. In February 1932, Pure Oil pumped some Dow Chemical acid down a Michigan well in two stages using 500 gallons of acid for each stage. Hydrochloric acid concentration was 15% by weight, and arsenic was added to prevent corrosion of the steel casing and tubing. The results were promising—the well's production went up from four to 16 barrels of oil per day.[27]

By the middle of 1932, it seemed certain that the acid process had real commercial possibilities, and Dow Chemical formed a subsidiary to handle the new chemical service. By the end of 1933, wells were being acidized by a number of oil companies throughout the areas where limestone was the producing horizon. These included Gulf Oil in Oklahoma, Shell in Kansas and Stanolind Oil and Gas Company—the newly established E&P subsidiary of Standard Oil Company of Indiana—in Kansas, Oklahoma and Louisiana. In just five years, acidizing replaced nitroglycerine for jump-starting oil production.[28]

Separating oil, gas and water

The more oil the young industry produced, the more urgent became the need to process what was actually coming out of the wells: a mixture of oil, gas and water. Separating oil and gas was important because of the oil purchaser's demand for pure oil and the usefulness of gas as fuel for the rig engine. In a first attempt at separation, at Oil Creek, Pennsylvania, in 1865, the produced fluids were passed directly from the well via a pipe into a tall, vertical barrel. Being lighter than oil or water, gas would rise to the surface and be led through a small pipe to the rig engine. This single-stage separator, or "gas trap" as it became known, barely changed for decades to come until US service companies such as The Ashton Valve Company, Oil Well Supply Company and The National Supply Company from various parts of the US gradually evolved the design.[29]

By the 1920s, oil-gas separators had become standard, but as well depths increased so did pressures. In the 1930s, wellhead pressures were reported as high as 2000 pounds per square inch (psi). Separation was now often accomplished in several stages, each stage removing gas and decreasing the pressure. In 1934, the Kettleman Hills oil field in California, for example, featured a three-stage separator.[30]

As the early fields started to deplete, water was increasingly produced along with the oil and gas, and this also had to be separated. The first technique, dating from the 1900s, simply used gravity, water being heavier than oil and also, obviously, gas. The produced fluids were led into the separator, and the water simply bled off at the bottom.[31] However, the gravity technique only worked if the water was "free"—in other words, not emulsified with oil. In 1936,

William Barnickel (1878-1923), who patented the first chemical oil and water separator on April 14, 1914, and branded it the Tret-O-Lite demulsifier. Three years later he founded Petrolite in St Louis, Missouri, a company that would become a pioneer in oil field treatment.

Jay Walker of Tulsa, Oklahoma, was granted a patent for separating water emulsified in oil.

Adding heat to the incoming oil–water stream was another method of separating the fluids. The addition of heat reduces the viscosity of the oil component, allowing the water to settle more rapidly. Another highly efficient method involved chemical demulsifiers. William Barnickel, a pharmaceutical chemist, invented this technique with a 1914 patent and three years later founded Petrolite in St Louis, Missouri, a company that would become a pioneer in oilfield treatment. Chemical demulsifiers became popular in the 1920s and 1930s and eventually were combined with gravity methods to separate all the fluid and gas components, or phases, of oil well production.[32]

Chapter 9: Exploring the World

INDUSTRIAL GROWTH following World War I and the proliferation of the automobile caused world oil consumption to increase dramatically in the 1920s. Oil companies began to expand their horizons and to do that they needed geologists with fresh ideas. As Wallace Pratt, the first geologist hired by Humble Oil, famously put it, "Where oil is first found, is, in the final analysis, in the minds of men."[1]

Stratigraphy

Since the 1880s, the anticlinal and salt doctrine had ruled rather sweepingly in the oil field. For example, in 1910 Frederick Gardner Clapp, a geologist working for the USGS, published an article entitled "A Proposed Classification of Petroleum and Natural Gas Fields Based on Structure." This touched on anticlines and salt domes, but that was all.

A decade later things started to change. In March 1919, Johan August Udden, the Swedish-born director of The University of Texas's Bureau of Economic Geology, spoke on "Oil Bearing Formations in Texas" at the AAPG annual convention in Dallas and expounded on two types of petroleum traps that thus far had received little attention—fault traps and stratigraphic traps, which he called "edged sands."[2]

Fault traps are created by a faulting movement of the earth, where two adjacent strata slip or slide against each other so that oil or gas gets trapped on the underside of the fault. Stratigraphic traps on the other hand are formed as a result of lateral and vertical variations in the reservoir rock that limit the movement of fluids. Udden was the first to recognize the importance of these two breeds of trap. Indeed, the discipline of stratigraphy developed into a central pillar of petroleum geology and spawned the discipline of petrology—the study of mineral content, grain size, texture color and fossil content.[3]

The idea of stratigraphic traps went back to pioneering petroleum geologist and engineer John Franklin Carll who was relatively unrecognized during his lifetime. In the 1880s, Carll first articulated how stratigraphy trapped oil. Drawing upon well samples and his own field work, he constructed and published maps of the Pennsylvanian Venango Third Sand, the formation at a very shallow depth where

Colonel Drake had struck oil in 1859, illustrating how oil accumulated in subsurface stratigraphic layers of sand and not for reasons related to structures such as anticlines or salt domes. Carll created some of the first cross-sectional maps, depicted from a side perspective and showing how strata are deposited and positioned relative to each other.[4]

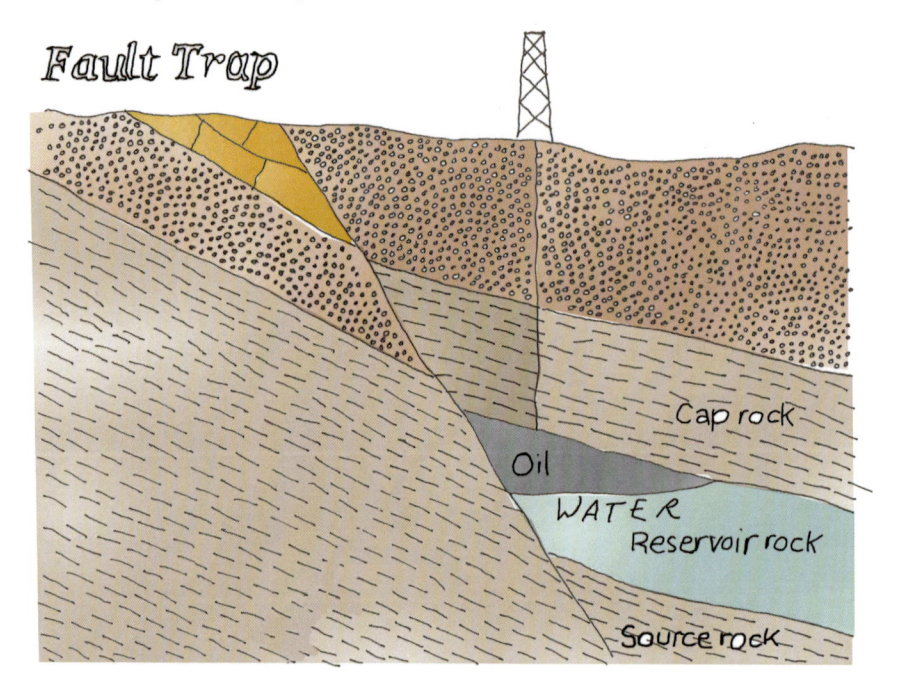

Fault trap. Faults are caused when tectonic movements result in an abrupt shift of adjacent strata. Where two pieces of the same strata slip or slide against each along a fault, a trap can form creating the right environment for creating an oil reservoir.

If Udden's 1919 address had planted the germ of an idea in people's minds, it was geologist Arville Irving Levorsen who gave stratigraphy the push it needed in his 1936 AAPG presidential address "Stratigraphic versus Structural Accumulation." Levorsen emphasized the importance of stratigraphic traps, ensuring that this terminology soon became as essential to the language of petroleum geology as the word anticline.[5] In February 1945, Stanford University appointed Levorsen professor of geology, an example of the rising importance played by petroleum geology in academia.

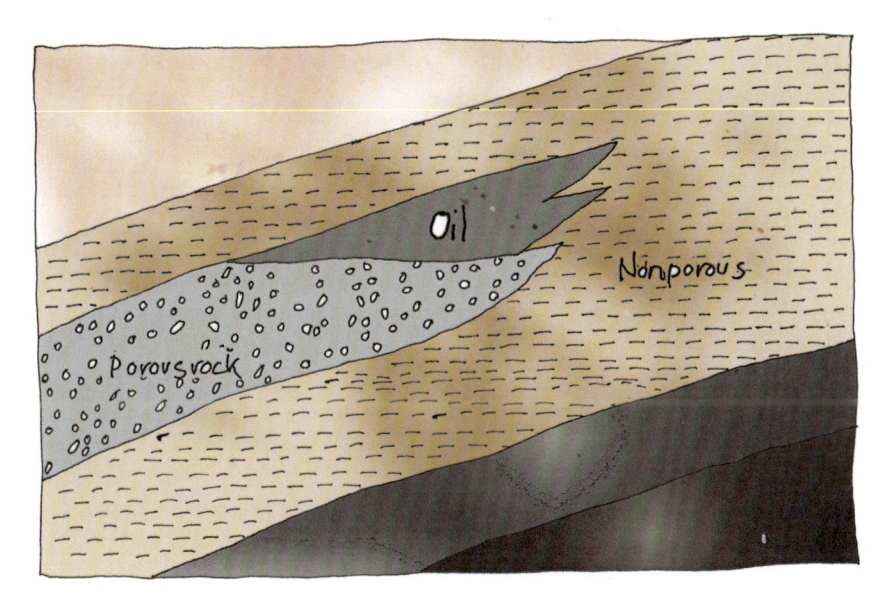

Stratigraphic trap, when a porous layer pinches out within non-permeable rock, for example sandstone in shale.

The first discoveries in Saudi Arabia

One of the early geology students at Stanford was Max Steineke, a son of German immigrants who had settled in Oregon.

Steineke graduated in 1921 and gained experience exploring for oil in California, Alaska, Canada, Colombia and New Zealand for Standard Oil of California (SOCAL, later becoming Chevron). In 1934, he was recruited to join a small team of geologists in Saudi Arabia. The previous year, King Ibn Saud had granted an oil concession to SOCAL, so it formed a new subsidiary, the California-Arabian Standard Oil Company (CASOC), the predecessor of Aramco, to explore for oil and gas in the vast desert country. The geology of Saudi Arabia could only be guessed at. Camel trips across the interior during World War I and the 1920s by Harry St. John Bridger Philby provided some insight. Collections made on these early travels indicated the presence of lower Kimmeridgian and Callovian carbonates, both respectively forming part of the late and middle Jurassic period. Geologists could surmise that these Jurassic rocks, were deposited in shallow waters because of fossil evidence and the widespread occurrence of evaporites, minerals caused by seawater evaporation. It was clear that at some point the sea had covered large portions of the Arabian shield.[6]

Max Steineke (1898-1952), US citizen and son of German immigrants, and a petroleum geologist who in the 1930s helped make the first oil discoveries for the California-Arabian Standard Oil Company (CASOC), the predecessor of Saudi Aramco.

Already in October 1933, the first two CASOC geologists in Saudi Arabia, Bert Miller and Krug Henry, had reconnoitred the limestone hills of Jebel Dhahran on the coast of the Arabian Gulf in eastern Saudi Arabia. They promptly found a favorable structure which they named the Dammam Dome, a geologic structure Miller and Henry recognized as a salt dome from their experience in the US Gulf Coast. The dome rose gently above the flat topography of the Arabian coastal plain and comprised a number of hills up to 350 feet in elevation.[7]

Early in March 1934, Richard Kerr—pilot, aerial photographer and geologist—arrived with an airplane specially designed and built by the Fairchild Aviation Corporation, New York, to conduct aerial reconnaissance to supplement the field work. Kerr took pictures of the Dammam Dome from the air while Henry and colleague J.W. "Soak" Hoover finished detailing the structure on the ground and staked a location for a test well in early June before the onset of the hot Arabian summer. Hoover noted in his diary for June 5, 1934, that they found organic-rich shale and marine fossils on one of the long, low hills of Dammam Dome, indicators that held out some possibility of an oil find. The first test well was spudded in April 1935 and drilled to a depth of 3,203 feet, resulting in some oil shows but nothing to get excited about.[8]

Despite this early work, the task in Eastern Arabia remained daunting. The CASOC geologists made gravity surveys, benefiting from a recently developed, much simpler and faster gravity instrument than the cumbersome torsion balance. Yet, the bulk of the exploration work consisted of ordinary field geology. When Max Steineke arrived, he crossed the Arabian Peninsula in both directions, carefully surveying the landscape as he went. The information he and his party gathered became the basis for all future geologic profiles of the country. In 1936, Steineke was made CASOC chief geologist.[9]

CASOC still believed Dammam to be the most attractive structure since it was close to a productive zone in the neighboring island of Bahrain. In December 1936, following five more dry wells at Dammam Dome, Steineke urged the drillers to go deeper with the seventh well. Throughout 1937 and into early 1938 there were still no positive results, and Steineke had to go back to San Francisco to persuade management to continue funding the Arabian operations. A meeting was scheduled for March 4, 1938, and on the very same day Dammam no. 7, now at 4,727 feet in the Upper Jurassic Arab Formation, came in at 1,585 barrels of oil per day. This discovery later became known as the "Prosperity Well."[10]

CASOC management were now willing to risk more investment around Dammam and decided to focus on a nearby area where a test well, Abu Hadriyah no. 1, was being drilled. In October 1938, the seismic company GSI was contracted by CASOC to obtain seismic reflection data around Abu Hadriyah, and their results indicated the potential for a vast reservoir below. However, the report was met with skepticism, and the CASOC production department wanted to abandon the well. Some even saw this as a good opportunity for getting rid of the geophysicists. But parent company SOCAL's management had earmarked cash for the Abu Hadriyah well. By the early days of 1940 the well had reached 8,656 feet, and still there was no hint of oil. The bit cut its way down past 9,000 feet, and the hole remained dry.

About this time, George Cunningham, exploration manager of SOCAL, and Cecil Green, who supervised GSI field crews worldwide, arrived for a visit of the Saudi operations. Green watched the drilling and later recalled, "You might go that deep at home, but certainly you wouldn't go all the way to Saudi Arabia and drill 9,000-foot holes looking for oil. It was too expensive." In mid-February of 1940,

Cunningham and Green proceeded eastward to visit SOCAL operations in India and Indonesia. While they were in the Indus basin, two telegrams arrived. Cunningham read the first: "Please return to Saudi Arabia before going on to Indonesia for the express purpose of making a post-mortem study of the loss of US$ 1.5 million on the dry hole at Abu Hadriyah." Cunningham was crestfallen and Green didn't feel any better. "Why don't you read the other one?" suggested Green. Cunningham opened the second telegram. It was from Steineke and read: "Abu Hadriyah well just came in at 15,000 barrels per day at 10,115 feet." GSI's reflection survey had been vindicated, and Green remembered the advice Everette Lee DeGolyer had once given him: "Use all the best geology you can and all the best geophysics. But be sure to carry a rabbit's foot in your pocket."[11]

Steineke died prematurely aged 54 in 1952, having discovered in 1948 the largest oil field in the world, the incomparable Ghawar field that produces today as abundantly as it did when first put on production.[12]

Of doodlebuggers and roving geologists

In the 1940s and 1950s, seismic exploration was still a job that demanded great personal flexibility. "It was common for my father to come home on Friday and say, 'We're moving on Monday,'" recalled Leon Thomsen, a geophysicist at the University of Houston, "So, we packed up and moved. It never occurred to me to complain, we just did it. I moved fourteen times before I left home."[13]

Like all geophysicists of that era, Leon's father, Erik Thomsen, was known as a "doodlebugger," a term borrowed from the dowser who uses a simple divining rod to locate subterranean water, minerals, oil or gas. In 1951, Erik Thomsen's wife wrote, "The doodlebugger of earlier times claimed occult powers in the matter of locating oil by the twitch of his stick. The most modern equipment has not been able to erase the old name, and seismic doodlebuggers cling to it with stubborn affection."[14]

In contrast, the job of the exploration geologist was well established. Covering large areas in the field, examining outcrops and collecting rocks characterized their daily routine. Many oil companies had reconnaissance geology groups. The most famous worked for Standard Oil Company of New Jersey and was called the "Rover Boys," a nickname derived from the official "Roving Geological Assignment Worldwide." The Rover Boys were an elite group of geologists, carefully selected and trained. From the late 1940s to the mid-1960s, they were dispatched by Standard Oil management into the unknown sedimentary basin areas of the world to make assessments of oil prospects.[15]

Standard Oil of New Jersey's Rover Boys, here in Timbuktu, Mali in 1959, an elite group of geologists who roved all over the world searching for promising oil prospects. From left: Dick Murphy, Dave Kingston, mechanic George Voutopoulos, and Shelly Eddington who appears to be aiming for some food which they often had to hunt themselves.

Dave Kingston, who was party chief of the Rover Boys for 15 years, conducted surface field work and regional studies in the frontier areas of South America, Europe, the Middle East and Africa. Kingston was hired into the group because of his experiences in the 1950s, performing a geologic reconnaissance of the unmapped Yukon Territory in Canada, where they even had to hunt for their own food. From this proving ground, Standard Oil did not hesitate to send him anywhere in the world. On top of this, a Rover Boy also needed the ability to learn new languages and have the knack for getting along with the locals. As Kingston remembered, "The first year I was in Turkey I was arrested five times for espionage. But after that, I learned to speak enough Turkish to be able to talk directly with the police without going through my interpreter."[16]

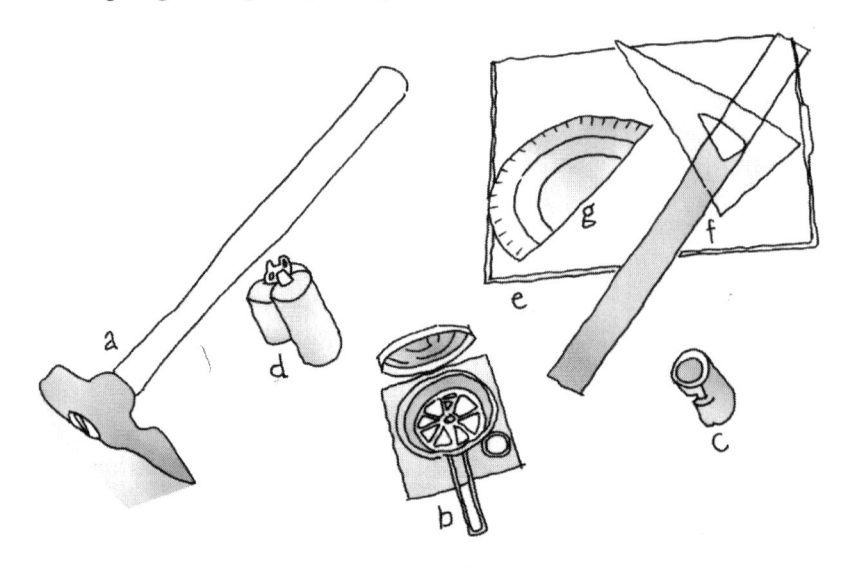

Basic tools for the geologist working in the field: a = hammer; b = compass; c = magnifying glass; d = hydrochloric acid bottle (the "acid test" of rock can signal the presence of carbonate minerals such as calcite, dolomite); e = field book; f = miner's level and triangle; g = protractor.

The Rover Boys explored basin areas to locate the size and shape of key structures and then work out stratigraphy from outcrops. At the very heart of the field work was measuring the angle of dip—in other words how steeply the strata were inclined to the horizontal. This allowed explorationists to estimate how deep the strata might appear in different parts of the basin. Dip was measured by a clinometer, but the measurement had to be combined with a compass reading to provide the direction the strata were dipping in.[17]

All that changed when both dip and direction could be measured using the Brunton compass made in Riverton, Wyoming. This instrument, patented in 1894 by a Canadian-born Colorado geologist named David W. Brunton, was originally used in mining and introduced into the oil field in the early 20th century. The Brunton compass measured dip more accurately than the old hand-held clinometer and was easy to use. In the late 1940s, Bert Bally, a geologist with Shell and then Rice University in Houston, recalled the importance of the sophisticated compass: "As a student I worked in Sicily and was hired by Gulf Oil to map the whole of southeast Sicily. They hired me because I spoke Italian, so I had to map and rank various prospects in the hot summer there. Initially there was no other technology except a Brunton compass. With that and subsequently a rather limited geophysical survey, the Ragusa oil field was found."[18]

Another fine example of the hardy exploration breed was Augusto Gansser-Biaggi, a Swiss geologist whose early work included trips to Greenland and the Himalayas. Disguised as a pilgrim, he once circumnavigated the holy mountain Mount Kailash in Tibet, discovering at the southern foot of the mountain marine sediments, a sensation at the time. After a spell working in Colombia, Venezuela and Trinidad for Shell, he joined the National Iranian Oil Company in the mid-1950s as chief geologist. Working from relief pictures taken by the Iranian Air Force, Gansser located the largest known wildcat oil gusher in the country, north of Qom, which produced 80,000 barrels per day before it caught fire August 26, 1956, and eventually collapsed on itself.[19]

If geologists were unable to access territory from the ground, they could still be dropped there from the air. As David Jenkins, BP chief geologist from 1979 to 1982, recalled, "When I went to Papua New Guinea in the 1960s we used helicopters to take us to clearings in the forest. We then walked off with one or two assistants collecting rock samples and later got picked up again."[20]

From aerial photography to satellites

Help from the air had been coming from the earliest days. Union Oil geologists tried to define the Santa Fe Springs and Richfield prospects in California using aerial photography soon after the close of World War I. They assembled 400 photographs, which pilots of the company's airplane took with an ordinary camera, into a mosaic that represented an area of some 6,250 acres. Most early aerial photography supporting geology was rather haphazard and lacked any technology to fit many individual photographs into a coherent picture.[21]

World War II changed that. In one development, called "trimetrogon" photography, British and Americans Forces operated reconnaissance airplanes, most notably the P-38 "Lightning" that had a special camera installed in its nose. The camera took vertical shots and at the same time a sideways shot on each side. The three shots gave a better view of objects on the ground as Dave Kingston remembered: "I worked up in the Canadian Rockies and these trimetrogon photographs were the first aerial photographs that we had up there; they were a big help to give the big picture and location of outcrops for further studying and mapping the surface geology."[22]

Another breakthrough in aerial photography was the wide-angle stereoscopic camera, combined with the technique of flying back and forth over an area taking overlapping strips of pictures. Then, equipped with a stereoscopic viewer, geologists could prepare detailed topographic maps without even visiting the territory. Martin Ziegler of Shell, one of three Swiss brothers, all of whom contributed to postwar exploration geology, recalls a trip to Nepal together with party leader Ken Glennie to study the foothills of the Himalayas in the early 1960s: "We looked at the overlapping parts of the aerial photographs stereoscopically and you'd see the ridges, you'd see the rivers and you could see how the beds were dipping . . . and then you jumped to the next photo and eventually you'd build up a whole sequence of data you're interested in."[23]

Drawbacks to aerial photography included cloud cover and vegetation, which, if sufficiently dense, could mask what the geologist was looking for. In areas such as deserts with neither impediment, the technique was a godsend. Recalls Dave Kingston: "They did a lot of aerial photography, for example in North Africa. All of a sudden we could map from the air in a very short time structures that previously had taken us years to do with surface field parties." Glennie, working in the Middle East in 1964, also has vivid memories of the benefits of aerial photogeology: "We drove on south into Oman where I managed to persuade the managing director of Petroleum Development Oman to hire a plane for me, so I could take photographs from the air. It cost me 500 pounds for a day's flying and I spent the day taking hundreds of photographs. I've never spent 500 pounds of Shell's money more usefully."[24]

The prevalence of aerial photogeology started to wane towards the end of the 1960s when NASA sent the first satellites into orbit able to take pictures of the earth. In 1972, the US space agency launched its Landsat program with satellites capable of repeatedly photographing practically every spot on the globe. These pictures, covering 115 miles on a side, showed immense detail about the earth and its geologic features. Dave Kingston recalled, "Satellite photographs gave us a totally new perspective on the world's onshore basins, a massive help for international exploration."[25]

A natural extension of aerial photography was aeromagnetic surveys. During World War II, the US Navy developed the first airborne magnetometer for antisubmarine warfare. Gulf Research and Development Company modified and adapted this for limited use in petroleum exploration work during the war years, and by 1946 several airborne magnetometers were available and got licensed to Aero Service Corporation and Fairchild Aviation.[26]

Magnetic data received from the airborne magnetometer, being plotted at Gulf Oil Company in 1947.

In 1947, full-scale commercial use of aeromagnetic surveys started, with most of the work conducted outside the US. The quality of aeromagnetic data was better than that obtained by land magnetometers because measurements at altitude averaged out minor surface and near-surface variations of no interest to petroleum exploration. Furthermore, in aerial magnetic surveys, it was possible to fly over the same area at different elevations in order to delineate better the depth of any anomalies that might be present. Another success factor was cost. Because aeromagnetic surveying usually cost a fraction of seismic surveying, the airborne magnetometer offered a rapid, cheap method of obtaining an initial assessment of the petroleum potential of large, unexplored areas. In 1964, the Soviets were the first to perform magnetic surveying from satellites; NASA followed suit the next year.[27]

Sedimentology makes an impact

Both in academia and in the oil companies, sedimentary rocks were now being scrutinized to an unprecedented degree. This was triggered by the search in the years following World War II for stratigraphically trapped oil. An early example was the API Project 51 that started in 1950. API Project 51 was a multidisciplinary study of sedimentation in the northern Gulf of Mexico and the largest single project that the API had ever sponsored. It was hosted by the Scripps Institute of Oceanography in La Jolla, California, under the direction of Francis Parker Shepard.

From his earliest years Shepard had been good at questioning accepted geologic knowledge. Based at the University of Illinois in 1923, he was on the New England coast collecting seafloor samples and found something that astonished him: "To my surprise, I found almost everything brought up from the seafloor seemed to clash with the antiquated ideas that we had been taught in our geology courses. For example, we were told that sand is found along the shores and this, in turn, is replaced by finer sediments outside, and only mud occurs on the outer continental shelf. That is not what I found at all. Mud often occurred right near the shore and was replaced by sand in deeper water outside; then sometimes by gravel still farther out." Shepard later became a renowned expert on submarine deposits and a key figure in the emerging discipline of sedimentology.[28]

Determining the age of the sediments was a perpetual challenge. Observing and analyzing the biostratigraphy of rocks was rather difficult and cumbersome work, but a new method derived from the early days of nuclear physics significantly improved the ability to date rocks correctly. The idea, which came from two renowned physicists, Ernest Rutherford and Frederick Soddy, in 1902, was first used to date rock samples by the British geologist Arthur Holmes in 1911. The technique is based on the fact that all rocks contain tiny amounts of certain radioactive elements that are unstable and eventually decay. The decay rate is constant and known, so the age of rocks may be determined by measuring for a chosen radioactive element present in the rock, how much has decayed and how much of the original remains. This works perfectly for igneous and metamorphic rock but must be used with care for sedimentary rock because its evolution through weathering and transport is so complex. An alternative dating for sedimentary rock was already available by measuring the direction of the rock's remnant magnetism and correlating that to the earth's known and numerous magnetic reversals.[29]

Another major area of research was carbonates. Unlike the basic chemistry of quartz, a compound of silicon and oxygen that originates deep in the earth and is progressively weathered and ground to form sand grains, carbonate's main constituent calcite is a compound of calcium, carbon and oxygen. Carbonates either precipitate from calcite-rich environments or originate from living organisms, for example coral reefs. In the world of sedimentology, carbonates make up 10 to 15% of all sedimentary rocks and contain some of the world's largest oil reservoirs such as the Ghawar field in Saudi Arabia. Worldwide, carbonate reservoirs contain at least 60% of all known oil and 40% of the gas. In the Middle East, those figures rise to 70% and 90%.[30]

In 1950, Shell recruited its first sedimentologist with a chemical and biological background, James Lee Wilson. From 1952 through 1966, Wilson applied himself to the new discipline of carbonates in Texas and New Mexico. He was followed in close order by Robert J. Dunham, who by 1962 introduced the first classification of carbonate rock in terms of its depositional fabric. This was a seminal piece of work because carbonate rocks vary in structure and appearance to a far greater extent than sandstones, and the industry badly needed some way of understanding and dealing with this important rock type. Dunham's classification with modifications is still in use today.[31]

Petroleum geology was slowly coming of age, and the training of young geologists became a priority for oil companies. Martin Ziegler was an instructor in carbonate geology for Exxon in the late 1960s and early 1970s and describes what it took to create the right mindset: "Annually we had training sessions for geologists, geophysicists, production engineers and others. We took them on a month-long training course, starting out at the lab in Houston to get them acquainted with various types of limestones. Then we took the group out to Florida and the Bahamas—they were in small groups, no more than about 14 people or so. And there we showed them where different types of limestones were formed in the Florida Bay; we showed them how a shoreline looks, what a reef looks like. We took them in a plane over to the Bimini and Andros islands in the Bahamas, and there we showed how an island develops. They had to take cores and they had to collect fossils. They had to make cross-sections and discuss what was going on."[32]

Understanding oil migration

Success in exploration depends crucially on finding the right geologic environment for a reservoir, which is to say a porous sedimentary rock capable of containing the oil and gas. But equally important is ensuring that somewhere nearby the right conditions existed far back in geologic time for oil and gas to be created and then migrate to its resting place.

Up until the early 20th century, geologists believed oil and gas were created where they were eventually found—in the reservoir. But in 1909, Malcolm J. Munn of the USGS surmised that the oil might be squeezed out of shale source beds and then swept up by water toward a reservoir formation.

Thirty years after Munn, Vincent C. Illing, head of the petroleum department at Royal School of Mines in London, tried to reproduce this idea in the laboratory. Illing's experiments used long glass tubes filled with alternate sections of coarse and fine sand. The sands were saturated with water and then a mixture of 90% water and 10% oil was introduced at one end. He observed that oil would fill a coarse section easily but get temporarily stopped at a coarse-fine interface. The water, meanwhile, would travel without hindrance through both coarse- and fine-sand sections, eventually providing enough impetus for the oil to break through each fine-sand section and build up in the next coarse sand, and so on.

Illing measured pressure changes across the coarse-fine interfaces and tilted the tubes to simulate oil accumulating against the forces of buoyancy. These simple experiments corroborated the idea of migration and provided a foundation for explaining accumulations not only in structural traps but also in stratigraphic traps. They demonstrated that oil-water contacts in reservoirs need not be horizontal if dynamic water conditions are present during migration.[33]

Yet, still in the 1940s and most of the 1950s, many geologists clung to the idea that oil-water interfaces were systematically horizontal. It took one extraordinary geologist named Marion King Hubbert to change people's perceptions forever. Many just called him King Hubbert, because in many ways he acted like a king. Hubbert became well-known for the ultimate put-down: "Our ignorance is not so vast as our failure to use what we know." He was a person of great authority and scientific rigor. He not only studied geology but also mathematics and physics and may be credited for firmly pulling geology into the quantitative sciences. In 1943, Hubbert joined Shell in Houston, where he later directed the Shell research laboratory. After retiring from Shell in 1964, he worked at the USGS until 1976.[34]

Marion King Hubbert (1903-89), a brilliant yet sometimes irascible Shell geologist, who from the 1940s through to the 1960s made significant contributions to exploration geology and petroleum engineering.

King Hubbert was clearly difficult to work with. When Martha Lou Broussard, a 1957-Rice University geology graduate and the first female geologist graduating from Rice, went for a job interview, he asked her if she intended to have children. When she said yes, Hubbert told her to go to the blackboard and calculate at exactly what point the world would reach one person per square meter.[35]

Only a cantankerous mind such as Hubbert's could shake up the geologic community about oil migration. In a 1940-paper on "The Theory of Ground Water Motion," Hubbert had already published some important work discussing underground movement and segregation of two waters of different densities. At Shell, he extended this work to movements of oil, gas and salt water, and in 1953 presented a paper at the spring meeting of the AAPG entitled "Entrapment of Petroleum under Hydrodynamic Conditions."

In this paper, Hubbert demonstrated that the presence or absence of hydrocarbons in all types of traps can be explained better in almost every instance by the existence of dynamic water conditions rather than static water conditions in the subsurface. As Illing had noted, if the water is at rest, the oil and gas interfaces will be horizontal, typically the case in anticlines. If the water is in motion, the interfaces may be tilted in the flow direction, sometimes the case in stratigraphic traps that would not otherwise trap the oil or gas in static conditions.

As an example, Hubbert referred to the Rocky Mountains oil province. Here, water is typically flowing through subsurface layers from the mountains toward the Wyoming basin. He observed that in some Rocky Mountain anticlines, the seal appeared not to provide closure. Yet, because the flowing water tilted the oil-water contact, the oil did get trapped.[36]

The importance of plate tectonics

Meanwhile, at the largest possible length scale, petroleum geologists were beginning to harness the new ideas of plate tectonics.

In 1897, the English geologist Richard Dixon Oldham used a seismograph to monitor a huge 8.1 earthquake in Assam, and this led to his proposal that the earth consisted of three major components: core, mantle and crust. Twelve years later, the Croatian meteorologist and seismologist Andrija Mohorovičić, following in Oldham's footsteps, postulated the existence of a boundary surface between the mantle and the crust, which came to be known the M-discontinuity or simply "Moho." In passing through the rocks immediately above this surface, earthquake waves reach a velocity of about 7.2 kilometers per second, whereas below the M-discontinuity the velocity suddenly jumps to above 7.9.[37]

These results provided the basis of a wild idea first propounded by Alfred Wegener, a German meteorologist, geophysicist and polar researcher. The idea was that the continents and ocean floors forming the earth's crust were sufficiently detached from the mantle to be able to "float" and move around. He published these ideas in 1915 in the now seminal work "The Origin of Continents and Oceans," claiming that all the continents had once formed one supercontinent and then, approximately 200 million years ago, split up and drifted away from each other to reach their present form. Wegener provided plenty of evidence for his so-called Continental Drift theory, but was met with universal skepticism. The problem was that Wegener hadn't been able to propose a valid mechanism behind the drift. "A lot of people pooh-poohed it because they wanted to know how it worked," said Ken Glennie recalling his student years at the University of Edinburgh during the late 1940s. John McPhee, the popular writer about geology, remembered when he was a graduate, "Nearly all the faculty at Princeton thought continental drift was sheer baloney." Former Exxon-geologist Walter Ziegler, the oldest of Martin's brothers, remembered, "When I came to Calgary in 1955, the research department head of Imperial Oil told me that continental drift was European bullshit!"[38]

Swiss geologist Walter Ziegler, who worked for Exxon from 1956 to 1983 overseeing the company's exploration activities in many places around the world, including Canada, Greenland, the North Sea, Turkey, the Middle East, Libya and offshore west Africa.

By that time, clues were beginning to emerge in a discipline closely related to geology: oceanography. This discipline took form thanks to the extraordinarily ambitious British Challenger expedition, which from 1872 to 1876 sailed more than 70,000 nautical miles around the world, taking depth soundings, describing seafloor sediments and identifying thousands of new species. Fifty years later, depth surveys in the Atlantic and Caribbean were revealing a highly irregular seafloor. Especially intriguing, a line of underwater mountains seemed to dot the mid-Atlantic. The picture sharpened after World War I, when Fessenden's echo sounding measurements revealed a long, continuous mountain chain. In the late 1940s and the 1950s, ocean surveys conducted by many nations filled in more detail.[39]

In 1947, Maurice Ewing of Columbia University in New York led an expedition on the US research ship *Atlantis* and found sediment layers on the Atlantic seafloor to be far thinner than expected. Many scientists believed that the oceans had existed without much change for at least four billion years, so the sediment layer was expected to be thousands of feet deep. The seafloor therefore appeared to be much younger, in the range of 200 million years or less.

Nine years later, Ewing, together with Bruce Heezen, also from Columbia, noticed that earthquakes in the ocean floor predominantly occurred along midocean ridges. In 1962, Harry Hess, a Princeton professor of geology, associated the earthquakes with the idea that ocean crust was forming at the ridges, with molten material such as basalt oozing up from the earth's mantle along the midocean ridges and spreading new seafloor away from the ridge in both directions. Although his theory made sense, Hess lacked convincing evidence to support it.[40]

This was to come from another science hot spot, the University of Cambridge in England. Geoscientists Frederick Vine and Drummond Matthews started looking at the magnetic patterns of the ocean floor. It was well known that the earth's magnetic field had reversed 171 times in the past 76 million years, so it seemed reasonable to observe this record in the vicinity of the Mid-Atlantic Ridge. What they found was crucial. Mirror image records appeared on either side of the ridge, suggesting that the seafloor was not only spreading but also documenting its age.[41]

Matthews and Vine published their results in *Nature* in September 1963, making history. But other geoscientists had had similar ideas. In January 1963, the Canadian geophysicist Lawrence Whitaker Morley submitted a paper including almost identical ideas to the *Journal of Geophysical Research*, which rejected it summarily. Morley's paper came back with a note telling him that his ideas were suitable for a cocktail party but not for a serious publication.[42]

The Canadian geophysicist John Tuzo-Wilson was also skeptical of plate tectonics but eventually became one of its most famous supporters. He resolved many unanswered questions, in particular the idea of the transform fault in which plates slide past each other without any oceanic crust being created or destroyed. The most famous example is probably the San Andreas Fault between the North American and Pacific plates. For Walter Ziegler of Exxon the transform fault concept was a key turning-point: "I remember having dinner one evening in the early 60s at the house of Professor Bob Folinsbee at the University of Alberta, and Tuzo-Wilson was there. Tuzo-Wilson was involved in all sorts of oceanographic studies and was the inventor of the transform faults. And there he was sketching it at the dinner table, explaining how it worked. Slowly the American scientific community and the oil community began to wake up."[43]

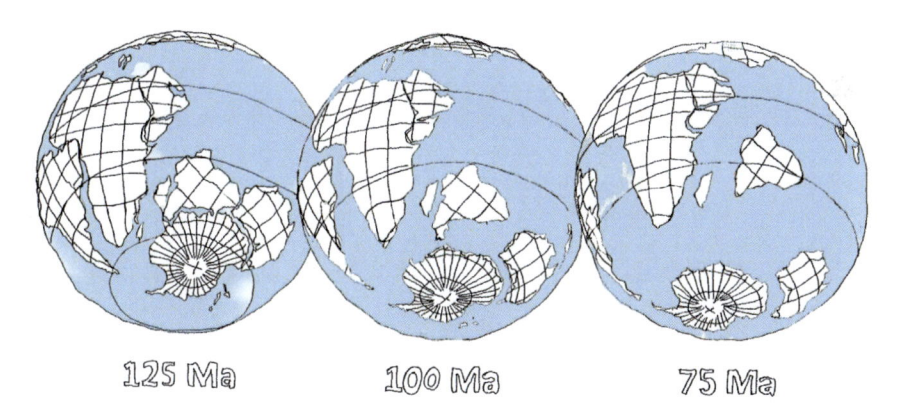

Plate tectonics, a key concept for petroleum geology because it explains long-term movements of the earth's crust. Here the first 50 million years of India plate's northward movement is shown, before it slams into the Asia plate.

In the US, at the Exxon research center in Houston, plate tectonics was finally becoming mainstream. A young geologist, Pete Temple, who had studied under Harry Hess at Princeton, was an early adopter. Dave White of Exxon remembers, "One day Pete and another member of the group, Tom Nelson, started looking at a seismic section from the Otway Basin off South Australia and they envisioned that what they were seeing was a pull-apart feature, where Antarctica had pulled away from Australia. Applying the theory, they then postulated that there should be a transform fault within a certain area near Tasmania at a right angle from the pull-apart. The fault turned out to be right where they thought it would be. I remember Pete coming into my office waving the seismic section in the air. He thought that was pretty neat, and I did too."[44]

Plate tectonics had begun its slow march into the minds of oilfield geologists.

Chapter 10: The Proliferation of Well Logging

WITH THE rapid expansion of the oil industry during the 1930s and 1940s and accompanying increase in world oil production, it became urgent to understand better the shape, size and nature of the reservoir. Downhole measurements were the key, and so began a golden era for the young science of well logging.

The beginning of log interpretation

The new electrical resistivity log and SP measurement had proven quite successful for correlating between wells and predicting which zones would produce hydrocarbons. But the interpretation was still only qualitative.

Jabiol's logs from Sumatra in 1930 had shown that high resistivity related to oil-bearing formations, and another Schlumberger engineer, Raymond Sauvage based in Grozny, Chechnya, had observed the same phenomenon and tried to take the observations a step further by looking for some kind of quantitative relationship. In Azerbaijan, Azneft officials at the Baku Oil Trust proposed to Sauvage that he determine the relationship by measuring the resistivity of partially saturated core samples in the lab. Azneft could provide work space and assistants but, alas, precious little equipment. Sauvage looked around and tried to improvise, but work in the Baku oil fields called. In 1931 alone, the Schlumberger team logged 1,200 wells just in Baku. The oil saturation project was forgotten.[1]

Three years later, geophysicist Vladimir I. Kogan at the Petroleum Institute of Azerbaijan picked it up. He packed a vertically mounted tube with sand cores obtained from the Baku oil fields, filled the sand with brine and then placed the bottom end of the tube in a beaker of oil. Air pressure was used to slowly force the oil upward into the sand and displace the brine. Kogan knew the water saturation from measuring the sample's weight and measured the sample's resistivity with a Wheatstone bridge, an electric circuit used to measure electrical resistance. Numerous measurements on two sand packs, respectively with 20% and 45% porosity, each containing a mixture of oil and water in varying proportions, confirmed a clear increase in resistivity as more oil saturated the cores. By 1935 Kogan had expanded his experiments to two sets of cores from the Surakhany oil field near Baku and then projected the results to estimate the field's reserves from well logs made by Schlumberger. By comparing this calculation with a similar one made using logs, run in late 1932 and early 1933, he was able to determine that reserves had dropped 8% in the previous two years. This was a first and major step toward reservoir surveillance.[2]

Between 1936 and 1939, three groups of US researchers made similar experiments: John Jakosky at International Geophysics in Los Angeles and Richard Hopper at the University of California; Ralph Wyckoff and Holbrook Botset from Gulf Research and Development Company in Pittsburgh; and Miles Leverett of Humble Oil, Houston. However, none of them was able to pull the data together and come up with a definitive relationship between electrical resistivity and rock properties.[3]

That fell to Gustavus Erdman Archie. Gus Archie, as he came to be known, joined Shell as a petroleum engineer in 1934. From early training assignments examining cuttings and cores, Archie learned to appreciate how difficult it was to determine formation porosity and permeability when cuttings were the only source of data. It convinced him that downhole logging measurements such as electrical resistivity were the key to quantifying reservoir properties. With Shell also convinced, Archie was told to lead the charge.

Gustavus Erdman "Gus" Archie (1907-78), a Shell engineer who in the 1940s almost single-handedly created the discipline of petrophysics. Modest as well as ingenious, Archie was regarded by many of his peers as "Shell's all-time best engineer."

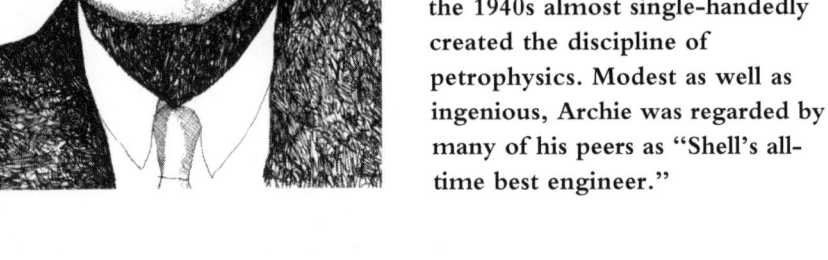

As a first step, Archie undertook a systematic investigation of every Shell well in the US Gulf, comparing each well's electric log with core analysis, mud log and test data. He continued this study as wells were drilled over the next two years. During this time, Archie had numerous discussions with the Schlumberger staff to review the physics of the measurements on offer. In the lab, he saturated numerous sandstones with water of different salinities and measured their resistivities. The experiments suggested to him that resistivity was proportional to water resistivity and inversely proportional to rock porosity raised by an exponent that came to be called "m", which seemed to be about 2.

Archie then combed the existing petroleum literature for any experiments where resistivity had been measured in rock saturated with a mixture of water and oil, and found the three sets of data from the US and through these an account of Kogan's work. The accumulated data led him to squeeze a water saturation term alongside porosity, raising it to an exponent that came to be known as "n", also around 2. Archie's equation was complete and rapidly adopted worldwide providing the standard for assessing reserves from well logs, although a cottage industry analyzing the exponents "m" and "n" to death has flourished ever since. In the short term, the Archie relationship reminded the industry how badly they needed a logging tool that could measure porosity.[4]

Gus Archie wrote up his work in four internal Shell reports, the gist of which was published in the transactions of the American Institute of Mining, Metallurgical, and Petroleum Engineers in his now famous 1942-article "The Electrical Resistivity Log as an Aid in Determining Some Reservoir Characteristics." But Archie did more than offer the industry his famous equation. He created an entirely new discipline that described how the physical properties of rock related to its solid components, its pore space and the fluid the pore space contained. In a 1950 *AAPG Bulletin* article he suggested "petrophysics" as a suitable term for this new discipline, the derivation being a combination of "petro," meaning rock in ancient Greek, and the self-evident "physics." The term stuck, and petrophysics soon became as indispensable to the successful oil company as geology and geophysics.[5]

The induction log

Concurrent with Archie's equation gaining acceptance in the industry, Schlumberger was fast developing more sophisticated versions of its original electric log. But one hurdle that stumped them was the introduction of oil-base drilling muds in the mid-1930s. The company's resistivity measurements needed the presence of a conductive fluid to transmit electric current to the formation; oil-base mud was nonconductive. The first idea was to use scratcher electrodes with sharp points, bristles or blades that were forced by springs onto the borehole wall in the hopes of establishing electrical contact. However, none of these devices was very successful, particularly in hard formations.

Meanwhile, Henri-Georges Doll had started working on a completely different solution that bypassed the drilling fluid altogether. This measured resistivity by inducing high-frequency alternating currents in the earth with a transmitter coil and measuring the resulting voltages with a receiver coil, rather like a transformer. This came to be known as "induction" logging. Already in 1940 Schlumberger had acquired two patents on this principle, granted respectively to California scientist Ralph Lohman in 1935 and University of Purdue professor Charles Aiken in 1936. The challenge was to develop a system that was rugged enough to survive the high temperatures and pressures in the borehole yet capable of detecting the tiny differences in the secondary currents generated in water- and oil-filled rocks.[6]

As with many other technology developments, it took World War II to move things along. Early in 1940, the French Ministry of Armaments had ordered Doll, who also was a reserve commander in the French Artillery, to work on a new system for detecting land mines. By coincidence, on the other side of the Atlantic, the US Army Corps of Engineers had received orders to start development of new mine detectors, including a detector that could be mounted on the front of a tank or a jeep. As the Nazis approached Paris, Doll escaped to the US where he immediately offered his services to the US Army for mine detector research and early in 1941 set up a subsidiary, Electro-Mechanical Research, to carry out the development work. Doll capitalized on his induction ideas and by 1944 was able to separate out the small signal created by secondary currents induced in the subsurface that might signal the presence of a mine. The key was developing a precision electronic circuit that had eluded him in his prewar efforts.[7]

Once the war was over and armed with his experience developing mine detectors, it didn't take long for Doll to build the first logging tool based on the induction principle. It was tested in a Humble Oil well drilled with oil-base mud in the Hawkins oil field, near Tyler,

Texas, on May 3, 1946, and in several other wells in the same field in subsequent months. The tests were promising but not entirely successful. The tool functioned as expected, and log interpreters were able to detect certain oil zones, but the log was much less sensitive to vertical changes in formation resistivity than the standard electrical log. Doll went back to the drawing board and in 1948, after some mathematical juggling, came up with a revised configuration of transmitters and receivers to improve the measurement's vertical resolution. The new induction logging tools could detect thin oil-bearing layers of only a few feet thickness.[8]

By now, Doll had opened and was directing the first Schlumberger research lab in Ridgefield, Connecticut, later to be named after him, and development of the induction tool continued there. In 1956, Schlumberger introduced a combined induction and electrical logging tool, two complimentary measurements that could be logged simultaneously. For a while, it became the logging tool of choice worldwide.[9]

Enter nuclear physics

One thing the Schlumberger tools could not do was measure through casing; the steel killed all electrical signals. One possibility was to use radioactive measurements of some kind, since some subatomic particles such as neutrons and gamma rays traverse steel as easily as they do most other materials. This idea had occurred to two physicists, William Green and Serge Scherbatskoy of Engineering Laboratories Inc., Tulsa, a spin-off of the young seismic company Seismograph Service Corporation. Green and Scherbatskoy built a specialized logging device with a small Geiger counter to detect gamma radiation and on October 29, 1938, ran it in a cased oil well, the Barnsdall Oil Company's Dawson no. 1 well in the Oklahoma City oil field. The results were promising. The gamma-ray logging tool, as it became known, indeed responded to lithology behind casing. The tool offered better depth control for perforating, by correlating the gamma-ray log with conventional logs made before the hole was cased.[10]

Encouraged by this trial, Green and Scherbatskoy joined forces with Socony-Vacuum Oil Company, the predecessor of Mobil, and in May 1939 founded Well Surveys Incorporated in Tulsa to commercialize gamma-ray logging services; the first survey was performed for

Stanolind in May 1940, at Spindletop. However, World War II soon threatened the fledgling company's survival, so in 1939 Well Surveys licensed its gamma-ray technology exclusively to Lane-Wells. While Well Surveys continued basic research, Lane-Wells built the gamma-ray tools and used them in their booming perforating business. In 1946, Lane-Wells developed the final piece of the jig-saw to make perforating on depth absolutely reliable. This was a simple electromagnetic device called a casing collar locator that was run with the gamma ray logging tool in cased hole before perforating and then again with the perforating guns. The sharp blips of each casing collar ensured perfect depth control.[11]

About this time during the summer of 1940, Scherbatskoy met an Italian physicist called Emilio Segré who had studied at the University of Rome and been a member of an elite group led by Enrico Fermi studying neutron transport, work that would lead to the building of the world's first nuclear reactor. Segré had emigrated to the US and was in the Tulsa area looking for a job. Scherbatskoy offered Segré a job, who declined. Instead he warmly recommended a colleague from Fermi's original team in Rome, Bruno Pontecorvo, who was now in Paris working with Pierre and Marie Curie. With Hitler about to enter Paris, Scherbatskoy's telegram offering employment was heaven-sent for Pontecorvo. He fled to Spain and sought aylum in the US.

Once installed at Tulsa, he signed on with Well Surveys. They were lucky to have secured Pontecorvo as other companies also desired the Italian physicist, including Schlumberger. When Doll got news of Pontecorvo's arrival in the US, although never comfortable with matters nuclear, he suggested hiring him to create a nuclear physics research department for Schlumberger, but it was too late.[12]

Pontecorvo's mandate was to bring his knowledge of neutron transport to the world of well logging. His first port of call was Shell engineer Folkert Brons based in Kilgore, Texas, who in November 1940 had patented a logging device that linked neutron transport to rock properties. Shell licensed the idea to Lane-Wells, and Pontecorvo at the sister company Well Surveys built the first neutron logging tool in 1941. The instrument consisted of a strong neutron source and an ionization chamber well shielded so it would see neutrons that had left the source, travelled through the formation and returned to the detector, rather than neutrons taking a straight line through the tool.

Because neutrons have the same mass as the hydrogen atom, a neutron hitting a hydrogen atom slows down considerably, whereas if it hits anything else it just bounces around maintaining its energy. Fewer neutrons reaching the detector therefore indicates the presence of hydrogen in the formation, and since hydrogen is a major component of both oil and water but not very much of rocks, the neutron log gave hope of not only measuring porosity, but also of differentiating between oil and water. The neutron log quickly became a bestseller.[13]

Bruno Pontecorvo (1913-93), the Italian-born atomic physicist and inventor of the first nuclear logging tool in 1941. A communist sympathizer, Pontecorvo defected to the Soviet Union in 1950 and was frequently accused of espionage but always stressed that he had only ever worked on the peaceful uses of atomic energy.

In 1949, Schlumberger decided to catch up, initiating an ambitious two-million dollar program in nuclear physics at their Ridgefield lab. The company hired a team of scientists led by consulting MIT professor Clark Goodman, who provided Schlumberger a cook-book of nearly all possible nuclear reactions that might find a logging application. It was bound in a black cover and became known in-house as the "Black Peril." Nearly every nuclear measurement that Schlumberger subsequently developed for well logging can be traced to Goodman's book.

An early member of the Ridgefield nuclear group, Jay Tittman, recalled, "In the first couple of years we were doing rather crude measurements of neutron moderation and diffusion in a big pit filled with water-saturated sand. We also had marble blocks cut from a rock quarry in Vermont to simulate zero-porosity limestone. The whole place came to be known as the rock farm. After a few years we had a pretty good handle on what was going on in neutron transport and also gamma-ray diffusion."[14] Simultaneously, in Houston, Schlumberger engineer John Dewan was hard at work building a neutron logging tool of his own design. By 1952, Schlumberger's version of the neutron logging tool was introduced in the US. From the very outset it proved comparable in quality to competitive services, and combined with their resistivity measurements reinforced Schlumberger's dominance in well logging.

The company also looked into logging techniques that used a gamma ray source rather than neutrons. The idea was that if you sent gamma rays into rock and the rock was dense, not many would scatter back to the borehole, but if the rock was not dense, then many would come back. Using gamma rays therefore opened the possibility of measuring formation density, and that was exciting because density could lead to porosity, providing an alternative to the porosity derived from the neutron log. Tittman began working on the idea in 1954 and was later joined by his colleague John Wahl. By 1957 the first density tool was ready and successfully tested near Midland, Texas, with commercialization following the year after. In time, the density log would become the gold standard for estimating reservoir porosity and, combined with Archie's law, the means of definitively assessing the total amount of oil in the reservoir.[15]

While Schlumberger caught up, Well Surveys lost Pontecorvo. In 1943, he joined the nuclear research institution Montreal Laboratory in Canada that was associated with the US Manhattan Project developing the atom bomb. Pontecorvo was barred from the Manhattan Project itself because he had earlier joined the Italian communist party in Rome in reaction to fascism and had remained socialist ever since. In 1949, Pontecorvo left Canada for the UK and worked for the Atomic Energy Research Establishment in Harwell, Oxfordshire. A year later, he moved on to the Soviet Union and joined the Institute of Nuclear Problems at Dubna, near Moscow, where he began a lengthy collaboration developing nuclear logging techniques for the Soviets.[16]

Acoustic logging

Similar to the beginnings of electrical logging, the origins of acoustic logging are to be found in surface exploration. The reflection seismic technique saw a sweeping success during the 1920s, but seismologists badly needed a technique for measuring the speed of sound at all depths in the subsurface. Without such data, they were unable to convert the recorded two-way time of seismic reflections to depth. Placing a geophone in a well was the solution, and in 1927 Amerada Petroleum Corporation conducted the first so-called velocity survey by setting off explosive at the surface and recording the acoustic arrival time at various depths in the well. Velocity surveys soon became a booming business, and in the mid-1930s, Schlumberger among others profited from it by renting out trucks, cables and personnel for the lowering of geophones in the hole while an oil or a seismic company supplied the geophone.[17]

In 1951, Humble Oil tried turning this technique on its head, planting a geophone at the surface, and using a modified perforating gun as a sound source. The gun could fire 24 cartridges at as many depths in the borehole, but the sound waves were too weak to be detected at the surface except in very shallow formations. Frank Kokesh, a Schlumberger engineer in Houston, tried a similar experiment in 1952, also without success. Another modification made in 1953 by The Carter Oil Company in Garvin County, Oklahoma, was to clamp the geophone to the borehole wall, providing a cleaner signal and less need for such a strong explosive source at surface.[18]

As early as 1934, Conrad Schlumberger had begun thinking about a continuous log of acoustic velocity and obtained a patent specifying how to use a source and two receivers to measure the speed of sound in a short interval of rock traversed by the wellbore. He then built an experimental tool and tested it at Pechelbronn the following year. The results were disappointing, but then the technology was very basic. The acoustic source was a horn from a Model A Ford modified to survive in the wellbore, and the two receiver signals were transmitted up the cable to either side of headphones that the engineer wore. Using this setup, the engineer tried to assess the time difference between odd noises reaching his two ears. The only record of the tests is of a practical joke. At a certain moment the engineer heard, instead of the usual cacophony, these distinctly spoke words: "Trilobites speaking. Go to hell!"[19] Trilobites are a fossil group of extinct marine animals.

After World War II, the Mobil subsidiary Magnolia took the lead. The key development was a new electronic circuit that could record tiny differences in the time it took a sound wave to travel from a source at one end of a logging tool to two geophones placed a short distance apart at the other end. An experimental tool built by Magnolia was successfully tested in 1954 and licensed to Seismograph Service Corporation and Schlumberger. In July 1957, Schlumberger marketed a version of a two-receiver system. But commercial success came for neither service company.[20]

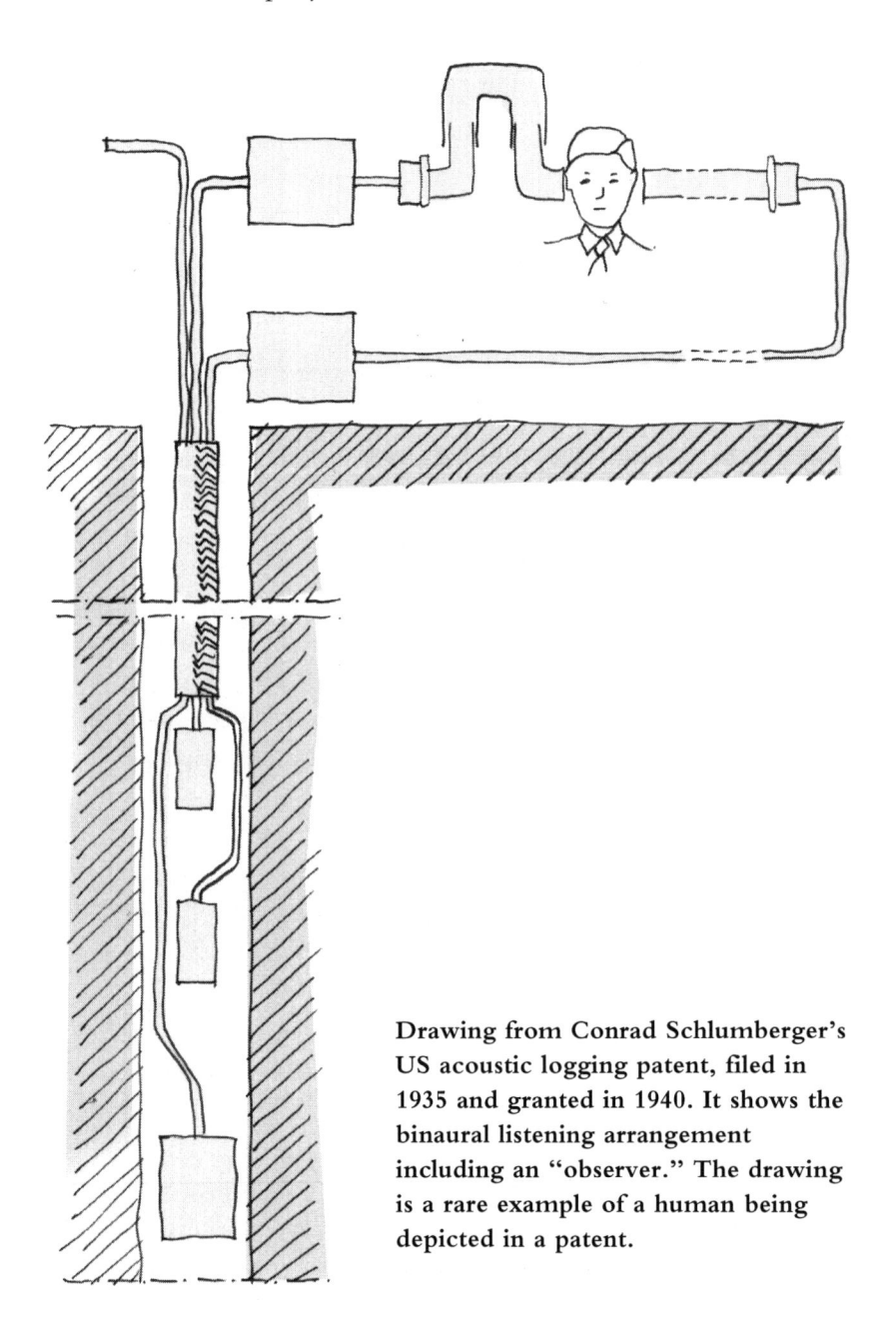

Drawing from Conrad Schlumberger's US acoustic logging patent, filed in 1935 and granted in 1940. It shows the binaural listening arrangement including an "observer." The drawing is a rare example of a human being depicted in a patent.

The early sonic logging tools with two receivers had one important drawback. The tool worked on the assumption that the part of the acoustic travel path in the mud was the same for both receivers. This was true if the borehole diameter was constant and the tool was perfectly aligned with the borehole, but it wasn't true if the borehole was of varying size or if the sonde was tilted. In the early 1960s, Shell researchers found a way using two transmitters and four receivers to get around the problem. The transmitters were placed above and below the receivers, causing variations in transit times to cancel out. Shell called the modified sonic log a "borehole compensated sonic tool" and licensed it to Schlumberger in 1964. "The borehole compensated sonic log became the standard for the industry for many, many years," recalled E.C. Thomas, a former Shell petrophysicist.[21]

During the Magnolia tests, geologist Tom Hingle noticed that the sound speed measured in different geologic layers appeared to correlate inversely with the porosity measured on core samples: the higher the porosity, the slower the sound speed. This was confirmed in 1956 by Malcolm Wyllie and his team at Gulf Research Development Corporation and packaged into a mathematical formula called the "time-average equation." The Wyllie time-average equation opened the door to using the new sonic log as a porosity measurement, and, more importantly, a new market for the technology.[22]

The following year, sonic logs serendipitously found another application. Schlumberger field crews at Lake Maracaibo, Venezuela, noticed that sonic logs showed unusual features, anomalies, at certain depths when the tool was run inside cased wells. The depths of the anomalies seemed to correlate with levels where there was a suspected poor cement bond between the steel casing and the formation. With a few modifications made in 1961, the sonic tool was quickly modified to provide a "cement bond log," superseding earlier attempts to find poor bond using temperature logging.[23]

Yet another possibility for the sonic log was permeability. A glimmer of hope centered around an acoustic wave called the Stoneley wave, discovered in 1924 by British geophysicist Robert Stoneley. Stoneley waves propagate along any solid-fluid interface, the borehole wall in an oil well being an example, and Shell reckoned they had observed the phenomenon in a borehole as early as the 1960s. The necessary theory had been worked out a decade earlier by Maurice

Anthony Biot, a Belgian-born professor of mathematics at Columbia University and consultant to Shell, who was already famous for unravelling the mysteries of acoustic propagation in porous rock.

What was interesting about Stoneley waves traveling along a borehole wall was their effect passing in front of a fractured or permeable formation. Theory said that some of the wave energy would initiate fluid movement between formation and borehole, resulting in amplitude changes that could be measured. In 1964, Schlumberger pursued this idea and started interpreting Stoneley waves to detect fractures, and in the mid-1970s, Shell went a step further and used the Stoneley wave to estimate permeability, initially in the Slochteren sandstone formation in the Netherlands. By the 1980s, Schlumberger had researched Stoneley waves enough to be sure of a permeability relationship.[24]

Chapter 11: Production Engineering Matures

AFTER WORLD WAR II, as logging techniques developed, production engineers were hard at work pioneering technologies to control and maximize each well's production. A key development was a radically new way of perforating casing. The new method came to be called jet or shaped-charge perforation, and as with many engineering breakthroughs, its origins belonged squarely with the military.

Of World War II bazookas and shaped-charge perforating

In 1935, chemical engineer Henry Mohaupt was completing his national service in the Swiss Army and incredulous at the ineffectiveness of the army's anti-tank weapons. Mohaupt thought he could do better and following his army duty established a laboratory in Zurich to develop a powerful anti-tank weapon that could be handled by the infantry soldier. Mohaupt's starting point was the "Munroe Effect." In 1888, the American chemist Charles Edward Munroe had discovered that a hollow or void cut into the surface of explosive would lead to a focusing and increase of the blast energy. In his own research, Mohaupt found that if the void was cone-shaped and, most important, the inside of the cone was lined with a thin layer of metal, the resulting jet formed by the melted metal was powerful enough to penetrate thick steel. By the late 1930s, Mohaupt, now working with the French, had created the first shaped-charge anti-tank device called a "bazooka," named after its resemblance to a crude wind instrument invented by Bob Burns, a US comedian who performed in the 1930s. But World War II was looming, so the French decided to share their research with the US, and Mohaupt moved there in October 1940 to direct the development and manufacture of the new devices. The bazooka was first deployed in earnest in the North African theatre, and being light enough to be carried by an infantry soldier transformed land warfare there and elsewhere.[1]

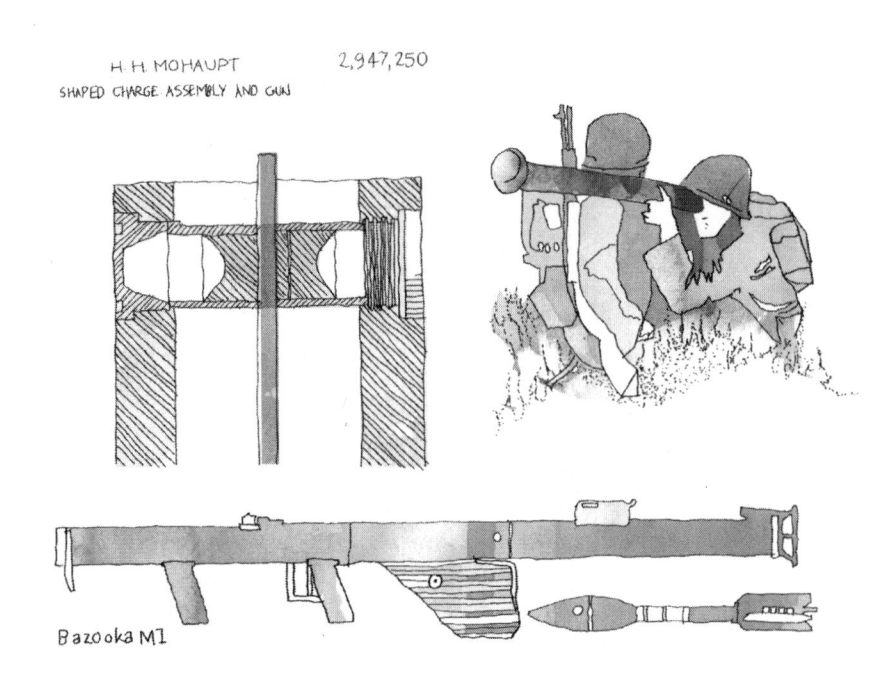

Drawing from Henry Mohaupt's patent "Shaped charge assembly and gun," filed in 1951 and granted in 1960. The detonating cord runs through the gun and shaped charges fire left and right. Mohaupt borrowed the shaped charge idea from the Bazooka used throughout World War II, right.

Meanwhile in Texas, oilfield hands were still saddled with the gun perforating technique that was palpably inadequate. Sometimes the guns didn't fire or, if they did, they would fail to perforate the casing. "We used to shoot little bullets into the casing down there and maybe only a quarter of them would penetrate the casing," recalled Bill Rehm, a retired Dresser engineer. One of the more successful practitioners was Ramsey Armstrong. In January 1945, Armstrong founded a perforating company called Well Explosives Company Inc. in Fort Worth, Texas.

Armstrong immediately acquired an exploding bullet patent, but this technique failed to provide any improvement. A year later, in 1946, two early Armstrong hires changed the company's fortunes. Robert McLemore, a petroleum engineer just discharged from the US army, and Eugene Tolson suggested developing a perforator using the same principles as the bazooka. But they needed people with the right know-how, so the company placed ads in newspapers in areas where DuPont, maker of the explosive end of the bazooka, had manufacturing plants. One applicant answering the ad told them about Henry Mohaupt, so Armstrong contacted Mohaupt and in October 1946 hired him.

In the beginning, the company lacked the funds to buy machinery for manufacturing the shaped charges, so the first few hundred charges featuring a copper liner were made by hand. Thanks to sponsorship from Stanolind, these first shaped charges were successfully tested in a piece of casing in a remote canyon near Pasadena, California. Although not yet commercially viable, the technology then caught the eye of the Byron Jackson Company of San Francisco, an oilfield equipment manufacturer dating from 1872; Armstrong and Byron Jackson knew each other from sharing the same patent attorney. Soon a deal was struck between the two companies: The Byron Jackson Company financed Armstrong's research at the rate of US$ 20,000 per month in return for 50% of Amstrong's perforating revenue.

The remaining pieces of the puzzle began to fall into place. In January 1947, Armstrong secured an exclusive license from DuPont for manufacturing the shaped charges. Next, the Timken Roller Bearing Company, founded by the German immigrant Henry Timken in 1899, was identified to provide steel with just the right hardness for the shaped charge carrier. Softer steels swelled when the shaped charges detonated, causing the gun to become lodged in the well, and harder steels shattered. By September 1947, Well Explosives had a working, commercial shaped-charge perforator.[2]

In initial tests in California and Texas, the perforators yielded a definite increase in oil production. Of particular significance was the ability of the shaped charge to penetrate through multiple casing strings and cement into the rock formation. "Those first jet perforators would penetrate several inches of steel," recalled George King, a chemical and petroleum engineer who started working for Amoco in 1971 and became a well-regarded perforating and stimulation expert. By January 1948, Well Explosives Company Inc., now renamed Welex to avoid the word "explosives," was offering its shaped-charge perforation services throughout the US Mid-Continent and Texas. Soon all major perforating companies were taking out licenses for the shaped-charge technology.

The success of shaped-charge perforation was reflected in 1957 by the Halliburton acquisition of Welex and the longevity of the original design, as George King remembered, "When I started in the industry in 1971, we were still using the basic shaped charge from the 1950s."[3]

The first well fraccing

The Stanolind Oil and Gas Company that sponsored the first shaped-charge tests also played a pioneering role in the evolution of hydraulic fracturing, or "fraccing" as it became known, a technique to stimulate oil production.

Stanolind already had experience in well stimulation with its acidizing activities, and after World War II the company accelerated research in this area. Already during the war years, Floyd Farris, a Stanolind engineer, had been studying the relationship between the pressure used during an acidizing treatment and the subsequent well production. He conjectured that high pressures might be fracturing the rock and enhancing production. He proposed dispensing with the acid and simply pumping highly pressurized liquid into the well to crack open the rock and then filling the cracks with sand to stop the cracks closing once the pressure was released.[4]

For the first test in 1947, Farris chose a low-productivity well in the Hugoton gas field in Grant County, southwest Kansas. The Keppler well no. 1 had originally been completed with a downhole acid treatment that hadn't been successful. The target was a gas-producing limestone at 2,400 feet. Assisting Farris was Robert Fast, later to lead Stanolind's fracturing operations. They used gasoline as the fracturing fluid and thickened it with napalm to carry sand—screened river sand from the Arkansas River—into the artificially induced cracks. Altogether, 1,000 gallons of napalm-thickened gasoline were injected. Afterwards, a so-called "gel breaker" was injected to reduce the viscosity of the fracturing fluid and allow the well to produce. This first test showed hardly any increase in production.[5]

Stanolind persevered, however. The company trademarked its hydraulic fracturing innovation as "Hydrafrac," and from 1947 to January 1949 performed 31 fraccing jobs in 23 wells in seven fields. Eleven of the treated wells showed a sustained increase in production. These first successes attracted the attention of the oil industry although many remained skeptical. One skeptic was Erle Halliburton, but his chief engineer Bill Owsley was hooked. Halliburton could easily adapt its cement pumping trucks for the heavy pumping needed for hydraulic fracturing. Owsley convinced Erle to take the plunge, and in March 1949, Halliburton bought an exclusive license for Stanolind's Hydrafrac technology.

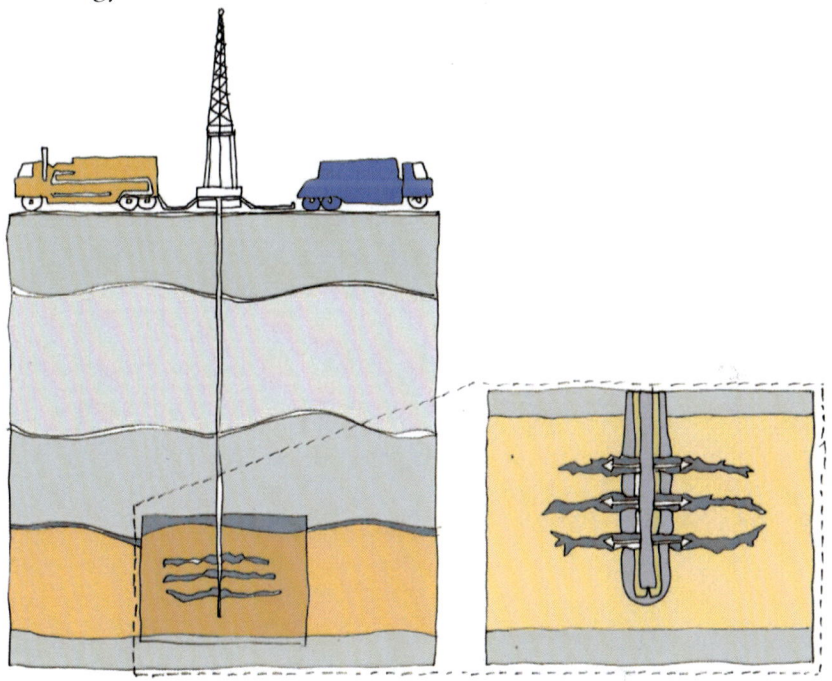

Hydraulic fracturing. Powerful trucks at the surface pump a mix of sand, water and chemicals under high pressures into a well to produce fractures deep in the rock.

On March 17, 1949, Halliburton conducted its first commercial fraccing operation, in a well 4,882 feet deep, in Alma, Oklahoma. Halliburton used a setup similar to Stanolind's: napalm-gelled gasoline, a breaking agent and 150 pounds of sand. The results were promising, and in the following twelve months Halliburton treated 332 wells, resulting in an average production increase of 75%. By 1952, Halliburton had performed 10,460 Hydrafrac treatments on 9,360 wells with even better results. In 1953, the Halliburton exclusive license lapsed, and the technique was open for all service companies to license. The number of treatments increased rapidly until a peak of 4,500 per month was reached in 1955. Fraccing increased the supply of oil in the US far beyond anything anticipated.[6]

Christmas trees

An absolute necessity for new production techniques such as shaped-charge perforating and hydraulic fracturing was a more sophisticated surface control installation. Following Drake's simple tubing gate-valve, the standard setup since the 1860s had been a wellhead accommodating one or at most two production tubing strings. This started to change in 1901 when Lucas's Spindletop well blew out and a new device had to be invented on the spot to cap the well. Lucas quickly built a gate-valve assembly which he dragged onto the top of the wellhead of the roaring well and finally managed to divert the stream of oil into a horizontal pipe. In the ensuing years, wells were equipped with a similar gate-valve assembly to control production better. During the 1920s, wellhead connections were developed to meet the growing challenge of deeper wells and correspondingly higher pressures.[7]

A modern wellhead, known as a "Christmas tree."

A number of specialized shops appeared in those years. One was Oil Center Tool (OCT), established in August 1927 by the brothers Arthur and Kirby Penick of Houston to produce pump liners and rods for the oilfields of East Texas. In 1929, OCT started manufacturing wellhead equipment and had immediate success with its casing and tubing heads. They soon noticed the disadvantages of the onsite assembly of wellheads which by that time had became known as "Christmas trees." The assembly of a typical Christmas tree, up to several meters high with two or more branches left and right, was not only expensive but also time-consuming and dangerous. The Penick brothers had a better and quite simple idea: assemble and test the wellheads in the factory before delivering to the wellsite. By the early 1930s, OCT was ready, and their pre-assembled Christmas trees were hailed as a landmark in efficiency, safety and cost-effectiveness.[8]

Chapter 12: Offshore Goes Deeper

BY THE late 1950s and early 1960s, the industry was moving into deeper seas, and rigs needed to be far sturdier to survive an increasingly harsh environment. Fresh ideas were needed, and size in terms of hardware clearly mattered.

Drillships and dynamic positioning

Several innovations derived from an ambitious scientific drilling project called Project Mohole, initiated in 1961 by the US National Academy of Sciences. The idea was to drill all the way through the earth's crust and penetrate the mantle below. The chosen site had to be in deep water, preferably more than 10,000 feet deep, the crust being significantly thinner under the oceans than under continents.[1]

When Project Mohole started, technical director Bill Bascom chose for the job the drillship *CUSS I*, named after a consortium of Continental Oil Company, Union Oil Company, Superior Oil Company and Shell Oil Company, that had been formed in 1946 to explore submerged acreage off California. The drillship had started life as a navy barge. The consortium cut a diamond-shaped hole, 32 by 22 feet, amidships and located a 98-foot derrick over the hole. The new vessel looked a bit strange for the era. The novelist John Steinbeck, who as a journalist reporting for *Life Magazine* sailed aboard *CUSS I* to the initial Mohole site near Guadalupe Island, 40 miles off the coast of Baja California, Mexico, noted that the ship had the "sleek race lines of an outhouse standing on a garbage scow." Offshore Guadalupe, the *CUSS I* successfully operated in 11,700 feet of water but only managed to drill 550 feet into the ocean floor before it hit hard, undrillable basalt rock. In the early 1960s, a new plan was made to reach the mantle, this time off of Maui, one of the Hawaiian islands. By 1966, however, mismanagement and lack of funding killed the project.[2]

But good came out of the Mohole failure. Engineers realized that a key challenge in deepwater drilling was keeping the drilling vessel in one place. Pete Johnson, a California-based engineer who worked on the Mohole Project, solved the problem by placing powerful thrusters at each corner of the ship and installing distance sensors that picked up radar and sonar signals from fixed buoys placed around the ship. If the wind or ocean current displaced the ship, the relevant thruster kicked in and compensated.[3]

Drilling contractors in the Gulf of Mexico jumped on the innovation. In 1970, Dillard Hammett at the Southeast Drilling Company (SEDCO) designed the first purpose-built dynamically positioned drillship, the *SEDCO 445*. Two years later, the *SEDCO 445* drilled a first well using dynamic positioning in 1,300 feet of water offshore Brunei. These early days of dynamic positioning were not without risks. Every so often, there would be a "drive-off", describing a malfunction in the sensors or computers controlling the thrusters, causing the ship to drift off in a random direction as drilling continued.[4]

Dynamic positioning matured after the success of another scientific drilling ship, the *Glomar Challenger* built in 1968 by Global Marine Exploration Company for a new scientific project called the Deep-Sea Drilling Project. The new ship was able to maintain position in the deepest of waters thanks to four "tunnel thrusters," huge propellers mounted in tunnels that ran the width of the ship, fore and aft. These units, each powered by a 750-horsepower electric motor, were capable of instantly producing up to 17,000 pounds of thrust. The *Glomar Challenger* was also the first commercial vessel to employ the US Navy's new satellite navigation and communications system. On Christmas day 1970, it completed a hole in 12,982 feet of water in the Atlantic Basin, 200 miles from New York City, a world record.[5]

Semisubmersibles

Moving deeper offshore meant that rigs were routinely subjected to stronger winds and higher waves. The major oil companies demanded a more stable rig design. Enter Bruce Collipp, an MIT-graduate hired by Shell in 1954 for his expertise on how floating structures respond to waves. At Shell, Collipp cast his eye on the *Blue Water I*, a massive submersible rig built by the Blue Water Drilling Corporation, which normally sat on a shallow ocean floor. His idea was to partially float it and hope that its mass and bulk would provide the required stability. Shell and Blue Water signed a five-year contract and began converting the rig in 1961. When Shell applied for an operating license, Collipp explained to a Coast Guard official that the rig, originally designed to sit on the ocean floor, would henceforth drill in deep water while only partially submerged. So to fill in the blank next to the heading "Vessel Type," the official had to be creative and wrote "Semisubmersible."[6]

Semisubmersible rig on the move. Semisubmersibles are usually towed to new locations, but it is sometimes more cost-efficient to transport these massive drilling units using a special carrier ship.

The converted *Blue Water I* could operate in 600 feet of water, and national headlines heralded the event. "Oilmen can now find and produce petroleum from the open sea regardless of depth of water or distance from land," reported the *Wall Street Journal*. Even today, the invention of the semisubmersible still impresses industry veterans such as John Thorogood, the former head of drilling technology for BP: "Putting a derrick on a ship was sort of obvious, but to produce a semisubmersible was a stroke of genius." With four stabilizing columns resembling monstrous milk bottles, the *Blue Water I* had the desired hydrodynamic properties. The hull pontoons could be partially filled, making them buoyant enough to keep the vessel afloat but heavy enough to sink below the lash of the waves. Alas, hurricane waves could still be too powerful. *Blue Water I* capsized and sank when Hurricane Hilda hit in 1964.[7]

Nevertheless, naval architects favored semisubmersibles for deep-sea drilling. ODECO, impressed by the performance of the *Blue Water I*, was the first offshore drilling contractor to build a semisubmersible from the keel up. In 1963, Alden Laborde, the brain behind the *Mr. Charlie* rig of the early 1950s, designed and constructed for ODECO a purpose-built, semisubmersible drilling rig, the *Ocean Driller*, that was shaped like a "V", viewed from above, with the derrick placed at the center.[8]

Other drilling contractors started copying the semisubmersible idea, one noteworthy rig being the *Pentagone 81,* so called because of its five massive legs. It was built in 1969 by the offshore drilling company Neptune that was working on a contract with the Institut Français du Pétrole (IFP), a French government-funded educational and research organization. André Rey-Grange, a Neptune design engineer, reflected on the difficulties designing and building the *Pentagone 81*: "The first unit was really a challenge because there was no experience of such construction. In fact, the ship building industry's experience wasn't very useful because the construction of these semisubmersible rigs had nothing to do with the construction of ships."[9]

The surge in offshore drilling soon resulted in the creation of an entire offshore marine industry. Large shipping companies were quick to engage, for example the Danish A.P. Møller Group. Its Maersk Drilling, established in 1972, hired Sam Lloyd of Bluewater Drilling Corporation to build an entire rig fleet from scratch. Gregers Kudsk, former chief drilling engineer of Maersk Drilling, recalled, "We started building quite a number of rigs of all different kinds, so we had semisubmersibles, we had platform rigs, and we had tender rigs. We had a whole range of equipment from day one so to speak."[10]

Sam Lloyd (1924-1978), former President of Bluewater Drilling Corporation that built the first semisubmersible drilling rig in 1961.

In the mid-1970s the most decisive player on the rig market was SEDCO with its 700 series semisubmersibles, pivotal for the North Sea boom. The most influential 700 series rig was the *SEDCO 709*, constructed in 1977. Frank Williford, a former SEDCO engineer, who managed the construction of this semisubmersible drilling unit, recalled, "The *SEDCO 709* was the most sophisticated drilling unit ever built at that time and had a design water-depth of 5,000 feet. Yet its first assignment was offshore Greenland in only 200 feet of water. The *SEDCO 709* had been chosen for the Greenland assignment because of its ability to disconnect while drilling, move away from an approaching iceberg, then reconnect and resume drilling in a matter of hours, all without any outside assistance."[11]

Offshore blowout preventers and marine risers

Ever since the first Cameron Iron Works and Shaffer BOPs were introduced in the 1920s, the technology to avert blowouts and fires had evolved rapidly. First, BOPs had converted from manual to hydraulic operation. Second, BOPs in the 1950s comprised a stack of three complementary functions: pipe rams that closed around a drillpipe but didn't obstruct the flow within the drillpipe, blind rams with no openings for tubing sealing the well and shear rams that cut straight through the drillstring or casing.

The idea of an annular BOP that closed around the drillpipe but still allowed the pipe to rotate was patented by Granville Sloan Knox of the Hydril Corporation in 1952. The packing element consisted of a doughnut-like rubber seal reinforced with steel ribs. The Hydril BOP soon became a standard in the industry and was typically placed on top of the traditional BOP stack.[12]

Offshore, BOPs were becoming a problem. By the late 1960s, BOP stacks had become such weighty monsters that they were becoming almost too large and heavy to fit on an offshore rig. Just the transport and installation of BOPs offshore could be a life-threatening experience, as David Llewelyn, later head of drilling for BP, recalls from his early years in the North Sea, "The BOP was just a land version with a lot of hydraulic hoists on it. It was so heavy it would swing around, and the vertical movement of the rig would make it very difficult to land."[13]

But a solution was underway. "The biggest change in offshore well control came when the industry started putting the BOP stack on the seafloor instead of the rig," remembered Ted Bourgoyne, a retired professor of petroleum engineering at Louisiana State University, who first joined the Conoco production engineering services group in the late 1960s. The CUSS consortium's pioneering *CUSS I* had already in 1955 featured a BOP that was lowered onto the seafloor via guidelines. But only in the 1960s did this practice become accepted for general application.

A blowout preventer (BOP) stack. An annular BOP is stacked on top of three plunger-type BOPs.

The crucial prerequisite was a secure connection to the floating rig that materialized in the form of a flexible marine riser—a large-diameter steel pipe that provides a barrier between the sea water and the drillstring, allowing drilling fluids from the well to flow back to the floating rig. In the early 1950s, the CUSS consortium had already experimented with a small suspended marine riser to allow mud return, but it wasn't before 1967 that George Savage and Robert Bradbury, two engineers working for The Offshore Company in Houston patented a "Marine Drilling Apparatus," which paved the way for the first stable marine riser.[14]

With the BOP sitting on the seafloor, a new problem arose. The BOP needed to be disconnected in case the rig shifted because of bad weather or any other technical reason. In the 1970s this was still a tricky hands-on maneuver. David Llewelyn, the engineer responsible for a well in 1,000 feet of water off the coast of Ireland then considered quite deep, remembers: "We had a diver called Jim who weighed half a ton. Jim would go down to maintain the BOP. He was so big we had to build a walkway around the BOP stack, in fact design the BOP around him. It was the only way to get down and work at 1,000 feet of water depth."[15]

Chapter 13: The Digital Revolution

WHILE THE oil industry was moving offshore, geoscientists were enjoying a new generation of seismic data acquisition and processing techniques. Since the 1930s, geophysicists had worked hard to improve the recording of seismic reflections, both in quantity and quality.

Magnetic tape recording and noise reduction

Until the 1940s and mid-1950s seismic signals could be recorded only on photographic paper, meter-long paper strips covered in reflection squiggles. This made interpretation painstaking and totally manual. An exception was a device invented by US geophysicist Frank Rieber of Continental Oil who in 1936 devised a sonograph capable of recording several geophone traces simultaneously in variable density on a moving 35-mm photographic film, a technique borrowed from the movie industry. This improved markedly the interpretation in areas of complex geology, which became ever more important. But the sonograph proved cumbersome and couldn't compete with what came next, magnetic recording.[1]

The foundation for recording magnetically was laid in 1929 when the Austrian-German engineer Fritz Pfleumer made the first magnetic tape by coating a long strip of paper with ferric oxide powder. The technology was further developed by Badische Anilin- und Soda-Fabrik (BASF), which manufactured the tape, and the German electronics company AEG, which manufactured machines to record and read the tapes. Because of escalating political tensions and then the outbreak of World War II, Pfleumer's innovation was kept largely secret. As the Allies invaded Europe near the end of the war, however, they captured some of the early recording equipment, and magnetic recording development soon began in the US.[2]

By the early 1950s the oil industry had caught up. In 1954, the Texas-based Magnolia Petroleum Company developed a 13-channel magnetic recorder and playback system for their seismic crews. And in 1958, Shell, while also adopting magnetic recording, developed the ability to plot seismic cross sections from tape. Meanwhile, GSI developed a system to record on magnetic disks, which had just become available.

The Rieber sonograph from 1936 that performed two novel functions: the first was recording ten geophone traces simultaneously in variable density form on film; the second was optically adding these traces with any desired increasing or decreasing time offset between them.

The versatility made possible by magnetic recording had a direct impact on seismic acqusition. In early 1950, W. Harry Mayne of Petty Geophysical Engineering was shooting seismic in south Texas and had problems with a massive surface layer of caliche, a hardened carbonate rock that produced large high-velocity reverberations, masking the desired reflections. High ground roll caused by seismic energy moving directly along the earth's surface also smeared the subsurface detail. As a result Mayne was not able to map a suspected series of small faults.[3]

Mayne decided to place the source and detector at progressively increasing and opposite distances from the desired reflection point in the subsurface, in what came to be called "offsets." For reasonably flat layering, the reflection point was then always near the midpoint between source and detector, technically allowing endless repeat recordings of the same reflection point in the subsurface. Mayne's new method was called common depth-point (CDP). The different recordings of the assumed reflection were then combined or "stacked"; a prerequisite was correcting for the fact that wider spacings created slightly longer reflection times, a phenomenon known as normal move-out (NMO). Stacking eliminated much of the smearing and gave a much clearer picture of the subsurface. In short, it dramatically improved the signal-to-noise ratio. Harry Mayne was granted a patent for his CDP innovation in 1956, and from around 1960 the technique saw universal acceptance. The real breakthrough for CDP innovation, however, had to wait for the move from analog to digital technology.[4]

The shift to digital recording and processing

On December 14, 1956, legendary geologist Everette Lee DeGolyer died by his own hand, unwilling to endure the pain of aplastic anemia, then an incurable disease. The geophysics industry that he helped create was about to be transformed. Just three days earlier, the Nobel Prize in Physics was awarded to US physicist William Shockley for the construction of the first solid-state transistor. The device Shockley and his team of scientists had developed at Bell Laboratories in 1947 was creating a digital revolution, in which the oil industry played almost as leading a role as the other most interested player, the military.[5]

The exploration industry's digital revolution was spearheaded by MIT. Between 1930 and the mid-1950s, MIT had transformed itself from a large and established engineering school into a major force in science research, in the process creating strong relationships with industry. Removing traditional barriers between science and industrial application, MIT gained a techno-scientific competence that moved easily from one problem to the next. In particular, Building 20 at MIT, designed in an afternoon and built within six months during World War II, was a ramshackle rabbit warren housing thousands of researchers that became a hotbed of new ideas and innovation.[6]

In the late 1940s, Mobil Oil began funding the MIT Chemical Engineering Department for research in refining that later would develop into programming digital computers to direct the operation of an oil refinery. MIT scientists were also active in exploration research and had initiated work on ways to eliminate unwanted reverberations between the seafloor and the sea surface, responsible for major distortions in the seismic image. In 1952, they successfully pulled in several major oil companies, including Mobil, to form an industry-academia collaboration called the Geophysical Analysis Group (GAG). It was to be the first of several such geophysics consortia.[7]

Enders Robinson, the GAG director, and his 18 graduate students had one main objective, as Sven Treitel, an early GAG student, remembers, "In this earliest stage of digital development, the digital revolution if you will, a key problem was processing seismic recordings to remove one of the most annoying sources of noise, namely the presence of multiple reflections caused by energy reverberating between strata."

One man who took the GAG ideas a step further was Milo Backus. He wasn't a member of the group but a "frequent guest in Building 20 and knew what we were up to," Treitel recalled. Backus joined GSI in 1956 and quickly became a key person for applying digital technology to exploration. To cope with the increasing sophistication of electronic seismic acquisition, GSI had earlier formed a Laboratory & Manufacturing division, which was spun off as Texas Instruments in 1951 with Cecil Green as its head. In a sort of reverse maneuver, GSI was then reconstituted as a division of Texas Instruments. Green would eventually own a large part of Texas Instruments, go on to become a noted philanthropist in the field of education and medicine, and live to the age of 103.[8]

Milo Backus, a former GSI geophysicist and later University of Texas at Austin professor, was one of the main driving forces behind the digital revolution in the seismic industry during the 1950s and 1960s.

GSI's hiring of Backus was part of the company's 1955 strategy to exploit the new digital seismic acquisition and computer processing to map out complex petroleum traps. Backed by Texas Instruments's strong position in the electronics sector, GSI launched a digital research program led by Backus who still remembers the risk his company took: "They put a huge amount of money into it and a huge amount of faith." At first GSI experienced a lot of headwind as the oil exploration industry hung on to their analog methods. Shifting from analog to digital required all the outmoded equipment to be completely discarded, and people were reluctant to do that.

The turning point was the invention of the transistor and the gradual miniaturization of computing power. Computer technology was so new that few fully grasped it, as Sven Treitel recalled: "There were those in the oil industry, generally people who grew up with vacuum tubes, who claimed that it was not possible to filter with a digital computer. So, it took quite some time for these ideas to be understood." Texas Instruments even did a survey of the industry to see if there was interest in digital recording and processing. "Except for a few majors, the answer was no," recalled Bob Graebner, a geophysicist working for GSI at the time, "The oil industry is a very tradition-bound business, I can tell you that."

The early 1960s saw several breakthroughs, mostly from Texas Instruments, including the first digital field recording system, the ability to record digitally in the field on magnetic tape and the first digital processing center. In a highly secretive GSI development, supported by Mobil and Texaco, the first transistorized digital field recording system was deployed in 1963. But in spite of everything, analog systems continued to dominate. When change came, it arrived in a totally new oil province, the North Sea.[9]

A few gas fields had been discovered in the southern North Sea, but the rather complex geology between the UK, Germany and Scandinavia put off the big players and demanded better exploration techniques. The first digital seismic lines in the North Sea were recorded by GSI in May 1964 and processed in Dallas the following winter. By 1965, several North Sea surveys had been recorded digitally, and many examples of processed data became available."[10]

This was the watershed GSI had been fighting for. Motivated by the success of these early surveys, six of GSI's top geophysicists and electrical engineers left and set up their own company, Digicon. Soon, other companies were pushing the digital envelope, including CGG and Petty Geophysical Engineering. Finally everyone joined the digital bandwagon. Amoco recruited Sven Treitel and Enders Robinson in the early 1960s, and about the same time Canadian geophysicist Roy Lindseth founded his Calgary-based firm Engineering Data Processors and soon was installing seismic computing systems for the major oil companies. In 1965, Shell got its first seismic system through the agency of Geosciences Incorporated, a company Robinson was now working for. "Many oil companies gradually came into it," recalled Milo Backus, "And of course before too long the whole industry was digital."[11]

Impact on seismic exploration

The digital revolution had a major impact on Mayne's CDP technique of stacking numerous reflections from the same point in the subsurface to create a cleaner, less noisy signal. The number of traces summed in a stack, or "fold" as it is known, increased rapidly, a good example being the first hydrocarbon find in the North Sea, the West Sole gas field discovered by BP in 1965. The first seismic data recorded there in 1962 was just a single-fold analog recording—in other words just one recorded reflection for each point in the subsurface. By 1964 it was four-fold, by 1965 eight-fold and by 1969 there was a huge jump to 48-fold. "All of this was enabled because of the introduction of digital recording and processing," recalled Jim Hornabrook, former BP Chief Geophysicist. "And of course," added Roy Lindseth, "the digital systems opened up scope for applying sophisticated processing techniques that had already been thought through, but which had been impossible with analog data."[12]

A key example was deconvolution, a signal processing algorithm that could sharpen the observed subsurface reflections. The underlying theory was developed during World War II by Norbert Wiener, the MIT-based eccentric genius, and adapted for reflection seismology in the 1950s by Enders Robinson and GSI's John Burg. Deconvolution is required because a typical seismic source spreads its energy over a period of time, albeit a short period of time. It has its own signal, as it were, and this interacts with the subsurface layers and smears the subsurface reflections. The mathematical term for this distortion is convolution, and removing its effect is achieved through deconvolution.

In 1961, GSI was ready with the first analog deconvolution method using electronic filters. Sam Evans, a retired GSI geophysicist recalled, "GSI devised a method in which they were able to filter data from a part of Lake Maracaibo that was a classic problem area." It gave a much clearer picture of the subsurface offshore. The North Sea was another oil province that profited from the deconvolution technique. "Data in the North Sea are almost useless without deconvolution," Bob Sheriff, former Chevron geophysicist and professor emeritus at the University of Houston wrote in his 1995 book *Exploration Seismology*.[13]

Canadian geophysicist Roy Lindseth who began his career in 1945 in the Venezuela oilfields. Lindseth co-founded Calgary's first digital processing company Engineering Data Processors (EDP) in 1964, and eight years later founded Teknica Resource and Development that commercialized synthetic sonic logs.

By the mid-1960s, deconvolution had come of age and completely changed the picture. Roy Lindseth remembers: "At one time the SEG published from time to time a map showing the US and other areas of the world with areas blacked out where it was impossible to get reasonable reflections. When the deconvolution method came along, these black areas disappeared. Fundamentally there was nowhere in the world that you couldn't see with seismic."[14]

Migration of seismic data

As deconvolution became indispensable for making sense of the acquired seismic data, yet another key technique in the processing chain emerged. This was migration, developed in the late 1960s and early 1970s. As Sam Gray, a former Amoco mathematician pointed out, "Deconvolution and migration have been the workhorse processes for the seismic industry ever since the advent of digital systems." Migration has evolved hugely over the years, but the objective has never changed. The goal of migration is to transform the vast set of acquired seismic reflections into as faithful an image of the subsurface as possible.[15]

For the subsurface is a chaotic place. Unlike neat mathematical models, sedimentary rock layers typically dip at all sorts of angles, and the whole subsurface structure may be crisscrossed by faults. In the early days geophysicists had to assume that reflections came from a point vertically below the shooting midpoint. If the formation is dipping, however, the reflection actually comes from a point in the subsurface updip from the midpoint. Not correcting for this difference makes the reflection appear deeper than it really is, and anticlines, for example, appear wider than they really are—not good for reserves estimation. The correction process is called migration since it involves a repositioning or migration of the assumed reflection point to its true physical position. Not doing this risks drilling in the wrong place.[16]

The need for migration was understood since the beginnings of seismic exploration. In 1921, Karcher at his company Geological Engineering performed the first hand migration of seismic data by pencil and paper on a section at Vine Branch Creek at Belle Isle, Oklahoma. Karcher applied a simple triangulation technique to backtrack the raypath to the reflection point in the subsurface and establish the contact between the oil-bearing Viola Limestone and the overlying Sylvan Shale. This was a laborious process. In the 1950s, a proliferation of mechanical migration devices appeared.[17]

The first digital migration was performed in 1967 by John Sherwood at Chevron in San Francisco using an IBM accounting machine. Then in the late 1960s, Sherwood hired a young scientist named Jon Claerbout as a consultant. Prior to this, Claerbout had been doing theoretical work in seismology for his master's and PhD studies at MIT and in 1967 was appointed professor of geophysics at Stanford University. "I had almost given up on seismology by that time," Claerbout recalled. The Chevron contract allowed him to put his theories into practice.[18]

Jon Claerbout, a professor emeritus of geophysics at Stanford University, who in the early 1970s discovered a key principle of seismic migration, the process that allows the geophysicist to transform acquired seismic reflections into a realistic image of the subsurface.

In 1970 and 1971, Claerbout published two seminal papers showing how the new computing power could finally harness the wave equation that describes exactly how sound travels through the subsurface. The key was his imaging principle, which states that reflectors may be correctly placed in an image of the subsurface by observing where a downgoing wave is exactly time coincident with an upgoing wave—sounds obvious, but it clarified the challenge. Thereafter seismic migration became forever rooted in wave-equation analysis, with increasingly sophisticated methods becoming possible as computing power increased. As Sven Treitel recalled, "In hindsight, Claerbout almost single-handedly created today's seismic migration."[19]

Wave-equation migration was so successful that it soon became the centerpiece of the second industry-academia consortium, the Stanford Exploration Project (SEP), begun in 1973. As Claerbout recalls, "In those days there was really no industrially funded research at American universities including Stanford. I was one of the first to break into that void." At the beginning Claerbout faced skepticism, but his novel approach to industrial research came at the right time. First, there was a growing need in the US for developing new techniques for resolving subtle and complex traps. Second, the problem of multiple reflections in deep water, recently observed in Eastern Canada, needed resolution.

During SEP's first year, no less than 18 operators and service companies subscribed. They also provided access to their field data. In return the SEP group gave back all the new ideas and concepts and, most importantly, the exclusive right to use them. The consortium structure was copied by other research groups at Stanford and became a model for many other industry-academia consortia in the exploration industry. The oil companies saw the consortium model as an ideal supplement to their own internal research efforts.[20]

By the mid-1970s, computer power had improved to the point where an alternative wave-equation method could be invoked that better handled steeply dipping beds. The method was based on a ray-tracing model and first presented at the 1976 SEG annual meeting by Bill Schneider of GSI. The new migration used a computational technique derived from work by the 19th century German physicist Gustav Kirchhoff. For a while, the so-called Kirchhoff migration became a standard.

But computer power continued to increase, and migration got even more sophisticated. A variation on Kirchhoff was the controlled-beam migration pioneered by Ross Hill at Chevron in the late 1980s and 1990s, and later commercialized by the company Veritas DGC. Controlled beam migration's forte was its ability to handle many arrivals coming from different parts of the same horizon and therefore improve imaging of steeply dipping or overturned events such as the all-important salt domes in the Gulf of Mexico. Even better was reverse time migration (RTM), conceived in the early 1980s by Dan Whitmore of Amoco and Dan Kosloff at Tel Aviv University, and first commercialized by CGG as recently as 2007. Reverse time migration has the potential of imaging formations with both great structural and velocity complications, such as sedimentary areas with steep salt inclusions. Yu Zhang, a former research manager at CGG and now a senior scientist at ConocoPhillips, commented, "RTM can easily image any kind of dips."[21]

Early detection of hydrocarbons with "Bright Spots"

Up to the early 1970s seismic surveys had the relatively modest goal of mapping subsurface structures and identifying possible oil traps. Operators still had to drill a well to see if the subsurface contained any oil and gas. What everyone sought was a technology to directly identify oil and gas without drilling a well. As BP's David Jenkins remembered, "If you were to say to somebody 'What's the innovation you'd really like to have and we've never managed to crack?,' then without fail the answer would be direct detection of hydrocarbons in the subsurface."[22]

The first glimpse of achieving this came in the 1970s. The technique was based on the principle that a rock filled with hydrocarbons displayed markedly different acoustic properties than similar rocks filled with water, particularly if the hydrocarbon was gas; rocks containing gas could exhibit abnormally low velocities. The idea had been proposed in 1960 by Carl Savit of Western Geophysical, but the crude state of seismic acquisition killed it. "The reason nobody paid attention in the early days," recalled Leon Thomsen, "was because the subsurface signal coming back to our receivers was extremely weak." The digital revolution provided amplification of the signal, but also crucial was better signal processing to preserve the true amplitude of the recorded reflections.[23]

Before this, however, geophysicists had already witnessed the hydrocarbon effect in the field. Stanolind's Erik Thomsen was one. Along the US Gulf Coast, he noticed that strong seismic amplitudes were invariably associated with wells that turned out to be good producers. Thomsen brought this to the attention of his management but was ignored. The Stanolind managers saw Thomsen as a field guy and told him, "We have PhDs working in the research center in Tulsa, they'll have the good ideas." He tried a second time to raise interest but was again rebuffed.[24]

Other companies's managements weren't as inflexible. Shell was interested enough to attempt hiring Thomsen. However, in the end, it wasn't Thomsen but Mike Forrest who made history. Forrest, a New Orleans-based Shell geophysicist, had come across some old geophysical abstracts translated into English describing how Russian geophysicists had detected hydrocarbons directly using strong reflectors in the seismic data. In 1967, he tried the technique in a new prospect offshore Louisiana, using some of the best seismic data available at that time. Forrest's observations of several strong seismic reflectors led to new discoveries for Shell.

Following more encouraging correlations in offshore Louisiana in 1969, Shell formed a combined operations-research team to study seismic amplitude changes and how they might be related to oil and gas in the subsurface. By now, the potential method had been christened the "bright-spot" technique because of the associated strong seismic signal that looked white and thus bright on a 2D black-and-white seismogram. In the December 1970 lease sale for offshore Louisiana, Shell exploration managers applied bright-spot evaluation for the first time in their bid. But Shell found itself outbid by another bright-spot player, Mobil. Since 1962, Mobil had been developing its own bright-spot interpretation, also in the Louisiana offshore region.[25]

By the early and mid-1970s, bright-spot technology improved considerably and became a reliable tool for many geologic provinces and at increasing depth. Both Shell and Mobil scanned almost every prospect in the Gulf of Mexico for bright spots and made bids based on quantitative measurements of seismic amplitude and sand thickness. Other companies eventually caught on. The bright-spot technique impacted exploration in similar environments around the world, most notably in the Niger Delta in West Africa.[26] More recently, bright-spot analysis has proved key for several deep offshore discoveries, for example Shell's Mars-Ursa fields and BP's Thunderhorse field in the Gulf of Mexico.

The air gun replaces dynamite offshore

Most of the seismograms Mike Forrest was looking at in the 1960s and 1970s were generated by dynamite explosions. Not only were quite a few fish regularly doomed to extinction when explosive was dropped in the ocean and detonated with a big splash, but the explosives were also rather dangerous to handle, as Jim Hornabrook recalled, "We were using 50-pound charges, which is really quite a lot of dynamite. With several charges strung out behind the boat, the danger was having a misfire. Several hundred pounds of dynamite sitting just behind your stern—and it could go off any moment." Sometimes undetonated charges would simply end up on the sea bed, obviously posing a problem near land. "At one point," added Hornabrook, "the Dutch authorities wouldn't allow any of the early seismic boats to dock in Holland."[27]

A solution in the early 1950s was to use safer detonating nitrocarbonitrate. But this didn't remove the basic environmental and safety issues of using explosives, particularly since more and more explosions were required for the new multifold CDP surveys. And another troublesome problem cropped up—the bubble effect. When dynamite detonates in water, the explosion creates an air bubble that oscillates below the surface, creating a secondary source of energy that confuses the entire picture. The only way to avoid this was detonating the explosive close to the ocean surface so the bubble vented to the atmosphere, but that made the whole operation even more dangerous.

What was needed was an entirely different kind of seismic source. The new source had to satisfy numerous requirements. It had to be capable of being towed and fired at frequent, regular intervals in a series of high-energy, uniform explosions, and to be deployable in a range of signal strengths. The answer could only come from the lab of a maverick inventor like Stephen "Bolt" Chelminski, a Columbia University engineering graduate teeming with new ideas.

Reportedly, he acquired the nickname Bolt when he was working with his brother on the engine of a Volkswagen and dropped a bolt into the engine. Upon discharge from Korean War military service, Chelminski used his mustering-out pay to equip himself with a machine shop and turn his hand to creating all kinds of devices, including pile drivers, a percussion drill and a controllable-thrust rocket motor. The Lamont-Doherty Geological Observatory at Columbia University learned of the young man and hired him as an engineer and mechanic in the late 1950s. Then on March 17, 1961, Chelminski's Lamont colleague John Hennion was killed when dynamite for a seismic survey detonated prematurely during an offshore survey. The accident spurred Chelminski to hunt for a nonexplosive marine source of seismic energy.[28]

By this time Stephen Chelminski had left Lamont and started his own company, aptly named Bolt Technology Inc. Ruling out any form of explosive, he soon came up with the idea of a pneumatic seismic source using compressed air. The device Chelminski constructed was essentially a steel container charged with very high-pressure air that on command decompressed rapidly to form a pressure pulse in the water. He called the device an "air gun."[29]

Chelminski's first air guns were tested out of Port Hueneme in California for GSI in 1966. Although GSI vice president Ken Burg was all for trying the Bolt air gun, his colleagues were lukewarm. "We just couldn't imagine something as simple as compressed air could replace dynamite as a seismic source," recalled Ben Giles, a long-timer with GSI. But the first test results proved very encouraging, and the Californian GSI team convinced GSI president Mark Smith to equip ten of the GSI vessels with Bolt air guns. The idea quickly caught on. By the mid-1970s, more than 50% of all marine seismic surveys used air-gun arrays as the preferred seismic source.[30]

Vibroseis

During the 1960s and 1970s the industry also saw new techniques for generating seismic waves onland. The traditional dynamite was becoming quite labor-intensive as the required number of shot holes increased, and explosives were anyway prohibited in urban areas. Another disadvantage was that dynamite explosions were rather uncontrolled and produced variable seismic signatures that were difficult to deconvolve.[31]

One of the first alternatives to dynamite was patented in 1956 by electrical engineer Burton McCollum. He had been a joint founder of Karcher's Geological Engineering Company in the 1920s and since worked for the US National Bureau of Standards. McCollum constructed a so-called "thumper" truck. A three-ton weight was raised by a hoist at the back of the truck to a height of ten feet and dropped to thump the ground. To achieve an adequate signal-to-noise ratio, records from up to 200 weight drops had to be combined into a single record.[32]

A Vibroseis truck in Libya, 2008. A heavy ground plate (center) is lowered to the earth and then vibrated through a range of frequencies by a powerful motor on the back of the truck, sending seismic waves down into the subsurface.

But in a parallel development, a more sophisticated idea took form in the minds of two Continental Oil researchers: chief of geophysical research John Crawford and research geophysicist Bill Doty. In August 1952, they attended a conference on seismogram analysis at MIT, to which all the leading US geophysicists of the time were invited. In one session, Enders Robinson at MIT suggested that "it might be possible to use a controlled signal in seismic work, as in radar." The two Continental Oil representatives started thinking.[33]

The first breakthrough came with the suggestion by John Crawford that a long-duration signal continually changing in frequency, either increasing or decreasing, would provide the desired high energy. The next step was developing a physical prototype. First, they attached a massive ground plate, which they called a "vibe," to the bottom of a heavy truck. Then they rigged up a 25-horsepower gasoline engine to hydraulically vibrate the ground plate for 10 to 15 seconds, sweeping through a broad range of frequencies from about 20 to 80 cycles per second.[34]

A patent was awarded in 1954 to the Continental team, and in 1958 the first field tests were successfully completed. Finally in 1961, the so-called Vibroseis was ready to be commercialized, the first licensee being Seismograph Service Corporation with that company's Nigel Anstey, a British geophysicist, making significant contributions to the subsequent development of the technology. Throughout the 1960s and 1970s, Vibroseis technology was refined and made even more effective thanks to the digital revolution. The trucks got to deliver more powerful vibrations, the sweeps got longer, and it became possible to control the signature of the sweep. By the mid-1970s Vibroseis had become the source of choice for land seismics worldwide.[35]

Chapter 14: Petroleum Geology Changes the Game

BY THE early 1970s, so many advances in geophysics had occurred that geologists had to learn quite a bit of extra geophysics to stay in the loop. What happened at the time was a sort of "amalgamation of geology and geophysics," recalled Martin Ziegler. Most major oil companies started to cross-train geophysicists and geologists, sending them on courses in the other discipline. Exxon was at the forefront of this skills management exercise, and geologist Peter Vail was one of the first to benefit.[1]

Sequence stratigraphy

In the late 1950s and early 1960s, Vail was studying at Northwestern University in Chicago where his professor, Lawrence Sloss, was establishing new ideas in stratigraphy. Sloss's area of interest was Paleozoic sediments within the US. Normally, rock units were subdivided simply on the basis of rock type or lithology, but Sloss was the first to recognize that numerous unconformities could be correlated across the continent and emphasized tectonic forces as the prime driver of unconformity generation. He then analyzed the sedimentary record in terms of the sequences of strata contained within these unconformities, creating a whole new subdiscipline called sequence stratigraphy.[2]

Peter Vail, an Exxon geologist, who became famous in the 1970s when he harnessed seismic data to the concept of sequence stratigraphy, creating the sub-discipline "seismic stratigraphy" that remains an indispensable tool for the explorationist.

Vail started in the oil business in 1956 in Tulsa, Oklahoma, in a mid-sized operator called Carter Oil, later acquired by Exxon. In 1965, he joined the new Exxon research center in Houston. Vail continued the Sloss program carrying out extensive studies in North America, dating sequences using fossils and other evidence from outcrops and drilled cores. Contrary to what his mentor Sloss had inferred, Vail believed that the unconformities were generated primarily by eustasy, a term describing worldwide variations in sea level. Crucially, Vail could now harness the increasingly prevalent seismic data.[3]

Vail's foray into seismics was regarded with suspicion by the geophysics community who considered geologists incapable of understanding the complexities of seismic technology. Traditional structural interpretation with thick colored pencils was fine with them, but subtle stratigraphic analysis using the fine detail and patterns within structure was considered a step too far. Fellow geologist Dave Kingston remembers the mood inside Exxon at the time: "We began to identify things, but the geophysicists wouldn't believe us because we were not geophysicists."[4]

By the mid-1970s, Peter Vail was identifying patterns within a seismic section that revealed the maximum and minimum sea-level events, the so-called high-stand and low-stand sequences of sediments created at the water's edge. This added a new dimension to the sequence stratigraphy thesis. The idea took the industry by storm. Vail and colleagues had effectively succeeded in building a solid bridge between geophysics and geology.[5]

In the 1980s, some of Vail's younger colleagues at Exxon, most notable Mac Jervey and Henry Posamentier, went on to develop a model that integrated the effects of tectonics and eustasy. Ever since, sequence stratigraphy and its subspecialty seismic stratigraphy have remained key working tools for the petroleum geologist.[6]

The concept of turbidite flow

The 1970s saw another key geology concept emerge. With an increasing focus on offshore exploration, one particular phenomenon was proving especially hard to explain. As Bryan Lovell, former chief sedimentologist at BP recalled: "In many places the geologic record showed a mixture of sandstones containing shallow water fossils alternating on a scale of a meter or less with mud stones containing fossils that were characteristic of deep oceans." Assuming the shallow-water sands were somehow redeposited in deepwater, the question on people's minds was what carried them there.

The first clues were surmised by Dutch geologist and oceanographer Philip Kuenen as early as the 1930s. A professor of geology at the University of Groningen, Kuenen proposed that earthquakes could trigger undersea flows of muddy sediments, effectively underwater avalanches that travel along the ocean-floor at considerable speed carrying all before them. Kuenen's evidence included experiments in tanks and field observations of the 1929 Grand Banks earthquake in the Atlantic Ocean off Newfoundland. He noted that material slumped off the continental slope was carried more than 450 miles beyond the continental shelf into the abyssal ocean plain, destroying several transatlantic telephone and telegraph cables in the process. Kuenen estimated the total mass of material to be an astonishing 100 cubic kilometers.[7]

The sedimentologist Arnold Bouma, Kuenen's compatriot and former student, analyzed the sequential breaking of the underwater cables and found that when the underwater avalanches came to rest, they formed a characteristic depositional sequence, later to be known as a "Bouma sequence." The bottom layer consisted of coarse sediments deposited under conditions of high depositional energy overlain by successively finer grained sediments such as sand and muds deposited under calmer conditions. The key was finding analogs on land, although many of Bouma's contemporaries were reluctant to accept that the phenomenon would be seen on land.[8]

One who embraced on-land analogs was Emiliano Mutti, professor of geology at the University of Turin. Mutti grew up in a small mountain village in the Northern Apennines, a mountain range that had been deposited millions of years ago in a marine environment but, which over time, had been uplifted. Mutti loved this region and as a geologist realized that the rocks in the valley of his hometown were similar to the offshore sequences that Bouma and Kuenen had described. In 1972, Mutti wrote a paper in the venerable *Memorie della Società Geologica Italiana* and coined the term "turbidite flows" for these murky, subsea avalanches.[9]

Oil companies took an interest in Bouma's and Mutti's work because if the turbidite theory was correct, there was every reason to explore in deeper water and even beyond the continental shelf. Another impetus came from Oxford professor of geology Harold Reading, who had started his career at Shell but left to pursue an academic career researching sandstones, including turbidites, which provided an immediate link back to the industry.[10]

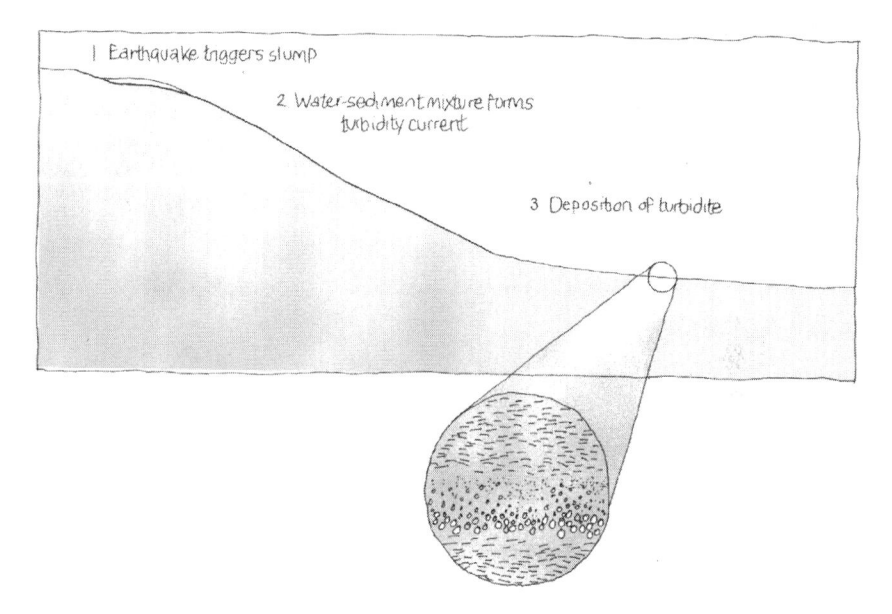

Turbidites, a sedimentary deposit found in many deep-water oil provinces. An earthquake triggers a slump of part of the seabed at a continental margin (1), setting off an "underwater avalanche" (2) that comes to rest creating a characteristic depositional sequence (3).

An early example of a turbidite field was the North Sea Forties field discovered by BP in 1970, a field with no less than five billion barrels of oil in place. Soon oil company geologists started looking for turbidites all around the world and especially the turbidite sandstones on the continental slopes of the Gulf of Mexico. Later, the turbidite hunt broadened to offshore western Africa and Brazil. The turbidite bonanza had begun and soon headed to deep water.[11]

Basin analysis comes into play

In the search for hydrocarbons, geologists always have to start at the largest scale by considering the entire sedimentary basin before zooming in on possible prospects. By the 1970s basin analysis had matured from fairly humble beginnings. One starting point was Lawrence Sloss's sequence stratigraphy that encouraged petroleum geologists to think big. An even earlier concept pioneered by Marshall Kay, a geology professor at Columbia University, was the geosyncline, supposedly a deep sedimentary basin that later evolved into a mountain range through multifarious geologic processes, but this was pre-plate tectonic days and an obvious weakness was not being able to explain exactly what these processes were.[12]

Inevitably, basin studies were transformed by the advent of plate tectonics. "This gave us a real idea of how converging and diverging plates affected basins," recalled Dave Kingston, "Plate tectonics gave us a real handle on how basins were formed." Soon new basin classifications were being postulated. For example, in 1974, Stanford professor of geology William Dickinson suggested five major types: oceanic basins, rifted continental margins, arc-trench systems, suture belts and intracontinental basins. In the mid-1970s, Dan McKenzie, a young geophysicist at the University of Cambridge, created a physical model of basin formation for which the dominant mechanism was stretching by tectonic forces. This was pure academic research. As Bryan Lovell remembers, "Dan wasn't specifically thinking about oil and gas at the time. He was simply trying to understand how the earth worked."[13]

In a 1978 paper, McKenzie presented the basic principles of his sedimentary basin model. As tectonic stretching causes the surface to subside, the model showed how the resulting low fills with sediments; the model included a number of factors controlling this sedimentation, ranging from the rate of subsidence and the amount of faulting, to compaction and even the heat flow and geochemical evolution of the sediments. McKenzie also postulated the broad principles that govern the all-important thermal history of a sedimentary basin, such as the heating of the rocks, the maturation of hydrocarbon and the release of oil and gas from the source rock. The model was the first to offer some kind of prediction on the probability of eventually finding oil and gas. "This paper explained much that the oil companies had learned drilling these basins," recalled Dan McKenzie. Even before publication, his model gained wide acceptance by word of mouth. Within two years, John Sclater and Phil Christie, also at Cambridge, showed that

McKenzie's model was consistent with published evidence of the North Sea crustal structure and its heat flow history.[14]

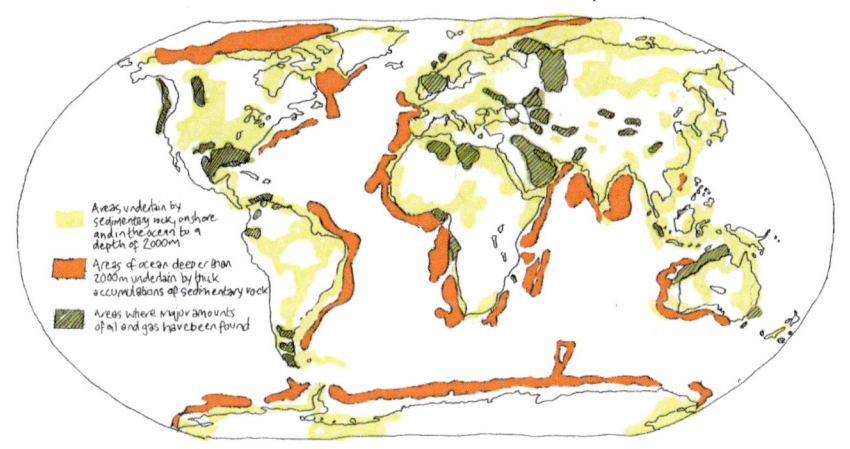

Sedimentary basins around the world. Yellow indicates all areas underlain by sedimentary rock, onshore and in the ocean to a depth of 2,000 meters; green shows the areas where oil and gas have been found by the early 2000s; red shows areas of ocean deeper than 2,000 meters underlain by thick accumulations of sedimentary rock.

By the early 1980s, basin studies had become key for every major oil company. BP formed a basin analysis group headed by geologist-geophysicist David Roberts from the Institute of Oceanographic Sciences in the UK. Exxon began cataloging all the world's sedimentary basin types, estimating their possible petroleum potential. Dave Kingston remembers: "In Esso we worked from satellite photographs, looking at them from a plate tectonic standpoint. How was each basin formed? By rifting, stretching, sagging and so on? We'd then predict the type of structure we'd expect to find in the basin and cross-check with well data once we started drilling."[15]

More recently, satellite gravity surveys have contributed to basin studies offshore. These satellites provide large-scale and precise measurements of the altitude of the ocean surface, data that can be correlated through models to the Moho structure and thickness of the earth's crust. The technique was first described in 1997 and has since become an essential part of the explorationist's toolbox.[16]

Petroleum systems

A companion to basin analysis is understanding how petroleum forms and matures, becoming liquid or gas. There was no doubt that temperature was important. However, in 1888 Hans Höfer, a geology professor at the University of Leoben, Austria, noted that the maturation of petroleum systems did not specifically require high temperatures, but rather the right combination of temperature and time. A Scottish-born geologist, Edward Cunningham-Craig would later rubbish this idea by saying it was equivalent to leaving a turkey in cold storage and expecting it to cook itself. Höfer's insistence on including maturation time would later prove correct.[17]

The mechanisms behind petroleum systems were still not well understood even in the 1950s, because exploration geophysics, petroleum geology and even specific subdisciplines were handled separately by oil companies, as former Shell geochemist Dietrich Welte recalled, "In the very beginning when I started to work for Shell in 1959, the focus was to make progress separately in different disciplines. There was no or very little direct communication between disciplines."

Dietrich Welte, a specialist in both geology and geochemistry, who worked at the German universities of Würzburg and Göttingen and the Technical University of Aachen, and from 1983 formed his own company Integrated Exploration Systems. Welte significantly advanced the petroleum systems concept in the 1970s and 1980s.

A key discipline was petroleum geochemistry, and an early pioneer was Wallace Dow, a geologist working for Amoco studying the Williston basin along the eastern edge of the Rocky Mountains. His aim was "to develop an understanding of the distribution of three major oil types in the Williston basin . . . and where each type was most likely to be found in the future." He presented his results at the 1972 AAPG annual meeting in Denver, Colorado, coining the concept of an oil system, later known as a petroleum system, which encompasses all the components necessary to create hydrocarbons, a migration pathway and a reservoir rock with its seal.[18]

Welte, who later became the director of the Institute for Petroleum and Organic Geochemistry at the Nuclear Research Centre Jülich, Germany, recalled the seminal impact of Dow's Williston basin studies: "From that moment on, I think, people understood that it is not enough to find a trap or an anticline and to drill them—the idea now was to understand the total process behind the cooking of hydrocarbon and its eventual resting place." Along with Welte, another key driver behind the petroleum system idea was geologist Bernard Tissot of IFP with whom he co-wrote the seminal textbook *Petroleum Formation and Occurrence*.[19]

Oil industry professionals studying a rock outcrop on the beach at Osmington Mills in Dorset, UK. The outcrop reveals an oil-rich layer at the height of their helmets.

The success of the petroleum system concept very much depended on computer modeling the basin, and this only became feasible in the

1970s. Probability methods also played a part. This allowed geoscientists to assess the uncertainty of each part of the chain, from source rock to maturation to migration. A key area of uncertainty was temperature history—key because this more than anything else determines the probable type of petroleum, whether it is heavy oil, light oil or gas. The next step concerned migration of petroleum, "When would migration occur, and which kind of migration avenue was available in the basin?" recalled Welte. "Finally, how was the petroleum finally trapped—how big is the trap, how much does it hold?"[20]

With increased certainty, exploration plays and prospects were increasingly developed in basins or regions in which a complete petroleum system could be proven to have some probability of existing. By the late 1980s, petroleum system analysis had become an absolute necessity.[21]

Soviet and Chinese contributions to geoscience

Many of the advances in geoscience in the 20th century originated in North America and Western Europe, but one should not forget the cradle of the oil industry, Azerbaijan and the Soviet Union. The Soviet oil and gas industry had been somewhat isolated since the revolution of 1917, and this led to parallel and different developments in many technological areas. "Even at a fundamental scientific level," recalled Bernie Vining, a former Exxon-geologist, "There were basic differences in geologic concepts used in exploration and production."

One difference related to the understanding of the origin of petroleum. Contrary to western opinion for which the organic or biogenic origin of oil and gas was established doctrine, Russian geologists generally believed in a theory that said hydrocarbons originated inorganically in the mantle. The concept of an inorganic origin of petroleum was heavily debated in the Soviet Union in the 1950s and 1960s and brought to international attention in 1974 by Vladimir Prorfir'ev, professor of geology at the All-Union Petroleum Scientific-Research Geological Prospecting Institute in Moscow. The idea was tested in the late 1980s and early 1990s when wells were drilled near the Siljan Ring in Sweden where a huge meteorite had fallen 377 million years ago, in the hopes of finding hydrocarbons emanating from the mantle. Nothing was found. Since then, the inorganic hypothesis of petroleum origin has fallen by the wayside.[22]

In other areas, Soviet geoscientists made significant contributions to hydrocarbon exploration. One such was Nikolai Lopatin who studied geology at the Northern Caucasian Mining Institute. After obtaining his PhD at Moscow University in 1968, he continued research in the chemistry department before joining the Coal Generation Institute at the Academy of Sciences of the USSR. His main work "Temperature and Geologic Time as Factors in Coalification" was published at the academy in 1971. Lopatin developed a method of using both time and temperature to calculate the thermal maturity of organic matter in sediments. His work was based on coal studies but applied equally well to petroleum and basin modeling. His time-temperature curve predicted which rocks were likely to be mature enough to produce oil and gas. "Soon everybody was calculating Lopatin curves and analyzing basins to find the sweet spots," recalled Exxon's Walter Ziegler, "Lopatin curves were way ahead of most Western thinking and became an all important part of basin and maturation analysis during the 1970s."[23]

In the Soviet Union, Lopatin's discovery was timely because numerous basins were in the process of being explored and exploited, most importantly the vast oil and gas fields in western Siberia discovered in the early 1960s. The West Siberian basin covers an area of 3.4 million square kilometers and is the largest continental sedimentary basin in the world.[24]

In comparison, seismic techniques in the Soviet Union suffered from a lack of ready computing power. But for those with the requisite skills, this created an opportunity as Bernie Vining remembers, "To make up for lack of computing power, Soviet mathematicians had to be much more creative in developing algorithms that did not require a high level of computing." Over the years, the Soviet theorists profoundly influenced the development of seismic processing in the West. As a consequence of the practical difficulties, alternative geophysical techniques such as magnetic and gravity techniques were preferred. Gravity surveys played a leading role in the early exploration of West Siberia in 1950 and especially in the discovery of the Tengiz field in 1979 on the northeastern fringe of the Caspian Sea in what is now Kazakhstan. In modern parlance, the Tengiz field is subsalt—its petroleum-bearing limestone reservoir sits below a massive salt layer. This is a classically difficult scenario for the seismic technique, but the density contrast between the reservoir and much lighter salt made it a natural target for the gravity technique.[25]

The Soviet geoscientists also influenced developments in China. Many Chinese geologists and geophysicists were trained at the leading universities and institutes in the Soviet Union. One was Zaitian Ma, who graduated from the geophysics department at the Leningrad Mining Institute in 1957 and later joined the Bureau of Geophysical Prospecting established in Beijing in 1961. During the 1970s and 1980s, Ma was involved in developing the first seismic processing system in China and became internationally known in 1981 when he published migration techniques to image highly dipping structures.[26]

Chapter 15: Making the Reservoir Work

UNTIL THE mid-20th century, oil and gas production was a simple matter of drilling wells and letting them produce. Oil or gas flowed up the well because it was pushed by the natural underground reservoir pressure, an initial phase of a reservoir's producing life that came to be known as primary production. As production declined, pumps and other lifting mechanisms could be installed, but in the long term a more fundamental shift was required. Pushed by rising energy needs, oil operators wracked their brains to find new ways to maintain reservoir pressure, resulting in a smorgasbord of techniques.[1]

Waterflooding

The first idea dated back to the late 19th century. John Franklin Carll, the Pennsylvanian geologist and engineer, knew that oil and water were immiscible, so in 1880 he raised the possibility that oil recovery might be increased by injecting water into the reservoir to push oil toward producing wells: "Some system might be devised by which water could be let down through certain shafts, and the oil forced towards certain other shafts where the pumps were kept in motion, and thus the rocks be completely voided of oil and left full of water."[2]

The first practical application of Carll's idea came rather accidentally ten years later when many wells in the Bradford field, Pennsylvania, were abandoned following primary production during the 1880s. Some wells were abandoned by pulling casing; in other wells the casing was simply left in the hole where it slowly corroded. In both cases, fresh water from shallower horizons was free to enter the producing interval. Suddenly some wells started producing again. It wasn't exactly a planned injection, but from then on operators ripped out the casing in poorly producing wells and hoped that the pressure of groundwater would increase production. In 1907, this crude practice started to reverse a decade-long decline in oil production from the Bradford field that had fallen to two million barrels per year. As late as 1920, the field was producing 2.5 million barrels per year.[3]

The first systematic technique of waterflooding, called a circle flood, consisted of injecting water into one well until the surrounding production wells started producing water with the oil. The watered-out production wells were then converted to water injection to create an ever expanding circular waterfront. However, not all operators believed in the efficacy of waterflooding. Moreover, a Pennsylvania law requiring the plugging of all abandoned wells to prevent water from entering oil and gas sands was construed by many as a show-stopper. So when waterflooding was performed, it tended to be done secretly. Then in 1921, the Pennsylvania legislature legalized the injection of water into the Bradford sands. At the same time, the circle-flood method was replaced by the more effective line-flood, in which two rows of producing wells were staggered either side of an equally spaced row of water injection wells. Within 10 years the production rate of the Bradford field increased five-fold.[4]

Waterflooding was slowly adopted throughout the oil provinces of the US: in Oklahoma in 1931, in Kansas in 1935, in Texas in 1936 and in Illinois in the 1940s. By 1955 waterflooded reservoirs contributed 12% of total US oil production, but the real jump came a few years later as former Atlantic Richfield Company (ARCO) engineer Fred Stalkup remembered, "Waterflooding came in big time in the early 1960s with those huge carbonate fields in West Texas." During the following decade, in the 1970s, waterflooding provided more than 40% of US production.[5]

In the Soviet Union, water injection was instigated in 1944 when the Tuymazy oil field in the Volga-Ural region was discovered. After World War II, the Soviets systematically used water injection in all new fields, notably the giant Romashkino oil field in the southern Volga-Ural region, where oil production commenced in 1954.[6]

Thermal recovery

But even waterflooding leaves a large fraction of the oil in place or unproduced, estimated in the mid-1970s to be as high as two-thirds of the original oil in place—particularly the case when the oil heavy and viscous. More ingenuity was required to recover this large remainder. One idea was to heat the oil since this lowers its viscosity and eases its flow to the wellbore. A downhole heater had in fact been patented by George Parry and William Warner of Philadelphia back in 1865, but there was no demand then for such an exotic technique.[7]

The first thermal recovery method was in-situ combustion, known as fire-flooding. The principles, which sounded quite dangerous, were set out in 1923 in patents issued to Edson Wolcott of Los Angeles and Frank Howard of Standard Development Company, New Jersey. As the patent explains, the oil-bearing formation in the neighborhood of a well is set alight with a downhole igniter and then fed oxygen by pumping air down the well; as the fire develops, the burning front pushes the mobilized oil toward production wells.[8]

Wolcott and Howard's idea was left in the drawer, however, so it was the Soviets in the mid-1930s who first attempted in-situ combustion, in a pressure-depleted reservoir in Azerbaijan. Difficulties with igniting the oil meant that only a small amount of oil was recovered. However, they did demonstrate that combustion can occur within the porous structure of an oil-bearing sandstone. Publication of the Soviet work in English in 1939 spurred at least one independent operator in the US to try the technique. In 1942, Brundred Petroleum Inc. obtained combustion in sand immediately around an injection wellbore in Oklahoma, but the experiment was abandoned because it failed to increase the field's production.[9]

During the postwar years in the US, research and development into in-situ combustion intensified, particularly at the Magnolia Petroleum Company and Sinclair Oil and Gas Company, both in Oklahoma. By 1953, the principles of the method were verified in the laboratory and tested in the field. As the process was further developed and more field tests undertaken in the late 1950s, engineering difficulties began to surface. Problems included explosions in the production facilities, heavy corrosion and sand production.[10]

While some operators played with fire, others chose the safer option of injecting hot steam to heat the oil. As Mike Prats, former Shell research engineer and thermal recovery guru, remembered from the 1960s, "Shell had the position that if there was a process in which thermal processes were required, doing it with steam was preferable to combustion." Their first attempt can be dated to 1931 when steam was injected for 235 days into an 18-foot sand at a depth of 380 feet in the Wilson and Swain lease near Woodson, Texas; the results were not recorded. The breakthrough came much later in a Shell field in Venezuela. And just like the beginnings of waterflooding in Pennsylvania, serendipity played a key role. "The big innovation in steam happened by sheer accident!" recalled Mike Prats.[11]

The Venezuela story begins in 1956 with a series of lab experiments in Shell's Rijswijk laboratory to investigate how steam displaced heavy oil in cores. The results encouraged Compañía Shell de Venezuela (CSV) to field test a steam drive in the viscous tar sands of the Mene Grande field, close to Lake Maracaibo. CSV engineers injected steam into a reservoir, but two years into the project, steam injection into the shallow sands at 550 feet began to exceed overburden pressure, causing cratering and ultimately steam, water and oil eruptions at the surface. The company therefore discontinued the test by relieving pressure in the injectors. At that moment one of the injector wells blew back, and along with the steam came a lot of oil. By the time everything stabilized, the wells that had previously produced just a few barrels per day now started producing 100 to 200 barrels per day.[12]

Cyclic steam injection was born. The method that became known as "huff-and-puff" consisted of three stages. Steam is first injected into a well for a certain amount of time to heat the oil in the surrounding reservoir to a temperature at which it will flow. Injection is then stopped, allowing the steam to soak for a while, typically not more than a few days. Then the heated oil is produced, at first by natural flow since the steam injection will have increased the reservoir pressure, and later by artificial lift. Production decreases as the oil cools down, and so at a certain point the steps are repeated.[13]

Huff-and-puff was used to great effect in the heavy oil sands in California during the 1960s. And in the Soviet Union the technique was applied in 1965 in the oil fields of Krasnodar Territory, north of the Caucasus Mountains and east of the Azov Sea.[14]

Miscible displacement with gas

Another way of pushing the oil out could be equally attributed to serendipity. In 1888, James Dinsmoor was working as a roustabout on the William Hill property in Venango Country, Pennsylvania, which produced oil from the Third Venango sand. On an adjoining property the Oil Well Supply Company deepened one of its Third Venango sand oil wells to the underlying gas-bearing Speechley sand. They found abundant gas but, temporarily lacking pipe or packer to prepare for production, shut the well in for a while. An immediate increase in oil production was observed in three nearby Third Venango sand oil wells on the William Hill property. When the Oil Well Supply

Company finally got round to completing and producing their Speechley gas well, oil production from the Venango wells returned to normal. Dinsmoor figured correctly that the Venango and Speechley were connected and that gas had been pushing the oil out of the Venango.

Dinsmoor made a business out of his observation and started to acquire shallow oil leases with low production with the hopes of increasing their production by injecting some form of gas. Others followed his example, and by 1917, gas injection had been used in over 90 projects involving some 4,000 wells, mostly in southeastern Ohio, northwestern West Virginia and Oklahoma. In most cases, operators recorded up to a four-fold increase in production. By the late 1920s, it was well established that gas injection would systematically increase production. In the late 1930s, gas injection was successfully applied in the Bahrain oil field, discovered in 1932.[15]

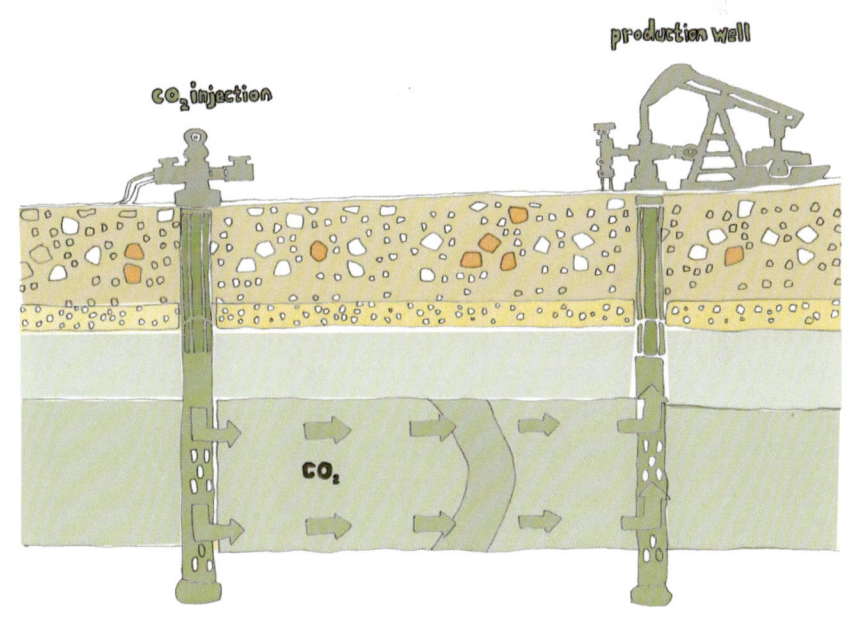

CO_2-flooding. Carbon dioxide is pumped into the reservoir through an injection well; the carbon dioxide mixes with the oil creating a lighter, less viscous fluid that flows up the production well.

The technique stepped up a notch in 1940 at The Atlantic Refining Company. Engineers at the company laboratory simulated gas injection using long samples of porous rock. The objective was to figure out, "What kind of oil recovery you could get by simply increasing reservoir pressure," as ARCO's Fred Stalkup recalled. The results were encouraging and surprising. "As they injected gas, they found that the higher the pressure the higher the oil recovery. Finally the pressure got high enough to push 99% of the oil out of the sample. It seemed too good to be true!"[16]

Further research by Atlantic Refining in the early 1950s led to an explanation of this phenomenon, which came to be known as "miscible displacement." In the Atlantic Refining experiments, the gas pressure had been raised high enough that the gas and oil began mixing, creating a lighter, less viscous fluid that could be produced more easily. During the 1950s and 1960s, Atlantic Refining as well as other oil companies such as Amoco developed many variations of miscible displacement, for example using liquefied petroleum gas. In the early 1970s, carbon dioxide was tried, the first time by Chevron in Scurry County, Texas in 1972, and eventually became the most commonly used miscible method.[17]

Chemical flooding

In the late 1960s, yet another recovery technique emerged, using special chemicals to push oil out of the reservoir. The first chemical used was polymer—long-chain molecules with very high molecular weight that occur naturally, as in starch and guar, or are made synthetically. Polymer science was born in the early 20th century in the laboratories of large chemical companies such as IG Farben, Dupont and Dow Chemical Company for the manufacture of plastics, rubber, adhesives and a host of everyday products.[18]

In 1964, David Pye of Dow Chemical and Burton Sandiford of Union Oil Company separately established that small amounts of polymer dissolved in water could significantly increase the viscosity of the water, and consequently when injected into a reservoir push more of the oil out. A number of oil companies immediately initiated their own research into polymers as well as other viscosity-influencing chemicals. "The major oil companies were doing lots of research in this area when I joined the industry in 1967," recalled George Hirasaki, a former Shell chemical engineer who worked on polymer flooding, and is now a professor at Rice University, Houston. One of the most popular polymer systems was developed at Marathon Oil Company's research lab in Littleton, Colorado, by Robert "Bob" Sydansk.[19]

There was also activity outside the US. The first tests of polymer technology in the Soviet Union were carried out at the end of the 1960s in the Mishkinskoye oil field, approximately 250 kilometers south of Tyumen, western Siberia. The tests were quite successful, increasing the field's recovery factor—a measure of how much oil can be produced versus the total amount of oil in place—to nearly 40%, from the 10% to 15% typically achieved with waterflooding.[20]

Ever since, however, the high expense of polymer flooding has limited its popularity, with the notable exception of the Daqing field, the largest oil field in China, situated in the northeastern corner of the country. The field was discovered in 1959 and had an estimated 2.2 billion metric tons of oil in place. Primary production levels started to decline during the 1960s, so a waterflooding program was started. Eventually in 1993, the China National Petroleum Company initiated polymer flooding, the first large-scale commercial application of the technique in China. This helped stabilize production and continues to control production decline today. Today, the Daqing oil field produces 40 million metric tons of oil per year.[21]

A couple of polymer injection pumps in the Daqing oilfield, part of an array of pumps housed in a warehouse for shelter against the harsh winter weather of North China. The polymer solution comes via the pipes behind the pumps.

Another idea was to add surfactant to the injection water. A surfactant molecule is attracted to both oil and water, reducing their interfacial tension to such an extent that the oil breaks into tiny droplets and is more easily drawn from the rock pores. Surfactant flooding was proposed as early as 1927 by Lester Uren, professor of petroleum engineering at the University of California at Berkeley and E.H. Fahmy from the same establishment, but it took 40 years before the oil companies and academia began playing with the idea, notably Marathon Oil, Shell, Exxon and The University of Texas at Austin. By the end of the 1970s, researchers realized that controlling slugs of surfactant injected into the reservoir was going to be a problem. Hirasaki recalled, "Does the surfactant preserve its properties when it mixes with the other fluids as it moves through the reservoir? We learned that it doesn't." A change in the 1980s was adding alkali to the surfactant to reduce the amount of surfactant required, always an expensive product. In the end, the 1986 oil-price crash temporarily killed interest in surfactants as it did for polymer flooding.

When the oil price recovered in 2003, so did research in surfactants. Hirasaki, who had joined Rice University in 1993, partnered with Gary Pope of The University of Texas in a US Department of Energy (US DOE)-sponsored project on surfactant flooding, resulting in new technology, improved surfactants and better knowledge of how to formulate them.[22]

Chapter 16: Logging Breakthroughs

ALL THESE reservoir flooding techniques required better understanding about the rock and its pore space. Since the early 1930s, knowledge about rock properties was revolutionized by the invention of electrical well logging, and as more sophisticated logging tools emerged, major improvements in log interpretation soon followed. However, two big challenges remained: unraveling the secrets of shaly sands and distinguishing between the three main sedimentary rock types.

Shaly sands and other rock types

Gus Archie, in 1942, had paved the way for log interpretation. But Archie's equations were established in the laboratory and specifically for clean sands, not for shaly sands. With the advance of exploration around the world after World War II, a greater variety of reservoirs was encountered, and shaly sands seemed to be the rule rather than the exception. In 1953, Henri Doll of Schlumberger wrote in a *Journal of Petroleum Technology* article: "The most important problem that has received so far no satisfactory solution is that of shaly sands." The problem was that shales appeared to conduct electricity in ways quite different from Archie's simple model, generally increasing conductivity and as a result invalidating any kind of water saturation evaluation.[1]

The first attempt to understand the electrical properties of shale was in 1950 by Malcolm Wyllie and his colleague Homer Patnode, both from the Gulf Oil Research and Development Corporation. They established experimentally that the conductivity of a shaly sand sample was proportional to the percentage of shale in the sample. However, this result did not explain how shales conducted electricity. That came in 1953 when Weldon Winsauer and William McCardell of Humble Oil realized that shales do not conduct electricity when dry, but only do so when wet because in this state positively charged cations can move along the shale-water interface, an interface that later came to be called the electrical double layer.[2]

The first modification of Archie's equation to account for the double-layer effect was developed in 1967 by two Shell physical chemists, Monroe Waxman, who worked at the company's research labs in Houston, and Lambert Smits, who worked at the research lab in Rijswijk near The Hague. Using measurements made on nearly 200 samples, they confirmed Winsauer and McCardell's findings and introduced a concept called the shale's cation exchange capacity in order to modify Archie's equation. In 1974, Waxman and his Houston-based colleague E.C. Thomas corroborated the Waxman-Smits modification in an experimental study of 12 moderately shaly sands. Gradually, the Waxman-Smits model became accepted despite a few inconsistencies with experimental results. However, Thomas remembered the main drawback: "Using the Waxman-Smits equation required that you develop an empirical data set from core analysis from each reservoir, so you could tie the value of this extra conductivity to some parameter that we could measure from logs." It was still rather complicated, in fact beyond the capability of most operators.[3]

The need for a more manageable solution brought Schlumberger into the arena. Developing a logging tool that could directly measure the all-important cation exchange capacity was judged impossible, so the company focused on establishing a simpler interpretation model. George Coates, a log interpreter working at the company's research lab in Ridgefield, Connecticut, made the assumption that one could determine the extra conductivity of a particular clay mineral from measuring electrical conductivity of the surrounding pure shales. That led to a reformulation of the Waxman-Smits equation, taking into account two different waters: clay-bound water to account for the shale cations and free water for the clean sand.[4]

This "dual-water model" was first presented by Coates and his Schlumberger colleagues Christian Clavier and Jean Dumanoir at the annual SPE conference in Denver, Colorado, in October 1977. The model was summarily rejected by industry experts, because it appeared to be a simple rehash of the ideas of Waxman and Smits and featured no new laboratory measurements. There was also Shell. "Shell simply didn't buy into it," recalled Coates.[5]

However, during the late 1970s and early 1980s, the dual-water model, the simpler of the two approaches, was seen to be the more feasible way to evaluate shaly sands and was successfully applied to thousands of wells. Robert Freedman, a former Shell physicist, recalled, "The two approaches are equally good, but Waxman-Smits is simply not as easy to use, because the shale parameters in the dual-water model can easily be estimated from logs whereas determining cation-exchange-capacity in the Waxman-Smits equation requires core measurements." The practical advantages of the dual-water model were also further underpinned in 1983 by Coates and Clavier's extensive study of log data from Texas, California, North Dakota, Oklahoma and Canada. The following year, the dual-water model was finally published in the *SPE Journal*, and it has since become extensively used for establishing water saturation in shaly sands.[6]

For reservoir understanding, however, shaly sands constituted only part of the story. More than 60% of the world's oil is found in carbonate rock, comprising mainly limestone and dolomite. The remaining interpretation challenge was figuring how logs could distinguish between the big three sedimentary rock types: sandstone, limestone and dolomite.[7]

The first clue came in 1963 from two field hands—Wayland Savre of Gulf Oil Corporation in Odessa, Texas, and Jack Burke of Schlumberger in nearby Midland, Texas—who were trying to unravel the complex lithologies of the West Texas Permian basin. Not only was establishing rock type a challenge, but so was estimating the rock's porosity. The two were obviously linked, and Savre and Burke realized that the right combination of the three available logs measuring porosity—the neutron log, the density log and the sonic log—could provide the answer to both conundrums. They were on the right track. Then in 1971, three Schlumberger veteran interpreters—Andre Poupon, Bill Hoyle and Arthur Schmidt—discovered that any two of these logs could do the trick, picking on the density-neutron combination to provide the best resolution for delivering both lithology and porosity. The so-called density-neutron crossplot became the basis for all lithologic log interpretation and was incorporated into computer-driven interpretation programs being developed at that time.[8]

Imaging the borehole

As these dramas played out, other advances in well logging followed the quite different track of actually trying to image the borehole wall.

Since the 1950s, several tools were developed to record optical images in the borehole, to see fractures in the open hole or to inspect completions and tubulars. The first was built by J.C. Dempsey and J.R. Hickey of the Tulsa-based oilfield service company Birdwell in 1958 and featured a black-and-white camera equipped with a 16-mm wide-angle lens. A similar camera was constructed in Russia in 1964 to see the distribution of fractures in oil and gas reservoirs.[9]

The same year, Shell patented an inspection tool with a wide-angle television camera that could take real-time black-and-white images; the technology was licensed to Oceanographic Engineering Corporation of La Jolla, California, which built the tools. The camera was mounted in a cylindrical housing with light sources and a mirror set at 45 degrees that could be rotated on command from the surface. This way, the camera could view the complete circumference of the borehole. A compass mounted in the field of view of the camera identified the direction in which the camera was looking. However, these early cameras saw limited use because they required a transparent borehole fluid or air-filled hole. In the case of the Shell tool, the quality of the coaxially transmitted video was compromised by high borehole pressure and temperature.[10]

These deficiencies inspired a group of researchers, led by Joseph Zemanek at Mobil Research & Development Corporation in Paulsboro, New Jersey, to seek an alternative imaging technology. In 1967, they constructed a logging device that imaged the borehole wall acoustically. Their "borehole televiewer" featured a rotating transducer that sent high-frequency sonic pulses to the borehole wall or casing and then measured the time for the reflected signal to return. The results were displayed on an oscilloscope and recorded by a camera attached to the oscilloscope. It was the first borehole imaging tool to successfully function in opaque liquids, clearly showing fractures, vugs and washouts in the open hole, a revolution at the time. In cased hole, perforations, splits and casing collars became clearly visible. In one case—a granite-wash field in Libya—the borehole televiewer was the only logging tool worth running since it alone could see the fractures through which the oil would eventually flow.[11]

However, the Mobil borehole televiewer had its limitations. Heavy muds would attenuate the acoustic pulses to the point they were undetectable, and the telemetry up to the surface could not cope with the long cables typically in use. Robert Hunt-Grubbe, a Schlumberger field engineer, recalled that "Mobil's borehole televiewer could only work on cables of 6,000 feet or less, and most of our cables were 18,000 feet! So, that was a problem." But the concept was sound, and following advancements in digital technology in the 1970s, the tool established its place in the oil field with other oil companies such as Amoco, ARCO and Shell building their own versions. By the 1990s, the service industry took control of the technology.[12]

Meanwhile, a more precise imaging technology was in development. Electrical resistivity logging had seen a rapid evolution since its early days in the 1920s and 1930s, but a perennial deficiency of the early tools was resolving very thin permeable zones. Then in 1950, Doll, pursuing an early idea of Conrad Schlumberger, built a logging device he later called the Microlog that could measure the resistivity of a small volume of rock just behind the borehole wall. The device featured three electrodes mounted on a flexible rubber pad that was kept in contact with the borehole wall. The spacing between the electrodes was only a few inches to ensure a shallow yet very precise resistivity measurement.[13]

The Microlog soon evolved into fancier variants with an increasing number of electrodes, but the true breakthrough in electrical imaging didn't happen for another 30 years. The basic idea came from Schlumberger theoretician Stan Gianzero and BP's Philip Threadgold, while the impetus to achieve a working tool came from imaging expert Mike Ekstrom, who was hired into the Schlumberger Ridgefield lab in the 1980s from the Lawrence Livermore Laboratory in California. Ekstrom was an expert in the relatively new field of digital imaging, so the solution he envisaged was hugely increasing the number of electrodes on the pad and using computers to process the data into a borehole image. The result was a logging tool called the Formation Microscanner, introduced in 1986, that had 54 electrodes distributed equally on two pads. It created quite a stir. Stefan Luthi, a geologist involved in the early trials and now professor of geology at TU Delft, remembers the breakthrough: "With borehole imaging you can actually see details in the well, you can very precisely determine, even without taking a core, structure at a large scale and the depositional environment at a small scale." Electrical scanning never looked back, and more sophisticated versions with hundreds of electrodes soon became available.[14]

Others had ambitions to image the cased hole, for monitoring casing corrosion and general wear and tear that potentially could compromise the entire well. One was Robert Hunt-Grubbe, who by now had left Schlumberger to found Sondex, a well-logging instrumentation company near Reading in the UK. In 1992, Sondex introduced its first well-imaging tool, a multifinger caliper with 40 fingers. Hunt-Grubbe recalled, "My dream was to provide an instrument that would deliver an accurate picture of the pipe surface in any well condition, and then find a way to present the data as a moving colored display as if the observer were traveling down inside the pipe seeing every crack and bump." The tool development was partly funded by a group of five oil companies led by BP, and was soon accepted as the most reliable method for inspecting the inside of casing.[15]

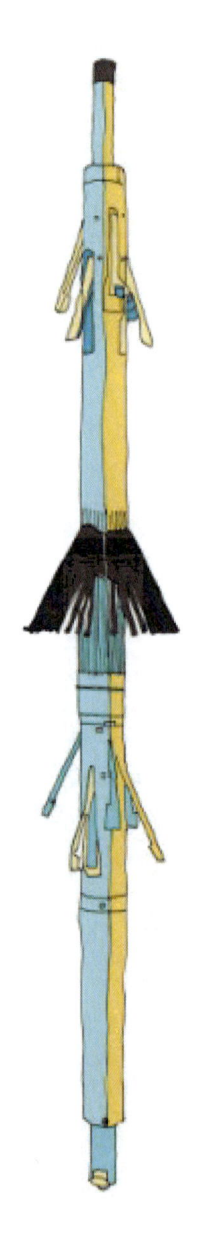

A Sondex caliper imaging tool from the 1990s, with 40 fingers, each capable of measuring very small changes in casing diameter. The individual measurements are transmitted up-hole to provide a real-time image of the inside of the casing.

Borehole geophysics

As it evolved, borehole imaging provided a very fine vertical resolution in the borehole, even if the measurement didn't extend much into the formation. Conventional logging could provide a deeper measurement, but somewhat at the cost of vertical resolution, which was typically six inches to a couple of feet. But the jump in resolution to seismic surveys remained vast. Seismics was unable to resolve features smaller than 30 to 40 feet. For the reservoir engineer, bridging these measurement scales with borehole seismics was of the utmost importance.[16]

The breakthrough came from the Soviet Union. In 1959, geophysicist Evsei Iosifovich Galperin at the Institute of Physics of the Earth at the Academy of Sciences in Moscow created virtually single-handedly the new discipline of borehole geophysics. Early checkshot surveys measured only the time of first arrival. Galperin understood the importance of recording the entire downhole waveform and at much closer vertical intervals than was prevalent at the time—25 meters was his suggestion. From complete waveforms, he showed how to extract the downgoing and upgoing energy, and began to provide a far more secure correlation with surface seismic data. The new method became known as vertical seismic profiling (VSP).

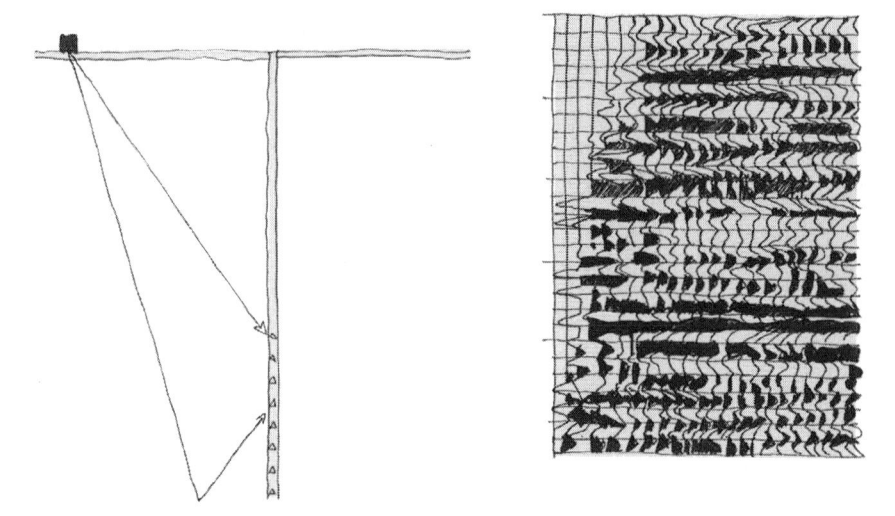

Vertical seismic profiling. Left, the black square represents the seismic source, and in the borehole receivers lowered on a logging cable record the arrivals. Right, an example of a seismogram with the horizontal axis showing the offset distance of the source from the well and the vertical axis showing the travel time of the recorded seismic waves.

The results of Galperin's work were published in Russian in 1971 and made available to the West in an English translation in 1974. One of the early adopters in the West was Nigel Anstey, now with Houston-based Seiscom Delta, who was granted a US patent in 1975 for a variant of vertical seismic profiling called broad-line seismic profiling. Anstey promoted the VSP method giving a special seminar on the subject in Oklahoma in 1979 for more than twenty US operators and several universities. By 1985, vertical seismic profiling had become standard practice in the West, as it had become in the Soviet Union. Its ability to close the gap between surface seismics and well logging was simply too important to ignore.[17]

Nuclear spectroscopy

Galperin was just one of a prolific scientific milieu in the Soviet Union of the 1950s and 1960s. Soviet scientists were active and successful in almost every technical discipline.

One area was nuclear spectroscopy logging, in which nuclear particles, for example neutrons, are aimed at the formation and then the resulting nuclear particles emitted by the formation in reaction are analyzed in terms of energy, concentration and timing. There were endless variants to explore, but the resulting spectra promised to yield DNA-like tagging of most of the chemical elements present in the formation, in both rock and pore fluid. And, crucially, such measurements could be made in cased-holes since nuclear particles, if chosen carefully, were comparatively untroubled traversing steel casing.[18]

With the Iron Curtain more open than it would later become, scientists and engineers in the West could keep up to date with Soviet work through Russian-language publications and their translations. One paper caught the attention of Pan-Geo Atlas Corporation, based in Houston. It described how to measure chlorine concentration in the formation, and by inference water saturation, using a chemical source of neutrons. In 1959, W.R. Rabson of Pan-Geo put this know-how into practice and introduced a nuclear-based "chlorine" log.[19]

Another Soviet article in 1958 aimed at measuring water saturation behind casing attracted the attention of Western companies. It described early experiments of a spectroscopy logging tool in which pulses of neutrons were produced at will, electronically. The technology for producing neutrons electronically had existed since the 1950s on both sides of the Iron Curtain for detonating hydrogen bombs, but of course it was highly classified. The challenge, at least in the West, was producing a version that could be used commercially for logging oil wells.[20]

The first to succeed was the Lane-Wells company, the California outfit that in 1955 had been acquired by Dresser and merged with Well Surveys. The leading physicist at Lane-Wells was Arthur Youmans, an Iowa farm boy who managed to educate himself during the Great Depression. Lane-Wells produced their first tool in 1964, followed by Schlumberger in 1966. The technical challenge was significant because the tool had to be slim enough to pass through production tubing, and the neutron pulses were generated by accelerating electrically charged ions through 120,000 volts, rather difficult to engineer safely in such a confined environment.[21]

Arthur Youmans (1922-2004), with Richard Conner and Erik Hopkinson, two members of his research team at Lane-Wells, and the first pulsed-neutron tool, in Tulsa, Oklahoma, 1963.

The door soon opened to using nuclear spectroscopy for identifying any number of elements in the subsurface. In 1980, Russ Hertzog at Schlumberger's Ridgefield lab reported on a prototype spectroscopic measurement using a pulsed neutron source that could identify no less than carbon, oxygen, silicon, calcium, iron, chlorine and hydrogen. It took a few years to make these measurements reliable, but the path was now clear except for one obstacle: converting elements to rock type. This might be straightforward for chemically simple rocks such as sandstone and limestone, but it was fraught with complexity for shales and clays with their endlessly variable chemical makeup. This provided the lifework of husband and wife team, Michael and Susan Herron, at Schlumberger's Ridgefield lab. Over the years, they obtained more than 400 core samples from numerous wells on four continents and analyzed them from both an elemental and lithologic point of view. The challenge was then finding a mathematical relationship between the two sets of data. Their results provided the missing link to move spectroscopy forward to the dominant position it holds today. Meanwhile, it became clear that nuclear spectroscopy was also possible with chemical source of neutrons—a throw-back to Rabson—and from early successes in Texas, this technology has also been progressively upgraded.[22]

The miracle of nuclear magnetic resonance

Although the oil industry seemed to have gotten its money's worth from measurements based on the principles of nuclear physics, the discipline could offer one more opportunity. In 1952, Harvard physicist Edward Mills Purcell and Swiss-born Stanford physicist Felix Bloch won the Nobel Prize for their 1946 discovery of nuclear magnetic resonance (NMR). It had been discovered previously that for elements with an odd number of protons or neutrons, the element's nucleus acts like a small magnet. Purcell and Bloch discovered that if you perturb the element with a strong magnetic field, the magnetized nuclei emit a characteristic and recognizable electromagnetic response.[23]

At first, the discovery was only exploited in academia for studying matter at the molecular scale. Then in 1948, Russell Varian, an electronics and aviation pioneer from California, started working on commercial applications of Purcell and Bloch's discovery. Varian proposed using the earth's magnetic field to stimulate proton NMR and in 1951 was granted a patent for an earth-field NMR magnetometer. At the same time, he had his eye on the oil industry and in 1952 filed another patent adapting his NMR magnetometer to well logging.[24]

By now, the oil industry was getting curious, particularly California Research Corporation in San Francisco, a subsidiary of the Standard Oil Company of California, today's Chevron. Beginning in 1953, the company had a research and development project headed by Robert Brown building an experimental earth-field NMR logging tool. By 1958, they had a functioning device that could detect hydrogen atoms, whose nuclei of a single proton have a relatively large magnetic moment. The interest in hydrogen of course was that it is present in both water and hydrocarbon, the two fluids filling the rock's pore space, so the logging tool had the potential to measure porosity directly. But there was more. In 1964, experiments linking NMR with fluid flow through porous rock had been conducted at the University of California, Berkeley, by a young Turkish geoscientist, Aytekin Timur, later known as just "Turk." Timur was soon working for Chevron where he extended his experiments to cover 150 samples from three US sandstone fields and developed the first theory positively linking NMR to permeability as well as porosity.

The first commercial logging service was introduced by Pan-Geo Atlas in 1962, using the Chevron design, but its commercial success would be dogged by a serious logistical drawback. The tool reacted to hydrogen in the borehole fluid, so to get any discernible measurement from the formation, the well had to be doped with magnetite to kill the borehole signal, an expensive and time-consuming procedure. Despite this potential showstopper, Schlumberger entered the fray, also using Chevron's design, bringing out a tool in the late 1960s and developing a second version in the late 1970s.[25]

Then, following the 1973 oil embargo, US President Jimmy Carter called on the US National Laboratories to become more involved in fossil fuel research. In February 1978, the Los Alamos National Laboratory formed a new geophysics group led by physicist Jasper Jackson, and one focus was nuclear magnetic resonance, but with a difference. This time, the NMR effect would be induced using a strong permanent magnet rather than the weak earth's magnetic field, and the field was directed into the formation to avoid stimulating hydrogen in the borehole. Laboratory tests of this technique were followed by a field test in 1983 and published in 1984. Building on "Turk" Timur's work, there was the tantalizing possibility that the industry might finally have a logging tool that would yield permeability.[26]

Although the Los Alamos design had serious limitations—such as low signal-to-noise ratio that prohibited continuous, nonstationary logging—the concept set the stage for the development of today's commercial NMR tools. The first company to jump on the Los Alamos bandwagon was a small firm called Numar, owned by Melvin Miller and based in Malvern, Pennsylvania. Miller had previously worked in the medical field and didn't hesitate to switch from one industry to another, but he didn't know a thing about the oil industry. Needing to talk to somebody with industry experience, Miller contacted Richard Bateman, a log interpreter who had worked with Schlumberger and then Amoco before becoming an independent consultant. Bateman recalled, "I get a call one day from somebody in Philadelphia who had been in the medical equipment business and had chanced across a patent written by Schlumberger about NMR in oil wells. He asked me to explain how logging worked and what was involved."[27]

In 1985, Numar obtained a license to produce a Los Alamos-type tool. Miller discussed the signal-to-noise ratio problem with Jackson and opted for a complete redesign. Their logging tool took four years to develop and was tested in 1989 in a Conoco test hole in Ponca City, Oklahoma. The magnetic resonance logs correlated well with porosity data from density and neutron logging tools and also water saturation measured on core samples. Numar hoped to partner with one or more of the three big service companies. "The expectation had been that after demonstrating technical feasibility," Melvin Miller wrote in 2001, "the service companies would be competing aggressively for rights to the technology." However, none of them took the bait. The price was not right, and at least one service company, Schlumberger, was already well advanced to achieving its own version of the technology, including a good deal of petrophysics research to interpret the new data.[28]

The race was on. Numar now started its own commercialization effort. In June 1991, it logged its first well for Mobil in Dewey County, Oklahoma, and within two years had logged 331 wells for many operators. The results were intriguing. "Measurement-wise it worked extremely well, very, very well," recalled George Coates who by this time had left Schlumberger and joined Numar as their interpretation guru. Its size was a disadvantage though. "It was a big tool, almost as big as the hole," recalled Coates, "Oilmen don't like to have logging tools in their hole in the first place, especially a big one." But the interpreted results promised huge improvements for both porosity and permeability. The industry suddenly waxed enthusiastic. For a moment, Numar's tool looked as if it was the measurement to end all other petrophysical measurements. Schlumberger now accelerated dramatically the commercialization of its own tool and was soon giving Numar more than a run for its money. For their part, Halliburton signed a cooperative agreement in 1996 with Numar for NMR logging services and eventually acquired the company in 1998.

Meanwhile, research into the connection between NMR and permeability continued apace, as scientists at Schlumberger's research lab in Connecticut provided a whole slew of new experimental data and advanced theorizing. By the time reliable logging tools appeared in the early 1990s, the stage was well set to ensure NMR's place as the preferred logging measurement for permeability. In the end, NMR was not to be the one logging tool to replace all others, but its place in the pantheon of logging tools was assured.[29]

From analog to digits

Logging, like its cousin seismic exploration, started life in the analog world, but by the 1960s was beginning a slow transition to the digital age. First, the recording of log data turned digital, the earliest manifestation being Schlumberger's attempt to record logs on punched paper tape in 1964, which proved to be rather noisy for the wireline engineer and anyway not capable of high data rates. The next and more successful attempt was recording data from the company's new high-resolution dipmeter tool on magnetic tape. The analog measurements made by the logging tool's sensors were transmitted to the surface using a new distortion-free telemetry and converted to digital format at the surface.

Not far behind came the use of computers for processing and interpreting log data. In those days, computers were scarce and so huge they filled rooms. In addition, logs were still mostly recorded in analog format, so every single curve on every log had to be manually digitized. Nevertheless, the industry soon realized the versatility of digital computation, but that didn't stop operators and service companies fighting about who had the right to perform computerized interpretations. A body of opinion among some operators held that since they had paid for the log data, and some of the data was highly confidential, they alone had the right to harness computers to interpret logs. The service companies disagreed, and after a few years, to the relief of both parties, the squabble petered out as an irrelevancy.[30]

The most dramatic step for logging's digital transformation was heralded by Intel's invention of the microprocessor in 1971. This accelerated the miniaturization of computers to the point that the control unit in the well logging unit could become digital. Previously, logging tools were controlled by the field engineer manipulating a stack of analog panels featuring a multitude of switches, connectors, adjusting knobs and meter dials. There was plenty of room for error. Computer control made life a bit easier, and in addition the computer could be programmed to make simple wellsite interpretations, cutting the time between logging and providing operators with usable results.[31]

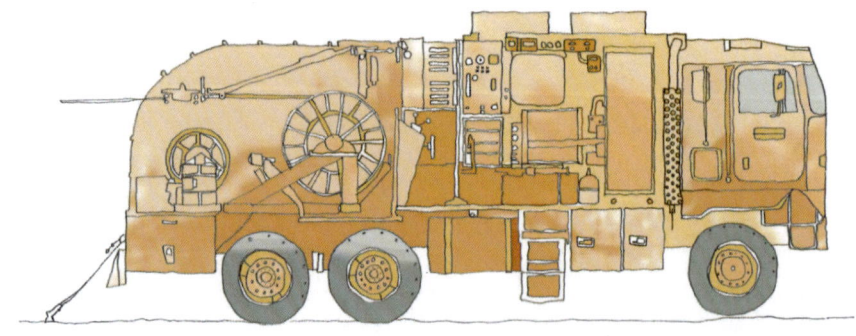

Cutaway view of logging truck from the 1980s, showing a winch with logging cable, the computerized operator's station and the driver's cabin.

In 1972, Gearhart-Owen Industries in Fort Worth, Texas, which had been founded in 1955 by Marvin Gearhart and Harold Owen, began developing a fully digital logging system. Gearhart hired Jack Burgen from Texas Instruments and Max Moseley who had 15 years of experience developing advanced digital systems, first for the military with Western Electric Company, then for seismic exploration in his last job at Teledyne Geotech of Garland, Texas. Gearhart-Owen's system was called the Direct Digital Logging (DDL) system and featured a ruggedized Raytheon computer and various peripherals. The first two

DDL logging trucks began their field trials in August 1975 and went into commercial service by the end of that year.[32]

The expectations for the new system were higher speed and better reliability. Oil companies were excited. A year later Schlumberger launched its own digital logging system, the Cyber Service Unit. Still today Marvin Gearhart recalls the thrill of beating Schlumberger to the market: "Schlumberger was the standard, the leader of the industry, so if you were ahead of Schlumberger, it made you look rather good. That really helped our business take off." By the early 1980s, digital logging had almost completely replaced analog techniques. But this didn't help Gearhart-Owen. When oil prices plunged in 1986, the company and its DDL struggled to survive, and two years later Halliburton bought them.[33]

Chapter 17: Controlling the Well

BOPs ARE an insurance against wells going out of control, but they don't absolve the driller from understanding and controlling kicks when they occur. "If I have seen further than others, it is because I have stood on the shoulders of giants", Bill Rehm, who worked for Dresser from 1955 to 1975, unpretentiously quoted Sir Isaac Newton during his interview for the *Journal of Petroleum Technology*'s 2008 special issue "Legends of Drilling."[1]

Well kicks and high pressures

Bill Rehm likes to refer to ideas put forward by T.B. O'Brian and W.C. Goins of Gulf Oil Research and Development Corporation from 1960: "The work by O'Brian and Goins on well kicks and well control was a major step because up to that point we had no idea how to handle high pressures. We did it with brute force and awkwardness. I mean you could always pressure up the bottle until something breaks, then hopefully it breaks a mile underground and you don't care. But they taught us how to do this in a controlled manner."[2]

Bill Rehm, a former drilling engineer with Magcobar, a division of Dresser. Rehm developed well pressure management and control, including the drilling choke that helps control kicks.

In 1962, Rehm was a technical service engineer for Magcobar, a division of Dresser, and posted to southwest Louisiana to look at the well control problem. The operator had assigned Drilling Well Control, based in Lafayette, Louisiana, to control pressure kicks, and they did so using a series of skid-mounted separators that progressively stepped down the pressure that was appearing in the annulus. This method was based in part on the ideas of O'Brian and Goins, and represented a significant step forward. But Bill Rehm proposed a simpler method, just using a well choke to restrict flow rate, effectively killing the kick. It worked, and well chokes to kill wells achieved rapid acceptance throughout the Gulf of Mexico.[3]

Some of the oil companies applied the new method deliberately to stretch their drilling envelope. In the Gulf of Mexico, Shell started drilling from kick to kick, controlling the mud density in such a way that the well kicks were small and easy to control. Other companies were not quite so lucky.[4]

Oilwell firefighting

In the early hours of February 20, 1962, the spacecraft *Friendship* 7 and John Glenn began their first orbit around Earth. As they flew over Africa at an altitude of 300 kilometers, Glenn noticed a fire in the western part of the Sahara and described it as "the devil's cigarette lighter." Glenn had spotted a well fire in one of the world's great natural gas fields, Gassi Touil, that had been burning in the Algerian desert since November 2, 1961. Every day, 16 million cubic meters of gas escaped from the well, billowing 140 meters into the air and releasing enough energy to power London for three months. Geologists warned that if the well wasn't extinguished it would burn for another 100 years.[5]

Red Adair did the job in just a few months. Braving the blaze that roared louder than a jet engine, the world's most famous firefighter detonated 340 kilograms of nitroglycerin on the wellhead and extinguished the

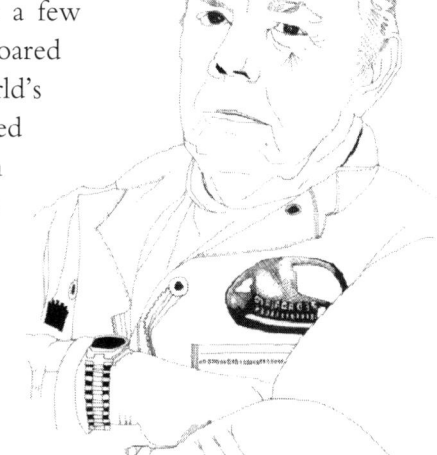

Red Adair (1915-2004), the celebrated oil well fire-fighter.

Flames; the explosion robbed the fire explosion robbed the fire of oxygen needed to support combustion. Red Adair then had to fix a new valve on top of the well before finally bringing the blowout under control.[6]

Adair had learned from Myron M. Kinley, who pioneered modern oilwell firefighting techniques. Myron Kinley's father, Karl, was one of the first daredevils to extinguish an oilwell fire with explosives. In 1913, Kinley Sr. walked up to a well fire near Taft, California, dropped a dynamite bomb into the well and ran. The subsequent explosion extinguished a fire that had been burning for several months. During the 1920s, shooting out oilwell fires with explosives became relatively common as a means of last resort, and in 1931, Myron Kinley extinguished a fire in Romania that had been burning for 890 days and was considered impossible to control. Shortly afterward he founded M.M. Kinley Company.[7]

Red Adair, who joined M.M. Kinley Company in 1946, can be characterized as fearless, painstaking and patient, particularly under the duress of a blowout. And he truly lived by his motto "Nothing is impossible." His authorized biography tells the story of his wife Kemmie giving birth prematurely to their first child Jimmy in the 1940s. The baby had difficulty breathing and after struggling for two days was declared dead. Adair stormed into the maternity ward screaming, "He isn't dead, I know he isn't, I can feel it. Don't let him go!" The doctors made a last desperate attempt to revive him, and miraculously the baby started breathing. Red shouted at the incredulous doctor, "Now you stay with him, because he's gonna make it." Jimmy did make it, and later in life became an oilwell firefighter like his father.[8]

Adair left Kinley in 1959 to form his own firefighting firm, the Red Adair Company, in Houston. The company developed numerous technological innovations refining the firefighting technique. Together with his lieutenants Asger "Boots" Hansen and Ed "Coots" Matthews, who would later form their own company, Red Adair tamed 42 blowouts in the 1960s alone. Adair's fame peaked in 1968 when John Wayne portrayed him in the Hollywood movie *Hellfighters*.[9]

The oilwell firefighters even became immortalized in a Lone Star State joke: "A Texan died and arrived at the Pearly Gates. Saint Peter showed him Heaven, but the Texan was unimpressed. So Peter showed him the fires of Hell and said, 'I bet you don't have anything like that in Texas.' The Texan thought for a couple of seconds and said, 'We sure don't. But there are a couple of boys in Houston that can put that out for you.'"[10]

Well integrity for deep, hot wells

As wells got deeper and hotter, there was renewed concern about well integrity, particularly the engineering behind the casing string and the process of cementing it into place. Keeping oil, gas and water zones isolated from each other and preventing fluids from migrating behing the casing to the surface was the bottom line. Leaking hydrocarbon lowers well productivity, threatens the stability of the entire well, and most important creates a mighty environmental and safety hazard.[11]

Quite different from the rather shallow wells of the 1920s and 1930s, wells now required several strings of casing with decreasing sizes before final depth was reached, rather like an extended telescope. The main suppliers of casing were Halliburton, Baker Oil Tools and Hughes Tool, but they were joined in 1941 by a company founded by Jesse E. Hall Sr. of Weatherford, Texas, a small town 25 miles west of Fort Worth. Jesse named the new company after the town.[12]

After World War II, casing saw several advances. Research into metallurgy produced superior steels, and a simple innovation made it far easier to install the large-diameter surface casing. As late as the 1970s, each length of surface casing had to be welded to the length already in the hole. Alistair Oag, who was working as a trainee engineer at the time and later rose to become head of drilling at BP, recalled: "This would take four welders about four hours. Then you had to radiograph it to make sure that there was no defect in the weld. Then you ran the string into the hole and picked up the next joint for another 4 hours welding." The change came in the 1980s, when casing manufacturers welded connectors on the pipe that snapped together in seconds.[13]

Hand in hand with casing, increasingly deep wells also required improved cement and cementing techniques. Erik Nelson, a veteran cementing expert from Schlumberger, recalled, "Portland cement is not stable at temperatures above about 110°C, so the challenge was finding stuff we could add to the cement to make it stable at high temperature and pressure." The most important of these stabilizing cement additives was pozzolan.[14]

Mark Mau and Henry Edmundson

Roughnecks at work on a traditional rig floor, about to fasten a large wrench to the pipe.

Pozzolan, the oldest known natural cement, was used for construction until the quicker-setting and harder portland cement was discovered. Pozzolan added to portland cement reduces the cement-slurry density and increases the net compressive strength. A lower density reduces the hydrostatic pressure during cementing, and this helps stabilize the drilling mud flow up and down the annulus and prevents lost circulation in case of weak formations."[15]

Already in the late 1940s, Halliburton started investigating the pozzolan and portland combination. In the early 1950s, oil companies began requesting that pozzolanic materials, such as diatomaceous earth and silica, be added to portland cement. It is estimated that by 1957 over two-thirds of cement jobs used portland-pozzolan mixtures. Two years later, Halliburton had developed a cement for temperatures as high as 230°F.[16]

Parallel to advances in cement came advances in the equipment used for performing cementing operations. In the early days of oilwell cementing, the pumps used to push the cement down the well could deliver up to 600 psi, more than enough for the comparatively shallow wells of the time. By October 1944, when Halliburton began research and development on their Super-Cementer, maximum pump pressures ranged between 6,000 and 8,000 psi. Their Super-Cementer, introduced in 1947, weighed in at 22 tons and provided 12,000 psi of pressure. But deeper wells soon required even more powerful cementing pumps. In 1957, Halliburton introduced a pump that generated 20,000 psi and deliver as many as 24 barrels of slurry per minute. This rise in efficiency went hand in hand with a significant growth in bulk cementing. During the 1950s, bulk cementing became the standard cement method, and by 1957 well over 75% of the portland cement used in US domestic oil wells was furnished from bulk plants.[17]

It is tempting to assume that once the hole is drilled, the casing has been placed in the well and cement successfully squeezed into the annular space between casing and the formation, then the job is done. It may be at the time, but the future is a different matter. In its liquid form before setting, the cement naturally withstands the hydrostatic pressure in the formation, but after the cement sets, the amount of hydrostatic pressure that it can bear slowly decreases with time. Its solid structure may become compromised and channels within the cement open up allowing oil but particularly gas to migrate up or down, causing loss of zonal isolation.

A Halliburton cement truck from the 1950s.

Service companies engaged in cementing therefore had their work cut out to develop cements that maintained their integrity over time. One solution simultaneously developed in the late 1970s in the Soviet Union and by Halliburton in the US was to add very fine aluminum powder to the cement, causing hydrogen gas to form within the cement structure, increasing its volume and blocking migration paths.[18] An alternative approach developed in 1985 by Dowell Schlumberger, a joint venture of Schlumberger and Dow Chemical, was to add latex to the cement slurry, the same chemical used in paint and comprising an emulsion of polymer microparticles. It was found that the latex particles would coalesce to form coherent, low-permeability plastic films blocking migration.[19] There have been many more advances in cementing ever since, and the cementer's recipe book is replete with additives for a wide variety of well conditions and circumstances.

Mud cleans up

As important as the integrity of the well was the drilling mud—ensuring that its many physical and chemical properties matched requirements at the bottom of the well. Since mud was continuously recirculated, cleaning the mud stream before it was pumped down the well again was critical.

Ever since the 1920s, a number of mechanical devices had come on the market to remove cuttings from the drilling mud, from simple shale shakers to more advanced desanders and desilters. But these machines were sometimes rather too efficient. If you cleaned barite-weighted mud too well, there was a risk of discarding the valuable barite. In

1970, Leon Robinson and his team at Humble Oil started working on a solution to circumvent this, securing a US patent for a new mud cleaning system.

Leon Robinson, an engineer who worked for Humble Oil and then Exxon Production Research, and developed numerous innovations in mud engineering.

First, the drilling fluid containing fine-sized weighting material and drilled solids was passed over a shale shaker as usual to remove some of the drilled solids. Then, a device called a hydrocyclone, working on the principle of a centrifuge, separated the remaining drilling fluid into low-density and high-density slurries. The low density slurry was returned to the drilling fluid system, whereas the heavy slurry, including most of the weighting material, was further processed through a second vibrating screen with a finer mesh to remove additional drilled solids but preserve any weighted material.[20]

During the 1960s and 1970s mud chemistry was becoming increasingly sophisticated, a key development being the use of polymers. The interest in adding polymers to drilling mud was twofold: Polymers provided extra viscosity to carry cuttings to the surface and they also controlled the amount of water seeping into the formation through the mudcake that builds on the borehole wall.[21]

At the same time, the industry slowly adopted oil-base mud, one of the movers being Martin Chenevert at Humble Oil Production Research. Chenevert had first started thinking about shales and stabilizing them while working on his master's thesis at The University of Texas at Austin: "My professor assigned me to work on the rock mechanics of shale rocks, which was interesting, but in fact I didn't get to graduate. In the middle of my studies Exxon hired me. The gist was 'Oh, you know about shale? Well, shale gives us real trouble, so we want you.'"[22]

In the following years, Martin Chenevert experimented with various formulations of oil-base mud adjusting the activity of the water component of the emulsion to match that of the shale. The net result was a huge uptake of oil-base mud throughout the industry pushing the envelope in deep and hot wells even further. Shell and El Paso Natural Gas Company, for example, used oil-base muds in 1974 in the no. 1 Benevides well, Texas, and set a new a new world record with a temperature of 555°F measured at a depth of 23,837 feet.[23]

When operators began using oil-base mud for offshore drilling, its environmental effects were not understood and the contaminated drill cuttings were simply dumped overboard. But it was soon realized that the aromatic chemical content in diesel and mineral oils was as toxic to marine organisms as it was to oilfield workers. The industry was slow to react.[24]

What forced its hand was action the US Congress took in 1972 when it enacted the first comprehensive national clean water legislation in response to growing public concern about water pollution. Administered by the Environmental Protection Agency (EPA), the Clean Water Act forced municipalities and major industries to implement effective pollution controls.[25]

It was only a question of time before the EPA went offshore, and in 1978, the Clean Water Act hit the mud industry. Oil companies wishing to drill exploratory wells off the US Mid-Atlantic coast had to agree, as a permit condition, to support a drilling mud bioassay program. A testing protocol using the shrimp species *Mysidopsis bahia* became the basis for assessing toxicity of drilling fluids and additives. In that same year, the state of Alabama moved aggressively to enforce discharge regulations on oil companies drilling in Mobile Bay, signaling that local governments would also look more closely at drilling fluids. Over the following years a number of bioassay programs were carried out in the Gulf of Mexico culminating in the outright prohibition of oil-base mud discharge in the Gulf in 1986.[26]

By then, the drilling industry had started to make a virtue out of necessity. Beginning in 1985, the industry initiated research on a biodegradable substitute for oil-base mud. Many took on the challenge, including mud companies M-I Drilling Fluids and Baroid, and the oil company Unocal. Researchers, most of them with a strong background in industrial chemistry, began developing synthetic-base drilling fluids to match the drilling performance characteristics of traditional oil-base mud while satisfying several environmental constraints. The fluid had to be nontoxic of course, but it also could not taint marine life, and it had to be readily biodegradable.[27]

They tried common vegetable oils, such as peanut, rapeseed and soy bean oils, and even fish oils. But all the oils suffered from various combinations of being too viscous, decomposing too quickly in contact with water, and becoming unstable at high temperature. Finally, they zeroed in on esters, which are liquids made by the reaction of a fatty acid with alcohol. In general, esters are exceptional lubricants, show low toxicity and have a high degree of aerobic and anaerobic biodegradability. However, there are a vast number of ways to synthesize esters.[28]

It took five years to find a recipe that satisfied all the environmental conditions yet provided the shale stabilization and superior lubricity that drillers needed. Phillips Petroleum Company Norway and Baroid undertook the first trials in the Greater Ekofisk basin in Norwegian waters in February 1990. Soon after, Baroid launched the first commercial product. M-I Drilling Fluids followed, providing synthetic-base mud for a well drilled in the Gulf of Mexico in June 1992. In 1993, synthetic drilling fluids were attracting keen interest, and at least 20 field trials had been conducted in various parts of the world. Cuttings could be discharged into the ocean, although a limit was imposed on the amount of synthetic material discharged.[29]

Another avenue was developing additives for water-base muds that would inhibit the reaction to shales and clays. The first such additive was the polymer PHPA—short for partially hydrolysed polyacrylamide-polyacrylate—introduced by Ronald K. Clark of Shell in Houston in 1976 and commercialized by Newpark Resources in Woodland, Texas. More efficient inhibitors followed, such as polyalkylene glycols, commonly known as polyols, introduced by BP and Anchor Drilling Fluids in the early 1990s.[30]

Overall though, synthetic-base drilling fluids have not replaced oil-base muds. Diesel and mineral oil muds are still used onshore if disposed of properly. Offshore in the North Sea, for example, operators have to inject oil-contaminated cuttings into dedicated disposal wells if they don't want to haul the cuttings on land for incineration. In the past decade, operators Total and Mobil have developed new technology to completely separate any remaining oil and water from cuttings. Their "hammer mill" process passes the cuttings through a barrel-shaped chamber in which hammers pound the cuttings, raising them to temperatures as high as 300°C. This causes the oil and water to evaporate, leaving the cuttings clean enough to be dumped overboard.[31]

Chapter 18: Going Horizontal

UNTIL THE 1920s, most drillers could not imagine or were not in the least bit interested in drilling wells other than vertical. It is believed that the first intentionally deviated well was drilled by Robert E. Lee for Big Lake Oil Company near Texon, Texas, in 1929. However, most of the early directional wells were probably illegal, drilled by unscrupulous operators who were trying to steal oil from their neighbors. In the 1930s, inspectors assisted by Texas Rangers found and shut down 380 deviated wells drilled across lease lines into productive portions of the Woodbine Formation in east Texas. An estimated US$ 100 million worth of oil was stolen in this fashion.[1]

Deviating from the vertical

The method of the slant-holers was rather simple. The illegal operators bought or leased low-producing wells from one of the major oil companies, and when production almost ceased they brought in a workover rig, pretending to carry out maintenance and improvements. They were actually running a "whipstock" down the shaft and slanting the hole toward an oil company's high-production lease. A whipstock is a long wedge with a concave face that is placed in the well. As drilling continues, the bit moves down the face of the whipstock and is pushed sideways into the formation.[2]

The whipstock was invented by H. John Eastman in 1929 though he developed the device for entirely legal reasons: reaching offshore oil sands from onshore locations without having to construct an offshore platform. In 1934, Eastman became famous for directional drilling when he drilled the world's first relief well to control a blowout in Conroe, Texas. Gradually, drillers learned the simple tricks of deviated drilling: exert more weight on the drillstring to push the drill bit upward; lessen the weight to let the drill bit drop. Playing with the bottomhole assembly with weights and stabilizers also helped.[3]

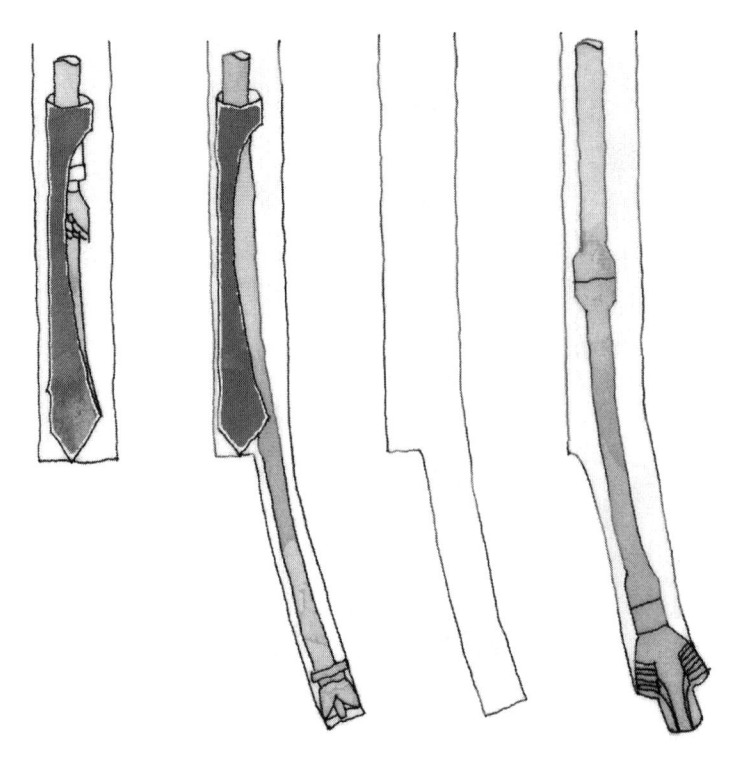

Kicking off a deviated well with a whipstock. From left, whipstock on bottom in oriented position; drilling the "rathole"; whipstock is removed; opening the "rathole" to full gauge with a hole opener.

In the Soviet Union, deviated drilling was approaching the horizontal. In 1937, the first experimental horizontal drilling took place in Yarega, Orenburg, and in March 1941, Alexander Grigoryan, a young Soviet driller and recent petroleum engineering graduate from the Azerbaijan Petroleum Institute, directed the first horizontal well drilling in Azerbaijan, at a depth of 1,700 meters.[4]

To drill horizontally, Grigoryan leveraged a key Soviet innovation, a turbomotor placed just above the bit and driven by the flow of drilling mud. Using a downhole motor to rotate the bit avoided the need to rotate the entire drillstring from the surface. The impact on directional drilling came immediately. With the drillstring stationary and a short, bent sub placed above the motor and bit, the motor powered the bit in a curve in whatever direction was desired. Once the hole was sufficiently deflected, the bent sub could be removed, allowing drilling to proceed along a straight line until more deflections were required. Turbodrills were used to drill experimental directional wells in 1939 in Grozny, and by 1941 Grigoryan claimed that turbodrilling was more efficient for directional drilling than using whipstocks.[5]

Turbodrilling in the Soviet Union had a long history. The first Russian turbodrill was built by Matvei Alkuvich Kapelyushnikov, a graduate of the mechanics faculty of the Tomsk Technological Institute in Siberia. In 1914, he was hired by the Baku Company of Russian Oil and after the Soviet takeover was appointed to the technical department of the Azneft Trust. This was a production association formed in 1922 to manage oil production in the Baku area by bringing together the oil wells, drilling rigs and other industrial facilities expropriated from their former owners. Despite reservations, Azneft allowed Kapelyushnikov to build a prototype turbodrill, and initial tests showed that it could outpace conventional rotary drilling. By 1925, the Azneft drilling department began using turbodrills in the Surakhany district of Baku.[6]

Turbodrilling suited the Soviets. At the time, the Soviet Union was unable to produce sufficient quantities of high-quality drillpipe necessary to withstand the torque applied in rotary drilling, so placing a motor downhole was timely.[7] In the US, meanwhile, only the Standard Oil Company of California took an interest in the new technology. In 1922, C.C. Scharpenburg, a Standard engineer, patented a multistage axial flow turbodrill that had many advanced features, including a sealed lubrication system, but it failed in tests in Elk Hills, California, and interest died with it.

In 1928, when a German journal ran an article describing Kapelyushnikov's turbodrill, the US drilling industry began to pay serious attention. Several American companies showed enough interest to invite Kapelyushnikov to the US to run demonstrations. At Texas Oil's Seminole oil field in Oklahoma, the Russian turbodrill drilled 60% faster than the rotary method.[8]

However, Kapelyushnikov's single-stage turbodrills had rather low durability, mostly because of erosion of the turbine blades by the drilling mud. Kapelyushnikov had to curtail his activities due to a car accident, but further developments in the Soviet Union were undertaken in 1934 and 1935 by Rolen Ioannesyan and Mikhail Gusman, two young engineers from Azneft in Baku. Their efforts resulted in a multistage turbodrill featuring special seals that prevented mud from getting inside the bearings. It delivered significantly higher power compared with the single-stage version.[9]

Enter the positive displacement motor

In 1942, Ioannesyan and Gusman were assigned to head a new department of Turbine Drilling in the USSR Oil Ministry with the goal of implementing turbodrill technology throughout the Soviet Union. A new low-speed, multistage turbodrill launched in the early 1950s proved reliable, and in 1955, turbodrilling managed to topple rotary drilling.

Nevertheless, opponents of turbodrilling in the Soviet Union continued to criticize the technique, saying it would compromise the development of high-strength drillpipe and drill collars, a continuing weakness in the Soviet oil and gas industry. But the rapid need for directional wells made possible by turbodrilling swung the argument firmly the other way. By the 1960s, turbodrilling was being used to drill 80% of all Soviet wells.[10]

Elsewhere, drillers began to reconsider downhole motor technology, primarily to drill directional wells for developing offshore fields. In 1956, the Soviets licensed turbodrill technology to four Western specialized drilling tools companies: Neyrpic (France), Salzgitter (Germany), Trauzl (Austria) and Dresser (US). Collectively these companies bought more than 100 turbodrills from the Soviet Union. In the late 1950s, Dresser extensively tested the Russian turbodrill and improved the durability of the rubber thrust bearings. In 1960, Dresser sold its turbodrill business to Eastman Industries that developed a modular bearing pack that facilitated the motor's maintenance.

Neyrpic made similar improvements on the Russian turbodrills marketing their own version of a drilling motor in the mid-1970s. The leading light at Neyrpic was French-Russian engineer Wladimir Tiraspolsky. During the 1960s, Tiraspolsky formed a company called Turboservice, and in the 1970s Turboservice was acquired by the Whipstock Company which itself later merged with Eastman. This flurry of corporate activity, however, did not secure the future of Soviet-based motor technology in the West.[11]

Back in the Soviet Union, Russian engineers, most prominently Mikhail Gusman and Dmitri Baldenko, were developing a completely new design that resulted in the so-called positive displacement motor. In this new design, the mud flow, rather than driving turbine blades, was directed onto a spiral shaft inside the motor housing causing the spiral shaft to turn and rotate the bit. The new motor failed to attract much attention in the Soviet Union, but in the US it caught on.[12]

Dyna-Drill Technologies, a Houston-based manufacturer of drilling equipment, was the company that ran with the idea. In 1959, it filed the original positive displacement motor patent, which would become a cornerstone in the company's success over the following years. Soon, the drilling industry took notice of the young and prospering company, and in the early 1960s Dyna-Drill was acquired by the Smith Tool Company, established in Houston in 1936. The new motor was marketed unsurprisingly under the brand name "Dyna-Drill."[13]

A big push initially for Smith's Dyna-Drill business was an order in the early 1970s from the US Atomic Energy Commission, which had a need to directionally drill into the massive cavities created by underground nuclear tests to sample contamination. This, together with the rising demand for offshore and directional wells, allowed Smith to eventually claim 90% of the US downhole motor market during the 1970s.[14]

Improving the roller cone bit

Meanwhile, the standard tricone bit was having a makeover. With drilling venturing into deeper, harder rocks, penetration rate was slowing and bits weren't lasting very long. A key area of research now addressed the materials the bits were made of. To the initiated, an obvious choice was to use tungsten carbide.[15]

Tungsten had been discovered in 1781 by the Swedish chemist Carl Wilhelm Scheele. It has a specific gravity close to 20, the highest melting point of all metals at 3,410°C and is extraordinarily resistant to heat and water. The breakthrough compound tungsten carbide was discovered during World War I and found to have a hardness close to that of diamonds. And a technique of adhering tungsten carbide to the cutting edges of steel tools was discovered in 1929. This opened up completely new opportunities for the drilling industry. During the 1930s, tungsten-carbide drill bits were tried in Germany, but development was curtailed by the time World War II began. In 1951, Hughes Tool developed a tricone bit with tungsten carbide inserts, as did the Smith Tool Company as soon as the Hughes tricone patent expired.[16]

Parallel to the tungsten-carbide innovation, Hughes Tool managed other improvements. One development in 1948 was incorporating mud jets in the tricone bit, though the idea of having the mud stream impinge directly on the bottom of the hole was tried variously from the early stages of rotary drilling. From the 1910s onwards, fishtail bits had been equipped with nozzle openings that forced the drilling mud out at great pressure, lifting cuttings away from the rock face.[17]

Then in the 1950s, Hughes engineers began studying the ball bearings inside each of the bit's cones. Over time, the cone bearings wore down due to abrasion from cuttings, especially in medium to soft formations. The only solution was to lubricate the bits on a regular basis, a time-consuming chore because the drillstring had to be pulled each time the bits needed lubricating. In 1959, Hughes Tool came up with a solution, in the form of self-lubricating sealed bearings.[18]

Next was the idea of replacing ball bearings with bush bearings, also called journal bearings. The balls in a traditional bearing distorted and failed during the intense flexural stresses experienced while drilling. Bush bearings coped better, and by the early 1970s, three drillbit companies offered a bush bearing that consisted of a cylindrical body around a rotating shaft. The Hughes Tool bush bearing, however, had the edge. The bush surface was made from sintered tungsten carbide to resist journal wear. In addition, the cone area was lined with strips of a silver alloy designed to lubricate and dissipate the extreme heat. Dan Scott, who was with Hughes Tool's metallurgical research group when bush bearings were introduced in 1969, noted: "It was the revolutionary feature in the 30 or 40 years from the time tricones were first developed. There were minor changes, but the bush-bearing rock bit lasted 4 or 5 times longer than other existing products."[19]

PDC bits

The bit renaissance continued. Enter Frank Christensen, a multitalent who played fullback for the University of Utah football team and later joined the Detroit Lions for which he was a member of the 1935 National Football League Championship team.

After returning home from Detroit, Christensen and a former Lions teammate, George Christensen (no relation) set up business in a corner of a machine shop run by Frank's family that sold mining equipment. George had previously worked with Koebel Diamond Tools, which marketed diamond-based tools to Detroit carmakers. In 1943, Frank and George christened their business Christensen Diamond Products and set about developing the first ever diamond bits for the mining industry. Ten years later, they were manufacturing diamond bits for the oil industry.

Frank Christensen (1910-2001) and George Christensen (1909-1968), not related, co-inventors of the polycrystalline diamond compact (PDC) bit, a technology that took time to make an impact, but from the 1980s revolutionized the drilling industry.

Diamonds had made an appearance as a cutting element as long ago as 1910 in specialized coring bits used to retrieve rock samples, mostly for mineral exploration. The drill bits introduced by Christensen in the early 1950s used industrial-grade natural diamonds crushed and processed to a specific shape and size. By the late 1950s, wells had been drilled with diamond bits in France and the US. In the 1960s, the Prikarpatburneft Drilling Company in the Soviet Union also began to adopt diamond bits, but drilling with diamonds nevertheless remained a niche business.[20]

Manufacturing bits with natural diamonds was very expensive and technically difficult. The game changed in 1971 when General Electric developed a process to make synthetic diamonds. This ultimately led to the development of polycrystalline diamond compact (PDC) bits. The industry was not slow to react. Christensen built a Diamond Technology Center in Salt Lake and other small companies ventured into the fledgling PDC market, most notably the Drilling & Service Company. General Electric tried to interest Hughes Tool in synthetic diamonds but got nowhere. Hughes had just introduced the bush-bearing bit, spent millions of dollars on a new factory and was not in the mood to invest in the new technology. This left the impetus firmly in the hands of Christensen Diamond Products.[21]

In 1976, Christensen engineers succeeded in bonding thin layers of synthetic polycrystalline diamonds to tungsten carbide inserts that were pressed into holes drilled in the body of a new type of bit, that unlike the roller cone had no moving parts. The cutters were fixed and designed to shear through the rock like a knife scraping through butter, but there were teething problems. The PDC bits were not able to drill anything other than slightly thick mudstones and soft shales. The mechanical construction of the bit was too vulnerable to the intense mechanical vibrations and stresses of the drilling process. And in hard formations, the bits whirled around the borehole and had a tendency to self-destruct.

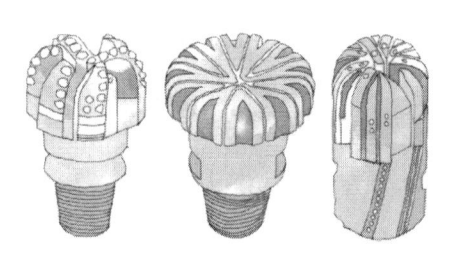

Fixed-cutter bits. From left, PDC bit, natural diamond bit, impregnated diamond bit.

By the mid-1980s, researchers, including Coy Fielder at Strata Bit Corporation, Houston, and Joe Kelly at Hughes Tool, were beginning to study bit vibration and drilling dynamics. At Amoco Drilling Research in Tulsa, Oklahoma, a team led by Tommy Warren was also researching cutter damage and bit performance, publishing several seminal papers and gaining key patents to improve PDC bit performance. Their findings were eventually licensed by Christensen to create an antiwhirl product line. By 1992, the antiwhirl PDC bits accounted for 9% of footage drilled worldwide. At this same time, through a series of corporate mergers and acquisitions, Christensen had become part of the newly established company Baker Hughes, formed in 1987 by the merger of Baker International and Hughes Tool.[22]

The high price of PDC bits still represented an effective brake for the new technology. In 1995, a single PDC bit could cost 5 to 15 times more than the equivalent roller cone bit. Then, manufacturers realized a new business model would help. They started renting and repairing PDC bits, reducing a high initial cost to a more reasonable long-term one. PDC usage took off, and in the 10-year period from 1995 to 2005 the PDC-drilled footage worldwide rose from 15% to 50%. Today, more than 60% of all bits are PDC.[23]

The arrival of topdrive

As bits became tougher and more reliable, the drilling rig was about to undergo dramatic change, and it would have been an unlikely bet that this would be triggered by an aerospace engineer named George Boyadjieff.

Boyadjieff's expertise was designing autopilots for commercial airplanes, but his new employer, the Abegg and Reinhold Company, a small drilling tools company founded in 1908, was precisely looking for somebody who didn't have preconceptions about the industry and who could see engineering problems from a fresh perspective. The new CEO of the company, Ben Reinhold, son of the founder Baldwin Reinhold, had a vision of a large international company and would later rename the firm with the shorter and punchier Varco International.

George Boyadjieff, inventor of Topdrive and numerous other drilling innovations.

George Boyadjieff was hired in 1969, and his task as the company's chief engineer was simply to develop new products. So Boyadjieff went visiting offshore oil rigs, the drillship *CUSS I* and a number of land rigs in Southern California. He talked to drillers about what needed improvement and then began developing new drilling technology for an industry slow to adopt anything new.[24]

The first tool Boyadjieff conceived was a device called the spinning wrench for making drillpipe connections. Prior to this, roughnecks on the rig floor wrapped a heavy chain around the drillpipe that spun the pipe together, a uniquely dangerous occupation that claimed many a finger. The spinning wrench mechanically spun the pipe using rollers and a motor clamped around the pipe. Boyadjieff's next development was the torque wrench, a hydraulic device that clamps on the pipe once it has been spun and uses hydraulics to make the connection tight.[25]

The spinning wrench and torque wrench typically hung on cables, which created problems for offshore drilling because they swung around and could potentially injure anyone on the drill floor. The spinning wrench and the torque wrench were therefore combined in a new large machine that was attached to the rig. Introduced in 1975, the Iron Roughneck could connect pipe automatically. Varco International began to be noticed.[26]

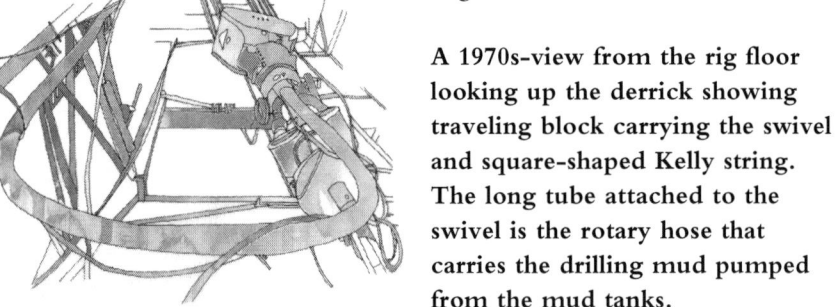

A 1970s-view from the rig floor looking up the derrick showing traveling block carrying the swivel and square-shaped Kelly string. The long tube attached to the swivel is the rotary hose that carries the drilling mud pumped from the mud tanks.

The next development, a kelly spinner, eliminated the rotary table, the usual means of rotating the square cross-sectioned kelly pipe that in turn transmitted the rotary motion to the drillstring. One of the first kelly spinners was installed on the Global Marine drillship *Glomar Challenger* as early as 1968. A few years later, Varco built a kelly spinner for a Global Marine project in the Arctic, but the use for kelly spinners was rather limited, predominantly for workover rigs.[27]

Global Marine and Exxon pushed Varco to develop a much larger device, a sort of power swivel that would hang directly under the traveling block, the huge pulley that moves the drillpipe up and down as drilling proceeds. One manufacturer of power swivels was Brown Oil Tools, founded by Cicero Brown in 1929 in Houston. In discussions with Brown, it became clear that even if a much larger power swivel was available, the real problem was ensuring smooth and efficient pipe handling.[28] On two jackups offshore Abu Dhabi equipped with Brown power swivels, as much time was lost tripping the pipe in and out of the hole as was gained with the swivel.

In 1980, SEDCO won a contract to build two new jackups with power swivels. Boyadjieff proposed that Varco build them a radiaclly different power swivel and started thinking.

One problem was attaching a torque wrench to the power swivel so the swivel could be disconnected from the pipe. The other was being able to reverse drive the swivel to backream out of the hole if the drillstring got stuck. Varco's Top Drive solved both issues and was marketed in 1982. First tests for SEDCO showed on the average a 20% to 25% decrease in drilling time.[29]

It took a while for topdrive technology to be accepted. A big selling point was being able to handle complete 90-foot stands of drillpipe rather than the conventional 30-foot singles, but even this benefit failed to convince some. Operators such as Maersk Oil, however, saw the benefits immediately. Maersk's experience pioneering container technology for their shipping business had taught them the benefits of automation. Eventually, from about the late 1980s, almost all large rigs started using topdrives.[30]

Topdrive, invented in the early 1980s. Topdrives have an in-built electric or hydraulic motor, here placed inside the rectangular box at the top end, just below the traveling block.

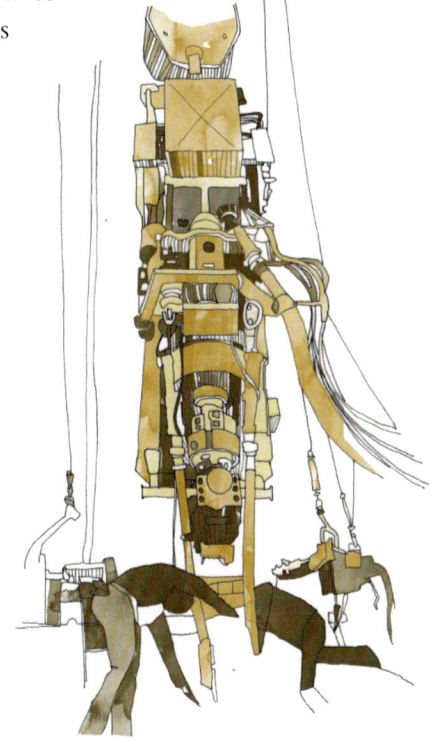

Horizontal and multilateral drilling

In time, these advances in drilling and bit technology would ease the introduction of horizontal drilling, but there remained many obstacles ahead, not least of which was the prevalent belief that horizontal wells were impossible to drill. It was scarcely in the collective consciousness of the industry, at least in the West, that horizontal wells would intersect more reservoir and consequently secure greater production, since oil and gas reservoirs are generally more horizontal in shape than vertical. The momentum remained with the Soviets.

By the 1960s, horizontal drilling in the Soviet Union had advanced considerably thanks to the continued innovations of Alexander Grigoryan. The Soviet Union's foremost driller had since been elevated to department head at the All-Union Scientific Research Institute for Drilling Technology in Moscow. The enabling technology was none other than the Soviet's turbodrill, it being far easier to steer a well horizontally using a stationary drillstring with a turbodrill on the end than rotating a drillstring from the surface. In 1968, the Soviets drilled a record horizontal well in Markovo in East Siberia that reached 2,507 meters total measured depth and had a horizontal section of 632 meters.[31]

But the results of horizontal drilling in the Soviet Union were disappointing, with production generally less than expected. The reason was failing to place the horizontal section in the reservoir. This was because they were not only uncertain about the exact location of the reservoir and also unable to control the well trajectory with any precision. For example, the record-breaking well in East Siberia was suspected of only intersecting 320 meters of the so-called "pay-zone," the portion of the reservoir that best produces oil and gas. In the 1970s, the Soviets cut back on horizontal drilling, concentrated on traditional vertical drilling and told their drillers to make as much hole as possible. At a time of steeply rising oil prices, maximizing cumulative footage was the best way to increase the production.[32]

Frustrated by restrictions imposed on his work by the Soviet regime, Grigoryan looked to the West and in the late 1970s started working with IFP, providing details of the technologies he had developed in the Soviet Union. Then in 1978, IFP and Elf Aquitaine launched a research project to investigate horizontal drilling named FORHOR (abbreviation for foragerie horizontale). In part, the project was motivated by a problem facing Elf in developing the Rospo Mare field, a heavy oil accumulation they had found in the Adriatic Sea. Most of the oil was contained in large vertical fractures, and the only way to produce the field was to intersect the vertical fractures with horizontal wells. At the time, Jacques Bosio, head of Elf upstream research and development, had been pushing to try the horizontal drilling technology in a few pilot wells.[33]

There was considerable resistance to horizontal drilling, not so much because of the technological challenges, but as a result of the industry's mindset. At the end of the 1970s, drilling engineers in the West still believed that the maximum possible inclination for a well was about 70 degrees from the vertical. In addition, reservoir engineers could not accept the idea that a horizontal well would increase the productivity of a well enough to compensate for the extra cost of drilling it. As Jacques Bosio once put it: "It proved far more difficult to change by 90° people's minds than the trajectories of the wells."

Using special drillpipe that could flex, Bosio finally succeeded in drilling two experimental horizontal wells in Lacq, an Elf field in southwest France, and by late 1981 managed to convince his management to risk horizontal drilling in the Rospo Mare field. A pilot project that included a drilling and production platform was deemed necessary, and in January 1982, Bosio's team drilled Rospo Mare no. 6 with a 370-meter horizontal section, at a cost twice that of a vertical well. But it produced 3,000 barrels per day, over 20 times more than a neighboring vertical well, boosting reserves from nearly zero to 450 million barrels, enough to save the Elf Aquitaine subsidiary in Italy. Elf Aquitaine's Rospo Mare field became the first oil field to systematically produce through horizontal wells.[34]

Grigoryan's influence didn't stop with horizontal wells. He was also the father of the multilateral well, in which several lateral branches are drilled from one mother wellbore. The world's first multilateral was drilled by Grigoryan in 1953 in Bashkiria in the Urals. Using a turbodrill, he branched the main hole into nine laterals spreading out like roots of a tree. This multilateral is said to have produced 17 times more oil than an average well in the field. Drillers in the Soviet Union later drilled more than 100 multilateral wells, of which Grigoryan himself drilled 30. But like the preceding horizontal well development, the Soviets abandoned multilaterals in favor of conventional, vertical drilling.[35]

The first multilateral well outside the Soviet Union was drilled by ARCO in 1980 in its Empire field, New Mexico, and from the mid-1980s horizontal and multilateral drilling steadily gained acceptance. In 1985, Amoco and Union Pacific Oil and Gas Company began exploiting the thin oil-bearing layer of the Austin Chalk Formation in Texas with a series of horizontal wells. In 1989, the first offshore multilateral in the Middle East was drilled by Arabian Oil Co. in the Khafji field, Saudi Arabia. The same year, Shell introduced horizontal drilling on a wide scale. In 1992, Maersk Oil drilled the first North Sea multilateral well. And in 1993, ADCO oil company in Abu Dhabi drilled the first multilateral onshore in the Middle East.[36]

In 2000, Saudi Aramco's then head of reservoir management, Nansen Saleri, proposed taking the multilateral idea one step further. His idea was the maximum reservoir contact (MRC) well that would have an aggregate reservoir contact in excess of 16,000 feet using ambitious multilateral configurations. From early 2002 to late 2006, Saudi Aramco applied Saleri's idea drilling 25 MRC wells in their Shaybah field. The MRC well idea translated into astonishing results: as much as a four-fold increase in well productivity and a three-fold decrease in unit development cost.[37]

Nansen Saleri, former head of reservoir management at Saudi Aramco. An authority on reservoir management, he pioneered the concept of maximum reservoir contact.

Chapter 19: Intelligent Drilling

AS WELLS went directional, knowing where the drill bit was heading became important. The traditional technique still in use in the 1970s was the single-shot survey. This comprised a small device lowered down the drillpipe that photographed a magnetic compass for direction and a form of plumb bob for the vertical angle. The measurement was slow to run and unreliable. Typically, a measurement was made every three or four hundred feet, and what happened in between was anyone's guess.[1]

Nevertheless, single shots were a welcome advance from earlier technologies. For example, one technique dating from 1874 used a four-ounce glass bottle, half-full of hydrofluoric acid and encased in a long slim-like cylinder called a bomb that was lowered through the drillstring to a point in the drill collar just above the bit. After the bottle remained at rest for about 30 minutes, the tool was pulled and the line etched inside the bottle by the acid revealed the inclination of the hole.[2]

The MWD struggle

By the 1950s, researchers were figuring how to measure inclination while drilling. The first to almost get there was J.J. Arps, chief engineer at British-American Oil Producing Company, Toronto, Ontario. His idea was to send the data to the surface using drilling mud as a transmitting agent. A shutter arrangement downhole would transform the data into mud pulses which would then travel up the well and be detected at the surface by a pressure transducer.[3]

It was clear that considerable research and development was needed to commercialize the idea. During the 1960s, several major oil companies, including Elf Aquitaine, Esso, Texaco and Mobil, became active. The first successful field test was made by Mobil in 1971, but problems persisted. Downhole electronics could not withstand the hostile downhole environment, and that together with a softening economic environment temporarily killed everone's interest. In 1973, oil prices rose and the development of reasonably robust microchips made it worth trying again. In addition, the resurgence of offshore drilling made it ever more urgent to accurately track well trajectory.[4]

Elf Aquitaine renewed its efforts by collaborating with Raymond Precision Industries, an aerospace company based in Middletown, Connecticut, that manufactured sensors capable of detecting mud pulses in the harsh drilling environment. In 1972, the two companies formed Teleco Oilfield Services with the specific purpose of developing mud pulse telemetry for directional surveying. By the time the technology became known as measurement while drilling (MWD), inventor Marvin Gearhart was on the scene. By the mid-1970s, the two front runners in MWD were Teleco and Gearhart-Owen. However, oil companies didn't want to be left behind, so some launched joint ventures with oil services companies. Mobil started a joint venture with Schlumberger, and Exxon worked with Teleco. In September 1978, Teleco Oilfield Services introduced the first commercial MWD tool after successful tests in the Gulf of Mexico, and the impact was immediate. By the early 1980s, all service companies were investing in MWD research and engineering.[5]

Marvin Gearhart, an entrepreneur who developed a number of oilfield technologies, for example measurement while drilling (MWD) in the 1970s.

Commercially, the acceptance of MWD was slow. Marvin Gearhart engaged in an SPE lecture tour at the time recalls the conservatism of the drilling fraternity: "I can still remember going to a town in Tennessee . . . I was giving a talk and after I'd finished, one of the guys in the audience, a local oilman, said: 'We're never going to do that kind of drilling here. Why did they send you here to tell us about technology that will never happen?'" Industry skepticism may have been rooted in the very slow rate of data transmission, since several minutes were needed to transmit each set of directional data.[6]

The telemetry challenge

Although the data rate had steadily improved over the years, the industry was always on the lookout for an alternative way of transmitting data up the drillpipe. One possibility was using optical fiber, a technology already attempted for wireline logging. Starting in 1980, Chevron had begun manufacturing a logging cable featuring optical fibers packaged inside the standard wireline's electrical conductors. Three years later, they had tested a prototype in a test well in Huntington Beach, California, achieving data rates of 10 megabits per second, phenomenally fast for that time. Schlumberger's first fiberoptic logging cable was tested in an 8,500-foot well in Marion County, Texas, in April 1985. Both developments suffered teething problems and it took a decade before fiberoptic logging cable saw widespread acceptance. For MWD, it was the fragility of optical fiber that proved a hurdle. In the 1990s, engineers at Sandia National Laboratories in Albuquerque, New Mexico, constructed a spooling system for running disposable optical fiber through drillpipe each time a new stand was added to the drillstring. The result should have been a data transmission of one megabit per second to an MWD tool at the bit, but the flow of mud invariably broke the fiber.

A recently emerging technology for MWD is wired drillpipe. In 2006, the first wired drillpipe system was commercialized by National Oilwell Varco, a company that had resulted from National Oilwell and Varco International merging the previous year. Each stand of drillpipe has built-in electrical wiring with connectors top and bottom. The system transmits data around 100 kilobits per second and has battery powered amplifiers every 3,000 feet. However, almost 10 years later, wired drillpipe still awaits general acceptance because of the system's complexity. But that hasn't stopped companies dreaming about the next logical step, which is equipping drillpipe with fiber optics. As Schlumberger's former CEO Andrew Gould commented: "If that works, it will be a real step change."[7]

Logging while drilling

The next logical step was extending MWD measurements to more than just simple directional information. Some visionaries even imagined somehow placing logging measurements in the drillpipe and dispensing with the wireline altogether. Instead of having to wait for the drillers to finish their job, it would then be possible to log while drilling, giving operators real-time information for making quicker decisions and the petrophysicist a first look at the formation while invasion was taking place. But this dream required far better mud pulse telemetry, and this was only one of many challenges. Just as serious was finding a way to measure formation resistivity, the key logging measurement, considered impossible in the presence of an electrically conductive drillpipe. In addition, logging sensors and their electronics were notoriously sensitive to shock and vibration, so how could they possibly stand up to the vibrations and jarring of a drillstring. In the end, it would take the industry more than half a century to make logging-while drilling (LWD) a reliable, commercial service.[8]

The first attempts at LWD occurred as early as 1932. Karcher, already famous for inventing reflection seismics, proposed a "continuous electric logging while drilling" by insulating the bit from the drillstring and passing current from the surface through the formation to the bit. The electrical return was to be up an insulated wire inside the drillpipe. A few years later, D.G. Hawthorn and J.E. Owen of the Amerada Geophysical Research Corporation in Tulsa had similar ideas, conducting field tests in half a dozen wells. Both systems showed that resistivity logs, and in the latter attempt the SP log as well, could be obtained while drilling. Colossal problems, however, prevented any of the systems becoming commercially feasible, for example the need for special drillpipe and the mechanical weakness of the insulated joint above the bit.

During the 1950s, Daniel Silverman of Stanolind thought up a method to eliminate the need for special drillpipe. He set up a winch on the rig floor that lowered sections of an electrical cable inside the drillpipe, with an electrical connector linking the sections each time a new stand of pipe was added to the string. Silverman obtained logs in the Seal Beach field of California, but there were obvious problems: The connectors leaked, and it was a tedious business connecting and disconnecting the cable sections as the pipe went in and out of the hole.[9]

The first steps towards a workable LWD system were taken by NL Industries, a Houston-based diversified conglomerate that had been formed in 1891 as National Lead Company and acquired Baroid, a leading mud logging company in the early 1920s. In 1978, NL asked

Baroid to focus on logging while drilling using the mud pulse telemetry from MWD as the conduit for passing data to the surface. Among several challenges remained the problem of making a formation resistivity measurement in the presence of an electrically conductive drillpipe.[10]

Two field engineers downloading data from a logging-while-drilling formation evaluation tool at the wellsite. The engineer in the foreground has attached a data transmission cable that feeds data to the laptop operated by his colleague.

NL Industries turned to Richard Meador of Texaco, an electrical engineer who since 1970 had been working on a tool to measure the dielectric constant of earth materials—a measure of the capacity of a material to store charge when an electric field is applied. The dielectric constant was interesting because water and oil have very different values, approximately 80 and 2 respectively. Meador, inspired by the work of Russian researchers Yuriy Antonov and Dmitri Daev, persuaded Texaco to sponsor the building of a dielectric logging tool. Meador tinkered with different frequencies and in 1973 found that a 2 MHz measurement virtually overlaid the traditional induction resistivity. In the mid-1970s, Texaco provided some of these 2 MHz tools to Aramco for logging an observation well in their Ghawar field.

NL hired Meador in 1978. He came with a bag full of expertise and was promptly made project manager of LWD research. "The first idea was to put the 2 MHz dielectric tool in a cutout in one side of a drill collar," recalled Meador. It worked, except that there was inadequate coupling between the transmitter and receiver coils to provide a usable measurement. So Larry Thompson, a recruit from Dresser Atlas, the result of Dresser acquiring both Pan-Geo Atlas Corporation and Lane-Wells, suggested wrapping the coil around the drill collar. This significantly improved signal coupling, and the NL electromagnetic wave resistivity (EWR) measurement was almost ready to go. In April 1979, the tool was tested extensively in a water tank and field testing began later that year. At this time, because mud telemetry was limited, data were recorded onto memory in the downhole tool, an option that remained an essential part of LWD, particularly when sensors started delivering large data volumes.

The whole development "was initially met with disbelief," recalled Roland Chemali, who had previously worked as a designer of electrical logging tools for Schlumberger. Experts at Schlumberger and in the rest of the industry believed that it was simply impossible to launch an electromagnetic wave from an antenna mounted on a metallic drill collar. More than just technical issues worried the company. Schlumberger was the undisputed leader in wireline logging, and in the late 1970s the fledgling LWD industry was seen as the enemy. Tom Zimmerman, a senior Schlumberger design engineer, remembered: "I know people who got sidelined when they tried to push LWD."[11]

In 1981, NL acquired directional drilling service company Sperry Sun, and soon after, the two companies merged to become Sperry Sun Drilling Services. Then the collapse of the oil price in 1982 nearly did it for Meador's project. "Management wanted to kill it," recalled Billy Hendricks, a member of the EWR sales team at Sperry Sun, "But a few people said just give him a little more time." In its final version, the EWR tool had a loop antenna in a recess around the drill collar for a transmitter and a couple of loop antennae for receivers on the same drill collar. Commercial operations started in 1983.[12]

These early LWD measurements were regarded as niche services restricted to simple resistivity curves, used mostly for well correlation. At Schlumberger, the interest for developing a tool remained within the engineering community. "The business people were totally disinterested", recalled Brian Clark, a Schlumberger LWD expert. "I remember in 1983 giving a talk at the Ridgefield lab in Connecticut and various wireline managers would say 'Stop working on this, it doesn't make sense.'"[13]

By 1985, Sperry Sun's EWR tool was becoming established along the US Gulf Coast and beginning to displace wireline logging. Schlumberger woke up, as Brian Clark remembered: "One day, in the spring of 1985, the Gulf Coast marketing manager came walking down the hall, agitated, waving a document and saying there was a company out there taking all the wireline business. I said let me see that paper. It was an early SPE paper on the Sperry Sun EWR tool."[14]

Clark read the paper and knew instantly what they were doing because he had written a patent memo in 1983 along very similar lines. "The paper got quite a bit of attention at the upper management levels," Clark recalled. By 1986, Schlumberger bit the bullet and initiated its own LWD development with generous funding despite the business downturn. Clark assembled a hand-picked team and within two years produced a tool that superseded the Sperry Sun tool in one important aspect. With its two transmitter antennae, the tool provided phase and amplitude measurements at two depths of investigation radically improving formation evaluation. But there was still deep skepticism about the physics. As Clark recalled, "I had an ongoing argument with a professor at the University of Houston who swore that it couldn't work. But I had done the modeling and the experiments." The tool was launched in 1989, christened the compensated dual resistivity tool and was an instant success. In 1994, Sperry Sun was eventually acquired by Dresser which four years later was taken over by Halliburton.[15]

During the 1990s and beyond, more and more wireline logging measurements were reborn as an LWD measurement, for example various forms of borehole imaging and almost the full spectrum of acoustic and nuclear measurements. In the process, the engineering and manufacturing specifications of LWD transformed the reliability of downhole equipment, whatever the conveyance. Logging while drilling tools have to be brutally tested for shock and vibrations, and as a result, more robust manufacturing techniques, such as soldering electrical components, had to be developed and would also get applied to wireline tools.[16]

The Schlumberger team that built the first compensated density neutron logging-while-drilling tool in 1989, pictured here in the Sugar Land facility near Houston.

In the end, every wireline logging company's nightmare, that LWD would kill the wireline logging market, never materialized. As with Numar's wild claims about NMR logging, the worst case scenario remained an illusion. However, as the drilling of highly deviated and horizontal wells proliferated, logging while drilling found its niche. It is always possible to push wireline logging tools to the end of a horizontal well with drillpipe, and indeed desirable if the measurement remains available only in wireline form, but drillpipe became a more obvious choice. In addition, the real-time advantage of the latest LWD measurements, such as being able to see ahead of the bit and around the drillstring to steer the well to reservoir sweet spots, leaves the operator with only one obvious option.

Geosteering

With so much information available while drilling and so much resting on the ability to hit the reservoir spot on, steering the well trajectory at will and accurately became crucial. The traditional way of steering was to attach a bent sub just above the downhole motor and bit. To steer the well, the string was rotated so the bend pointed in the right direction, and then drilling continued using just the motor, a technique called sliding. MWD measurements allowed the driller to monitor the well trajectory. But drilling a straight section then required pulling the entire drillstring, removing the bent sub, and going back in the hole, a long and tedious process.

The next step was to put the bent sub below the downhole motor, an idea that came from Robijn Feenstra of Shell and was commercialized in 1984 by Norton Christensen, the result of Norton, an abrasives manufacturer in Massachusetts, acquiring Christensen. In this configuration, the bent sub was sufficiently close to the bit that straight sections could be drilled by rotating the drillstring without removing the sub, and therefore without having to trip out of the hole. This would result in a slightly enlarged borehole.[17]

A problem with both configurations, however, was the slow pace of drilling in the sliding mode, generally 50% slower than regular drilling. This resulted from the drillstring being stationary. Cuttings accumulated on the low side of the hole, and the pipe's friction and drag along the borehole wall limited the ability to continue drilling.[18]

Surprisingly, the next innovation did not emerge from the mainstream oil and gas industry but from a German deep-drilling research program in Bavaria in 1985. This was run by the German Science Foundation and famously called the "Kontinentales Tiefbohrprogramm der Bundesrepublik Deutschland" or KTB project. With an objective of drilling 9,000 meters, hopefully to near the crust-mantle interface, the research well had to be kept within one degree from the vertical for its entire length to limit friction on drilling and to avoid exceeding the rig's hoisting capacity. To get help, KTB turned to a drilling technology center in Celle, Germany, that had been established in 1957 by Christensen Diamond Products as a manufacturing plant for diamond core heads and drill bits.

One of the employees at Celle was Volker Krüger who developed, built and operated a vertical drilling system (VDS) custom-made for the KTB program. This included a near-bit inclination sensor and a downhole processor programmed to activate steering ribs on a stabilizer just above the bit. Powered by a hydraulic system, the ribs continually nudged the well back to the vertical. The VDS was used to a depth of 7,500 meters at the maximum allowable temperature of 175°C for the electronics. In part, because the system controlled the well within a third of a degree from the vertical, the KTB project surpassed its target and reached a final depth of 9,101 meters, alas without reaching the mantle.

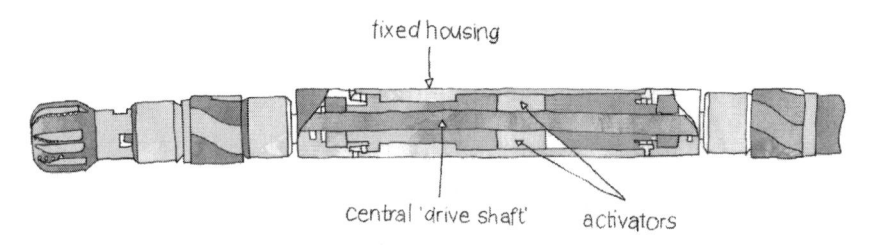

A modern rotary steerable system, with PDC bit on the left and, at center, a view inside the fixed housing showing activators that extend up and down to control drilling direction.

The VDS system was a stepping stone for a similarly conceived device for the oil and gas industry. The key backer was the Italian oil giant Agip, later a part of Eni, and their partner was Baker Hughes. In 1993, after four years of development, a new drilling system was launched that could steer a well in any direction while continuously rotating the drillstring. The era of rotary steerable drilling had begun.[19]

The introduction of rotary steerable technology eliminated several disadvantages of previous directional drilling methods. Because a rotary steerable system drills with continuous rotation from the surface, weight is transferred to the bit efficiently, increasing the rate of penetration. Rotation also cleans the hole by agitating drilling fluid and allowing the cuttings to flow to the surface. The decisive advantage, though, is not having to slide the tool, which becomes almost impossible at extreme horizontal distances. With rotary steerables, it became possible to drill some very long wells which soon gave rise to a new piece of terminology—extended-reach drilling (ERD) wells.[20]

BP was an early adopter of ERD drilling. Since 1984, it had been operating the Wytch Farm oil field in Dorset, England, close to Bournemouth. The field is mainly onshore, but significant parts of the reservoir stretch under Poole Harbour, a cherished and protected environmental area. Denied by local authorities from building artificial islands from which to drill, BP based their operations on land and used extended-reach drilling to reach the reservoir. In 1991, extended-reach drilling was still young —the 5-kilometer mark had been reached just once in 1989 in the Statfjord field in Norway by Statoil—but for BP it seemed a risk worth taking.[21]

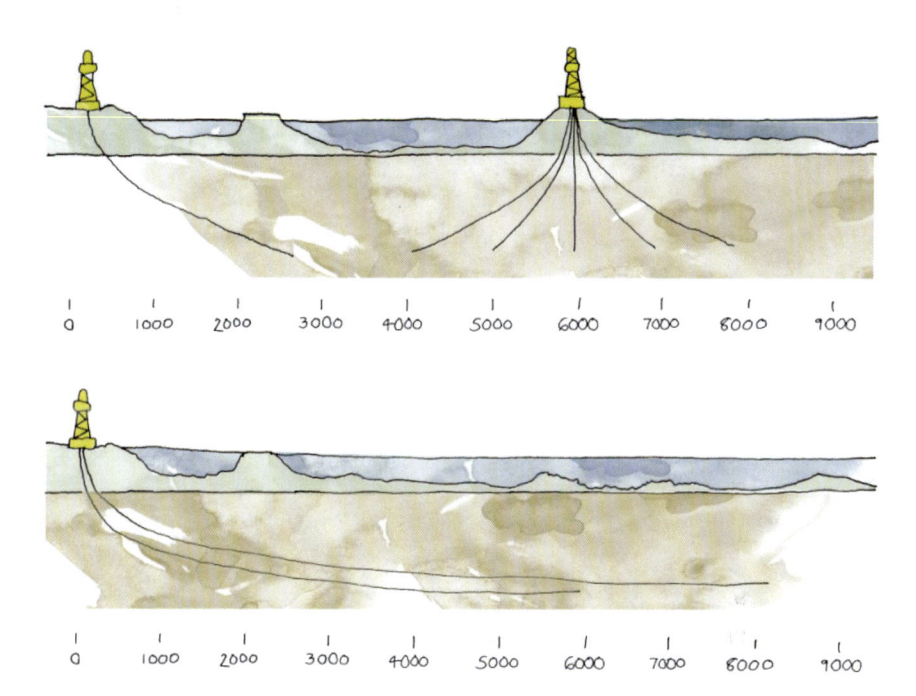

BP's Wytch Farm development. Top, the original proposal called for constructing an artificial island and drilling simple directional wells; bottom, final implementation in which extended-reach wells were drilled from the shore.

The rest is history. The first BP ERD well at Wytch Farm reached 3.8 kilometers. The company's main rotary steerable contractor was Camco Drilling Group Ltd., based in Stonehouse, England, a division of Camco International. The driving force at Camco was John Barr, who had been one of the pioneers in both PDC and rotary steerables while working for the Drilling & Service Company and later Reed Hycalog. The Camco rotary steerable prototypes successfully drilled three ERD wells and, in 1997, broke the 10-kilometer milestone.[22]

With two rotary steerable systems on the market, from Baker Hughes and Camco, complex wellbore geometries became possible, and the overall rate of penetration increased dramatically. Schlumberger was not slow to respond to this trend. In 1998, the service company acquired Camco and over the next years merged the Camco technology with its own.[23]

John Barr (1931-2007), left, a pioneer of rotary-steerable drilling in the mid-1990s for Camco International. Barr is shown with Roy Caldwell, President of Reed-Hycalog, a drilling tools manufacturer that collaborated with Camco.

The emergence of geomechanics

Drilling such sophisticated wells would have been impossible without some understanding about rock strength. To drill safely, the mud needs to exert enough pressure to hold back formation fluids. But it cannot exert so much pressure that it fractures the formation, resulting in loss of mud and the possibility of a blowout. Nor can the mud pressure drop to the point that the borehole wall collapses inwards. In short, balancing the mud pressure within these various limits needs much care, and it becomes increasingly difficult as wells become deviated and horizontal. The key to maintaining this balance, as drilling went horizontal, was provided by an emerging discipline called geomechanics, also known as rock mechanics.

The earliest understanding of why and how rocks fail is due to Charles-Augustin Coulomb. However, a practical method of predicting failure of rock would have to wait a century for Christian Otto Mohr, a railroad engineer from Hannover, Germany. Mohr recast Coulomb's ideas into a neat graphical technique later called Mohr's circle, to predict rock failure under three-dimensional stress. However, applying this to porous rock would require another pause. The complicating effect of fluid pressure in the pore space was provided in the 1940s by Karl von Terzaghi, an Austrian civil engineer.[24]

Thus far, the interest in geomechanics was confined to mining and civil engineering, particularly dam building. In neither case was it desirable to have a catastrophic rock failure. A more esoteric application of the discipline was precipitated by the development of nuclear bombs and their proliferation during the Cold War. Underground testing of nuclear weapons required a good understanding of rock failure in the vicinity of the test, and one man seized the intellectual challenge and, later on, a business opportunity.

Sid Green graduated from Stanford University in the late 1950s and joined a 900-person think tank hosted by General Motors that worked on advanced problems for the US Department of Defense. Green developed new ideas in rock mechanics to account for the incredibly high pressures present in an underground nuclear test, and in 1969 he founded Terratek Inc. in Salt Lake City to deliver rock mechanics services to the US government. As the Cold War petered out, Green turned his attention to another environment with very high stresses and pressures, the deep subsurface. Terratek shifted its sights toward the oil field and prospered.

Meanwhile, a small army of geomechanics experts was emerging from the few academic centers of excellence specializing in the discipline. One was the University of Minnesota, where UK-born Charles Fairhurst headed a rock mechanics department that produced an unending series of PhD students who today provide leadership in both academia and industry. Among these was Jean-Claude Roegiers, a Belgian engineer, who joined the cementing and fraccing company Dowell in Tulsa and created a team of rock mechanics specialists.[25]

The community was expanding. From the 1980s onwards, geomechanics matured from being a niche discipline few people in the industry understood, to being recognized as affecting virtually every phase of the exploration and production lifecycle. Most importantly, oil companies began to understand how much money geomechanics could save them, for example through better planning of the drilling process. The service companies were not slow to act. Schlumberger opened a geomechanics research lab in Beijing in 1998, taking advantage of China's abundant expertise in the discipline because of the proliferation of dam building in the country, and in 2006 acquired Terratek.[26]

Geomechanics appears in several guises, of which drilling is just one. It would soon be required for hydraulic fracturing, and also for the sanding problem when completing wells in unconsolidated reservoirs.

Chapter 20: Modern Completion and Production

AS THE early Azerbaijan oil producers discovered, large quantities of sand produced along with the oil provided a major headache, and sanding, as it became called, was soon experienced anywhere production came from shallow sandstones with little or no cementation to hold individual sand grains in place. Sanding can be catastrophic. Sand grains moving at speed are capable of eroding steel and can compromise an entire well, let alone the surface equipment. In addition, produced sand eventually fills the well, reducing or even killing production.[1]

Sand control becomes important

In the early 20th century, oilmen tried to combat sanding by using screens or slotted liners, rather than perforated casing. Having many entry points into the well reduced the forces tearing the sand grains from the formation, and the holes or slots in the screens and liners were designed to be of a size that encouraged the loose sand grains to create a bridge, blocking more sand from entering the well. The technique enabled longer sand-free production, but in the end the problem would always recur.[2]

Lester Uren at the University of California at Berkeley was the first to devise a better solution. In 1924, he conceived the idea of enlarging the hole a little and then filling the space between the borehole wall and the production liner with gravel, an idea inspired by techniques used in the water industry since the 1870s. The gravel was effective at causing the sand to bridge rather than produce but also supported the borehole wall from collapse. In an oil well, however, simply pouring gravel into the borehole caused too much damage to the formation. Uren therefore suggested mixing accurately sized gravel—large-grained sand about five to six times larger than the formation sand size—with water, oil or a clay-laden fluid, circulating it through the space between liner and formation and allowing it to settle from the bottom up. Uren's gravel pack worked, with only the finest sand particles able to move through the gravel.

Many in the oil field remained skeptical. An exception was the Layne-Atlantic Company, a regional affiliate of a company called Layne Inc. operating in the southern US states. In 1882, Layne had started out as a water well drilling company, and this origin may have triggered Layne-Atlantic's curiosity. Uren licensed the process to Layne-Atlantic who successfully applied the gravel-packing method in oil wells in the Gulf Coast in 1932. In the mid-1930s, the Texas Company, later Texaco, began experimenting with gravel packing in some of its wells in the Coalinga and Wilmington fields in California. The results were good enough that they bought Uren's patent outright and started licensing it to several oil and service companies. The Texas Company meanwhile developed a derivative method comprising a prepacked gravel pack, constructed by placing gravel of the right size in the annulus space between two perforated liners.[3]

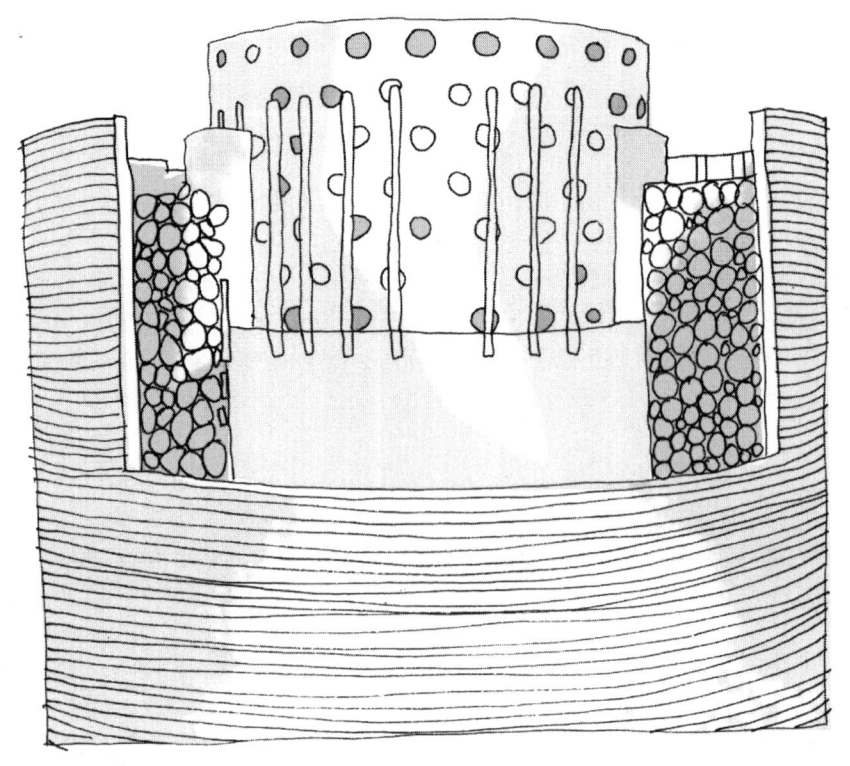

Cutaway view of a pre-packed gravel pack. The gravel is placed in the annulus space between two perforated liners.

Erle Halliburton came up with a completely different idea. The cementing company developed a plastic material from phenol-formaldehyde resin that, when injected into a poorly consolidated formation, bound together the sand grains for several feet around the wellbore. Halliburton's innovation was an outgrowth of the use of plastics for water exclusion, first performed in 1942 in an East Texas oil field to control water entry. Halliburton carried out its first commercial plastics treatment for sand control in January 1946 in the Gulf-operated Pierce Junction field in Harris County, Texas. The results were good, but the conditions were ideal: the formation had high permeability and easily accepted the plastic injection. On the whole, Halliburton's method failed to bite, as Harry O'Neal McLeod, a petroleum engineer who worked for Phillips Petroleum, Exxon, Dowell and Conoco recalled, "The problem was not being able to inject plastic through every perforation, so with time the plastic would be by-passed."[4]

During World War II, research continued in gravel packing. One major problem when circulating gravel in the well was the infiltration of carrier fluid into the formation, driving the oil away from the borehole. In 1945, Uren, who by this time had started consulting for the Texas Company, suggested lightening the carrier fluid with compressed air or some other gas to lower the risk of formation intrusion. In the following years, a number of other carrier fluids were introduced to decrease formation damage. The most popular was brine, which was mixed with the gravel and then pumped into the hole, but poor mixing often compromised the results.

In the late 1960s, research efforts by several companies, including Shell working with the Dowell Chemical Company, Continental Oil and the service company Byron Jackson, focused on ways to viscosify the gravel-transport fluid to achieve a better and sustained mix. A more viscous fluid could transport high gravel concentrations in a controlled manner yet remain in the well rather than infiltrating the formation. A particularly interesting viscosifier was hydroxyethylcellulose (HEC), and HEC-based fluids rapidly replaced brine as the transport fluid of choice and remained state of the art until the late 1980s. By then, HEC was seen to have problems. For example, it became common knowledge that HEC produced voids in the packs, and already in the mid-1970s, Continental Oil researchers had pointed out that HEC did not pack efficiently in deviated wells. Service companies therefore looked again at possible viscosifying agents. One alternative to HEC was crosslinked polymers, but these remained a niche product for highly deviated wells only.[5]

In the early 1990s, Baker Hughes in conjunction with sand-control guru Wally Penberthy at Exxon concluded rather surprisingly that brine could be the best carrier for gravel after all, especially for low-angle well deviations and relatively short completion intervals. This change of heart was due to an improvement in mixing hardware that allowed a more consistent mixing of gravel and brine. The introduction in the 1980s of improved filtration systems to filter large quantities of brine quickly at a reasonable cost was another contributing factor. By the late 1990s, most of the industry was back on brine as the preferred gravel pack carrier fluid.[6]

Hydraulic fracturing meets sand control

Combining technologies often brings dividends. This was the case in the early 1990s when operators started to embrace a novel sand control method called a "frac pack." "It's a gravel pack but with a frac on the other side of the pack," recalled Martin Rylance, a BP fraccing expert. "Frac packing dramatically improved the behavior and reliability of wells." The trick was to create short, wide and highly conductive fractures and stuff them with gravel. Once filled, the fractures offer more conduits for production, and the gravel-pack screen prevents flowback of sand. The result is improved productivity and less formation damage.[7]

In fact, the idea of combining sand control and stimulation in a single treatment began three decades earlier, in Venezuela in 1963. The Venezuelan Ministry of Mines and Hydrocarbons and the Creole Petroleum Corporation performed a well treatment that consisted of perforating the pay zone, followed by a small-scale fracturing treatment using viscous crude oil and sand, sized to control formation sand particles. This approach proved successful but failed to be adopted in other areas.[8]

Interest in frac packing resurfaced in the early 1980s, when BP, Amoco, Sohio and ARCO began to use the technique in the Prudhoe Bay and Kuparuk fields on the North Slope of Alaska and in the chalk formations of the North Sea. These results attracted attention in other areas but did not signal a step change because the treatments were designed and executed in a similar way to standard hard-rock fracturing. This resulted in long, narrow fractures, and any initial productivity gains were short-lived. The breakthrough for frac-packing came after 1985, driven by activity in the Gulf of Mexico where conventional gravel packs were proving significantly inadequate.[9]

A driving force was Bob Hannah who joined Sohio in 1983 with a reputation as an expert in hydraulic fracture acidizing, which he had gained working for the service company, The Western Company of North America founded by Eddie Chiles in Fort Worth, Texas, in 1939. Eddie was a classic Texan who advertised his company on TV with the slogan, "If you don't have an oil well, get one—you'll love doing business with Western!" His secret was to use tip-screenout fractures, in which pumping is continued after the fracture has filled with gravel, or screened out, causing the fracture to widen. This provided fractures with the right shape to increase production yet hold enough gravel to prevent sanding.

The technique revolutionized sand-control operations in the Gulf of Mexico. During the early 1990s, BP, Shell and Chevron all adopted the technique. Shell was also using it in shallow environments such as the Turtle Bayou field, in an inland lake in Louisiana. Later during the 1990s, Shell expanded the use of frac packing to the North Sea, Borneo and South America. By the year 2000, frac packing had reached the Middle East, West Africa and Brazil. Further advances by companies, including BP and Pennzoil, led to equipment and technique innovations that helped extend the length and width of the fractures to give much higher sustained production rates than were typically seen in gravel-packed wells. Today, many operators use frac packs as part of their standard completion.[10]

Advancing hydraulic fracturing

The frac pack had only become possible because of a number of advancements made in hydraulic fracturing. First was the issue of fracture orientation. By 1955, more than 100,000 well-fracturing operations had been performed worldwide, and nearly everyone accepted the hypothesis that the fractures created were predominantly horizontal. The theory was that the hydraulic pressure must have lifted the overburden vertically, causing the fracture to follow a horizontal path.[11]

Some oil company researchers were skeptical, but it wasn't before Shell asked Marion King Hubbert, its intellectual star, to make a critical review in 1955 that the horizontal fracture hypothesis would collapse under its own weight. Together with David Willis, whom Shell had just hired from Stanford University, Hubbert made an elaborate theoretical study of stresses around the wellbore. In all cases in relatively deep wells where the weight of the overburden was greater than any likely horizontal stress, Hubbert and Willis predicted that hydraulic fractures would form vertically, not horizontally.

To substantiate their theory, Hubbert and Willis performed a simple, makeshift experiment. They took a two-gallon flexible plastic bottle and filled it with gelatin. While it was still liquid, they inserted a glass tube to simulate a well. When the gelatin solidified, they withdrew the tube partway to simulate both a cased and openhole environment. In the first experiment, they simulated a horizontal least stress by squeezing the bottle between two clamped boards. In a second experiment, they simulated a vertical least stress by stretching the bottle vertically. For each stress condition, they fractured the gelatin by injecting a slurry of gypsum plaster, which solidified into a permanent record of the resulting fracture geometry. When the least stress was horizontal, the fracture surface oriented vertically, and horizontally if the least stress was vertical. They concluded in their classic 1957 paper "Mechanics of Hydraulic Fracturing" that fractures are "approximately perpendicular to the axis of least stress." Accordingly, in the tectonically relaxed areas of the US Mid-Continent and Gulf of Mexico coast, below a certain depth, fractures had to be vertical.

Their paper created a storm. Most in the industry refused to believe the results—horizontal fractures still sounded more plausible. Throughout the 1960s, however, evidence began to accumulate to give Hubbert and Willis's ideas new credence. Vertical-fracture imprints were reported on hydraulically inflated rubber packers, and before-and-after downhole photographs in uncased sections of gas wells revealed vertical fractures. By 1970 it was generally accepted that nearly all hydraulically induced fractures in deep wells were indeed vertical, and that fractures are horizontal only in shallow wells where horizontal stress exceeds overburden stress.[12]

Meanwhile in the mid-1950s, Soviet theorists including Grigory Barenblatt, who would later move to the University of Cambridge and finally the University of Berkeley, had been developing a theory of fracture propagation, but this failed to provide any significant practical breakthrough. That would come from Thomas Perkins, who joined The Atlantic Refining Company in Dallas in 1957. Engineers there were looking for ways to maintain fracture width, and one possibility was to use propping particles, generally sand particles, to keep the fracture open and conductive. Bigger propping particles "required that we had a good understanding of fracture width," recalled Perkins, "How they open, what factors influenced the width and how you then go and calculate the width in advance." In 1961, together with Loyd Kern, the leader of the Atlantic hydraulic fracturing group, Perkins published the seminal paper "Widths of Hydraulic Fractures." Still today, the paper is a key reference and ranks as one the most downloaded papers at OnePetro, an online library of technical literature provided by the SPE.[13]

Another advance also came in 1961, when Shell physicist Mike Prats published a paper that has enjoyed almost the same popularity as the Perkins and Kern paper. Similar to Hubbert, Prats went against a common belief, in his case that the bigger the fracture the better the results. Recalling his paper, Prats said, "There was an optimum length-to-width-ratio to get the best response, so for a given volume of proppant there was a certain shape that could optimize production. It wasn't just size." Prats found a mathematical formula to determine that optimal ratio.[14]

While Hubbert, Perkins and Prats revolutionized the understanding of hydraulic fractures, the fluids that transported the proppants saw a rapid evolution. The fluid used for the first fraccing job in 1947 was gelled oil, but an alternative from 1953 onward, was water, generally slick water of low viscosity. "At that time, we didn't really know how to gel water," recalled Larry Bell, a retired petroleum engineer who had worked for ARCO. The first water-base, gelled fracturing fluids surfaced only after 1960. One of the first was guar, used in the food industry for many years as a thickener primarily for ice cream—80% of the world's supply of guar is grown in India and Pakistan. Guar was first used in 1961 by Halliburton on a job in West Texas for The Atlantic Refining Company. Thomas Perkins noted, "Guar gum was and still is a tremendously important part of fracturing."[15]

Meanwhile, development of proppants kept pace, particularly during the 1970s when wells got deeper and hotter. "At that time hydraulic fracturing was marginal in wells as deep as 10,000 feet and these were primarily gas wells," recalled Claude Cooke, a physicist who worked for Exxon Production Research in Houston from 1954 to 1986. Sand, still the most commonly used propping material at the time, tended to be crushed by the stresses encountered in deep formations, producing fine particles or fragments that could drastically reduce the flow through a fracture. A novel idea for proppant was glass beads, which were intended to be high strength, but lab tests carried out by Cooke's Exxon team showed that they shattered. Cooke found a variety of other materials and tested them, both in-house and in cooperation with Carborundum Abrasives of Niagara Falls, New York, which eventually found a tough enough propping material in sintered bauxite particles.

Sintered bauxite is produced by crushing bauxite to a powder and then fusing it at high temperature into spherical beads that become extremely hard and durable. In 1977, sintered bauxite beads were successfully tested in twelve wells in south Texas and Mississippi, all deeper than 10,000 feet. Service companies were a little hesitant at first, because "pumping abrasives was tough on their pumps and pumps were expensive," recalled Larry Harrington, a former hydraulic fracturing expert at The Western Company of North America. But the gains in productivity soon redeemed any hardware repair costs. Claude Cooke recalled: "The application of sintered bauxite as a propping material really increased the efficiency of hydraulic fractures in deep wells."[16]

Workover and coiled tubing

During the life of a field, one of the production engineer's responsibilities is to monitor every well and make sure it keeps producing. If production drops, then the operator doesn't have much choice but to kill the well and bring in a small workover rig to pull the completion hardware and investigate. At least, that was the standard procedure until the 1960s. About that time, the first coiled tubing systems appeared, which allowed the operator to enter the well against pressure and to perform remedial activities without a rig. It had the makings of a production engineer's dream come true.[17]

The origins of coiled tubing can be traced to the 1944 Normandy invasion. A critical part of the D-Day planning was how to supply the forces landed in France with sufficient fuel. In a top-secret mission code named PLUTO—short for pipelines under the ocean—pipes were deployed across the English Channel to several points along the coast of France. George King, ex-Amoco and now working for Apache, explained: "In UK harbors, Allied engineers took 50-foot long sections of three-inch pipe, welded them together and spooled them onto an 80-foot diameter drum, until they had enough to go the width of the channel." The pipe was then unspooled and laid to rest on the seafloor. The plan worked and fuel flowed to the troops.[18]

The oil industry took note. In 1951, Houston-based Shell engineers George Calhoun and Herbert Allen patented an "Equipment for Inserting Small Flexible Tubing into High Pressure Wells." The concept included an injector head with two vertical, counter-rotating chains to move the tubing in and out of the well. The first application of the patent was not in the oil field but for the US Navy. Brown Oil Tools was contracted to adapt the idea to let submarines, submerged in depths up to 600 feet, deploy a radio antenna up to the ocean surface. Meanwhile, oil and gas operators sought ways to make practical use of the Calhoun and Allen patent. One was Jim Rike of Humble Oil. In 1959, Rike developed the key concept of "managed fatigue" to prove the feasibility and safety of repeatedly coiling and uncoiling steel pipe, inherent in coiled tubing deployment.[19]

Jim Rike (1927-2014), who worked 40 years for Exxon and contributed to a plethora of upstream technologies, including coiled tubing.

By 1962, Brown Oil Tools, together with The California Oil Company, had constructed the first functional coiled tubing unit for the oil field. The tubing string was fabricated by butt-welding 50-foot sections of 1.375-inch pipe to achieve a total length of 15,000 feet and spooling it onto a 9-foot diameter reel. During 1963 and 1964, it was tested in several wells, both inland and offshore in south Louisiana, for washing sand out of a well and fishing a storm choke. However, the low-strength steel and numerous welds meant that the tubing could not withstand repeated bending cycles during unspooling and spooling nor bear high tensile loads. Operators soon lost confidence. "But the companies that were supplying it loved it!" recalled George King, "I have several friends who in the 1960s made more money fishing broken coiled tubing than running it."[20]

A coiled tubing unit with, middle, its reel of flexible steel tubing. To deploy tubing downhole the operator spools the tubing off the reel and leads it through a "gooseneck", top right, which directs the coiled tubing downward to an injector head, where it is straightened just before it enters the borehole.

Throughout the 1970s, Brown Oil Tools and several other equipment manufacturing companies continued making improvements, but poor reliability continued to plague operations. It took improved steels and a new welding technique to change the picture. In 1980, Southwestern Pipe introduced a high-strength steel, and in 1983, Quality Tubing Inc. in Japan began using long sheets of steel to reduce the number of required welds. Then in 1989, Quality Tubing introduced bias welding, cutting steel sheets diagonally to spread the welding heat spirally around the tube. In 1990, suppliers began manufacturing coiled tubing with a 2-inch diameter, and in the subsequent decade, with diameters up to 4.5 inches. Today, coiled tubing has become an integral component of workover, and in certain circumstances it has been used for drilling, such as multilaterals from a mother bore.[21]

Pressure-controlled perforating

Shaped-charge perforating, developed in 1947 and used ever since for creating conduits for oil and gas flow through casing, was typically performed with the well slightly overbalanced, in other words with wellbore pressure greater than formation pore pressure. Unfortunately, this encouraged the shaped charge to pulverize the formation creating a zone of low permeability that impeded flow. Therefore, in the 1950s and 1960s, the oil industry began investigating ways to perforate underbalanced, impossible at the time because the heavy guns used to convey the shaped charges were deployed using a standard wireline, too large for any pressure-control equipment at the surface.

The first solution pioneered in 1952 by Schlumberger and Humble Oil was to complete the well, assemble the wellhead and then perforate through tubing. Schlumberger developed a new type of perforating gun small enough to pass through tubing and equipped it with shaped charges powerful enough to perforate the casing. They also deployed a single-conductor wireline small enough to allow pressure control at the surface. The gun was tested in 400 wells along the Gulf of Mexico coast. Through-tubing perforating became a standard procedure.[22]

The next leap in perforating technology can be attributed to Roy Vann, originally a Schlumberger field engineer. His solution was to mount the guns on the bottom of the production tubing and then complete the well before firing them. The guns were fired by dropping a bar down the well to trigger detonation and then left in the hole as part of the completion. The technique was called tubing-conveyed perforating (TCP) and promised to facilitate running very long guns.[23]

"Schlumberger initially didn't believe in TCP," recalled Larry Behrmann, the company's leading expert in perforating technologies, "Because they were a wireline company." Vann left Schlumberger in 1969 after 18 years of service and established the Vann Tool Company in Artesia, New Mexico, later renamed Geo Vann. In an interview with the *Tyler Morning Telegraph* in 2009, he was quoted: "My whole life has been looking and searching for what hasn't been done. I try to find something that's new." In 1970, Vann performed the first successful TCP operation for an independent operator in southeast New Mexico.

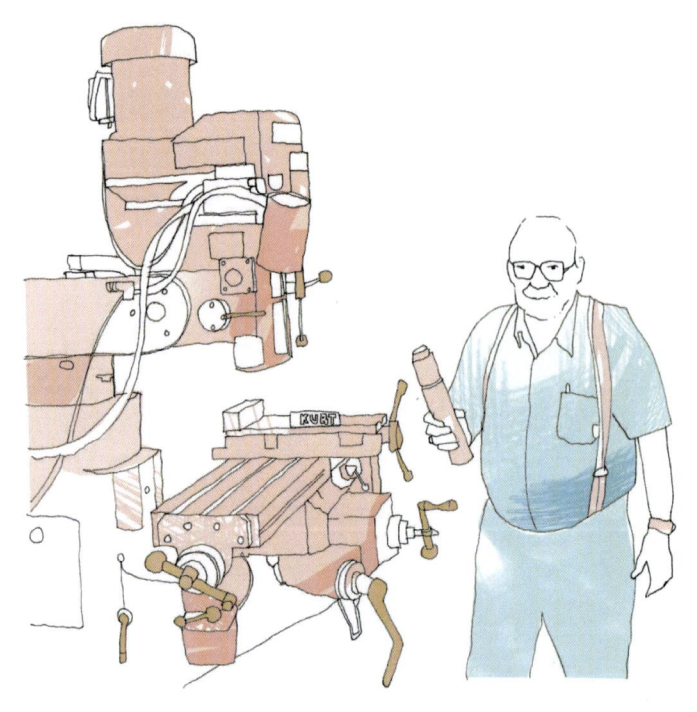

Roy Vann (1924-2013) in his workshop in Tyler, Texas. Vann was best known for his tubing-conveyed perforating innovations from the early 1970s. In his later years, he founded and ran Vann Pumping Systems, and in his lifetime accumulated no less than 37 US patents.

By 1980, Vann's company had tripled in size, as demand for TCP grew worldwide. By the mid-1980s, tubing conveyed perforating started to become mainstream. Soon the large service companies seized the initiative, and in 1986 Geo Vann, Inc. was acquired by Halliburton. Ten years later, TCP accounted for more than 25% of all perforating worldwide and was especially in demand for highly deviated and horizontal wells. Today, TCP strings have exceeded 8,000 feet in length.[24]

Modern pumping units

Until the mid-1950s, REDA Pump Company remained the only company producing and running electric submersible pumps (ESPs). As demand for oil grew in the 1950s and 1960s, demand for pumps of any kind picked up, and competition in the ESP market began to take shape.[25]

Armais Arutunoff, the irascible founder of REDA, demanded complete allegiance from his employees, allowing no credit for innovation or individual contribution. In the mid-1950s, several disgruntled employees approached Dresser in the hope that a large corporation with financial muscle could be persuaded to work with REDA. When Arutunoff found out, he forbade any cooperation with Dresser and systematically fired everyone he thought was involved. Two of those fired, Joe Carle and Eldon Drake, then approached the Byron Jackson company about developing an ESP product line. They were met with open arms, with the result that by the fall of 1957 Byron Jackson started developing its own ESP. The first prototypes were installed in the Shell Signal Hill and Round Mountain fields near Bakersfield, California.

In the early 1960s, more competition came from manufacturers of water well electrical pumps. Goulds Pumps of Seneca Falls, New York, and Franklin Electric Company of Bluffton, Indiana, combined forces but struggled to develop a reliable design for oil wells. Then in 1966, Franklin Electric hired Joe Carle from Byron Jackson to spearhead a redesign of their first attempt. The following year, Goulds Pumps and Franklin were able to establish Oil Dynamics Inc. (ODI), becoming a third major player in the ESP market. The first ODI pump was run in February 1968 in a Kewanee Oil Company well in Burbank, Oklahoma. A fourth ESP manufacturer appeared in 1965 when the Peerless Pump Company from Los Angeles introduced its Hydro Dynamics product line.

For Armais Arutunoff, the combination of losing two of his best engineers and having to face three strong competitors was more than he could handle. In 1969, he sold REDA to TRW Inc., a US engineering conglomerate that had already started producing ESP power cables in 1966. Now aged 76, Arutunoff went into the hotel business but nevertheless had the satisfaction of seeing ESPs become a worldwide success. During the 1970s, bulk ESP sales to Russia and China resulted in a significant growth for the industry. Russia soon caught up manufacturing their own ESPs.[26]

The Achilles heel of early ESPs was the electricity supply, a faulty cable or some electrical connection being the culprit. Adoption of ESPs in the North Sea in the early 1970s pushed these issues to the forefront. At that time, the lifetime of an ESP was around 20 months at a unit cost of around US$ 350,000, so any small improvement was worthwhile. Roger Hoestenbach, a Shell facility engineer responsible for the company's ESPs over 34 years, recalled how he was forever spotting problems, suggesting solutions and reporting back to his bosses, "One time a VP of Shell said to me: 'If I hear one more thing about your damn pumps, I'm gonna fire you!'"[27]

By the late 1970s, ESPs started to interest the large service companies who progressively bought their way into the business through a series of mergers and acquisitions. In 1980, Hughes Tool bought Byron Jackson's ESP division, and in 1986, Baker International acquired the Peerless Pump ESP, renaming it Bakerlift. Ten years later, Baker Hughes took over ODI and consolidated its ESP business under the Centrilift brand. Meanwhile in 1988, TRW sold REDA to Camco International, and in 1998, Schlumberger acquired Camco and with it the legacy Camco ESP brands.[28]

Since then, electric submersible pumps have evolved into an efficient and reliable artificial lift method for lifting moderate to high volumes, with at least 15 separate brands available on the market. Worldwide, more than 100,000 wells have electric submersible pumps installed, with Russia being a prolific user and manufacturer. New varieties of ESPs include a water-oil separator, which permits reinjection of the water without having to lift it to the surface.[29]

All this time, there was a continuing demand for traditional sucker rod pumps, but here too, changes were in the works. The key innovation in 1930 was the progressive cavity (PC) pump, also called the Moineau pump, named after the inventor René Moineau, a French pioneer in aviation engineering. Moineau's pump comprised a housing with a wave-like interior and an inner spiral-like shaft that when rotated, caused a cavity and any fluid within it to move continuously from one end of the housing to the other. As early as 1936, the PC pump was introduced into the upstream industry for pumping oilfield fluids at the surface.[30]

Joe Dunn Clegg, an industry leader in artificial lift. He worked for Shell from 1952 to 1991 and remains a consultant, teaching at the University of Houston. Clegg was the editor of the second edition of the Production Operations Engineering volume of the SPE Petroleum Engineering Handbook, published in 2007.

Except for limited field tests, it was not until the late 1970s that PC pumps began to be used downhole for artificial lift. By this time, the pumps were more durable thanks to the development of synthetic elastomers in the late 1940s, which were crucial for manufacturing the housing. A drive mechanism also had to be developed, usually a surface motor coupled through a right-angle drive and then via rods to the downhole pump. Among those marketing PC pumps for artificial lift today is the company Moineau had founded in 1932, Pompes Compresseurs Mécanique SA, that with help from IFP commercialized their first downhole product in 1985. In the decades since, PC pumps have enjoyed increasing success, in no small part because of enthusiasts like Jim Lea of Amoco Production Research in Tulsa. The pumps have proved especially suited for pumping heavy oil. By 2006, it was estimated that more than 50,000 wells worldwide were being produced with progressive cavity pumps.[31]

Analyzing production

As demand for oil and gas continued unabated, oil companies began to wonder if every well they owned was producing to its true potential. What was lacking was a bit of theory that could describe how to optimize any individual well.

In 1954, W.E. Gilbert of the Shell subsidiary Asiatic Petroleum Corporation proposed the first systematic analysis of well production at an API meeting in Los Angeles, analyzing the chain of pressure drops as oil flows into and up the well. Although Gilbert presented methods for optimizing any given well, the calculations were rather cumbersome in the precomputer age, and few operators yet saw the need for such fine-tuning of their wells. A widely used simplification was later developed by Jack Vogel, one of Gilbert's protégés, and provided the basis for the next step.[32]

The change came in the 1970s when the Arab oil embargos forced US producers to squeeze more oil out of their wells, and computers started to make an impact. Another follower of Gilbert's work was Kermit Brown, a professor of petroleum engineering at the University of Tulsa and already a world renowned gas-lift authority. Brown and Joe Mach, one of his students in Tulsa, redeveloped Gilbert's ideas and turned them into an industrywide success.[33]

Mach first worked for Gulf Oil Corporation but in 1976 was hired by the Schlumberger gas-lift subsidiary Macco. Schlumberger Macco was interested in Gilbert's approach and joined forces with oil company Tenneco to create a modern well production analysis. The new analysis broke the problem into distinct stages, each stage connected to the next via a so-called node. By monitoring pressure losses and fluid rates between consecutive nodes—as oil moved from the reservoir to the wellbore, then up to the surface and finally to the gathering station—the analysis could identify bottlenecks and compare the actual performance of the well against a theoretical ideal. The difference between the two, the performance gap, allowed the analyst to decide how to make the well more productive, for example whether to reperforate, stimulate or change the pump. From the outset, this nodal analysis was programmed on a computer.

Joe Mach, a member of the Schlumberger team that developed the Nodal production analysis method in the late 1970s. He rose to fame 20 years later when he successfully applied Nodal analysis to the Yukos company fields throughout Russia.

In 1979, Joe Mach, Macco colleague Eduardo Proano and Kermit Brown published their breakthrough work through the SPE: "A Nodal Approach for Applying Systems Analysis to the Flowing and Artificial Lift Oil or Gas Well." By now, "Nodal" had become a brand of some significance, and was soon trademarked by Schlumberger. Petroleum engineer Harry McLeod recalled the importance of nodal analysis: "One of the most significant things in the early 1980s was the introduction of nodal analysis. The idea was not totally new, but it had never really been applied because the calculations were so tedious. Nodal analysis simplified everything and allowed the engineer to evaluate performance quickly and make changes to the well." By the early 1990s, Schlumberger was offering a computerized nodal analysis that could identify the best wells for remedial treatment.[34]

A striking success for nodal analysis occurred in Russia in the late 1990s and early 2000s. The collapse of the Soviet Union in 1991 had taken its toll on the oil industry. From its peak in 1988 to the worst year in 1995, production in Russia fell by 40%. During this chaotic period, however, a young man named Mikhail Khordokhovsky had managed to cobble together several old Soviet production companies into one large conglomerate called Yukos. Khordokhovsky was ambitious and pro-Western. In October 1998, Yukos signed a five-year strategic agreement with Schlumberger, in which Schlumberger not only provided services for Yukos's ailing wells but also seconded personnel to jump-start Yukos's financial and technical management. One Schlumberger secondee was Joe Mach, who eventually became responsible for Yukos's total production and later a full-time Yukos employee and vice-president.[35]

In early 1999, Mach started implementing his nodal analysis approach, identifying performance gaps and selecting candidate wells for treatment. He bought large numbers of laptop computers and trained more than 1,000 young Yukos engineers in nodal analysis. "In 1999, I put members of my team in each production unit in the company," Mach said. "These people had to look at each well, calculate its performance gap and sort them in descending order. The well with the biggest gap went on top and we worked on that well." He then ordered a massive shutdown of Yukos's poorest producing wells. From 14,000 active wells when Mach arrived in early 1999,

Yukos's stock of producing wells dropped to 7,000 by the beginning of 2004.

In most cases, poor performance could be traced to the ESP, the lifting mechanism in virtually all the wells. Mach replaced old, inefficient ESPs with new and more powerful ones. And the analysis showed that existing ESPs should be placed much lower in the well, even to their temperature limit and beyond. This met with strong resistance from the Russian operators, but Khodorkovsky backed Mach up, recollecting: "When Joe first arrived and said set the pump lower, our guys basically said get lost, probably something less polite! But Joe's equally down-to-earth Oklahoman drawl won the day." From 1999 to 2004, Yukos's oil production rose from 44 to 86 million metric tons per year.[36]

Production logging and downhole tractors

Critical for improving any well's production is knowing how the well is behaving downhole, for example knowing which zones are producing, how much they are producing and what fluids are being produced. This is the role of production logging, generally made with small-diameter tools so they can fit through production tubing.[37]

Production logging can trace its origins to the logging tool invented by Schlumberger in 1937 to record temperature versus depth. Temperature is particularly suited for detecting gas, since gas entering a well expands and cools, so the tool found some application although a formal interpretation of temperature logs was seen to involve some tricky thermodynamics theory. The first logging tool to measure flow was invented in 1948 by Stanolind Oil and Gas Company for use in gas wells. It featured a hot-wire anemometer—basically a thin wire that heated up with increasing gas flow, displaying an increasing electrical resistance that could be measured. The first flowmeter for oil, constructed by Humble Oil in 1955, employed a spinning helical rotor whose speed of rotation was measured by a clock-controlled 16-mm film recording.[38]

Production logging surfaced again in 1957. That year, Shell was experiencing water encroachment in its oil field in Brunei, North Borneo, and the company needed to establish the depth at which water was entering their producing wells. They started building a production logging tool locally, calling it the "water witch." When the Shell engineer in charge fell sick, Schlumberger wireline engineer Simon Noik decided to jump in. Noik redesigned the tool to be able to measure holdup, the proportion of oil to water. Although the measurements were rather crude and not always reliable, Shell was enthusiastic.[39]

The early 1960s witnessed a steady development of production logging tools. Schlumberger introduced a tool to measure the density of produced fluids and a water-cut meter to establish holdup. By 1971, the company had launched its production combination tool, featuring six sensors that recorded flow rate, pressure, temperature, hole size, fluid density and casing collar depth. But a constraint was the telemetry—the data had to be transmitted sensor by sensor. Only in 1980 did the telemetry catch up, making simultaneous measurements possible, as Schlumberger demonstrated for the first time during a campaign in the Amoco South Swan Hills miscible flood project in Alberta, Canada. Production logging was maturing, and indeed graduating from helping oil companies diagnose single-well performance to optimizing whole fields.[40]

During the 1980s, high-angle and horizontal wells suddenly posed totally different challenges, not only for tool designers but also for the production engineering fraternity. First was the realization that flow in horizontal wells differed dramatically from that in vertical wells. A mixture of oil and water flowing up a vertical well was more or less understood, although the flow regimes could be complex and hard to model. Flow in horizontal wells, however, looked quite different. For a start, oil would always flow on top of water because it was lighter, making flows stratified rather than mixed. Then it was realized that a horizontal well was never precisely horizontal but undulating, so water could get trapped in the down part of an undulation, called a sump, and actually stop the oil flowing altogether. The complications were endless.

All this came to a head at the BP Wytch Farm field, being developed exclusively with extended-reach wells. A completely new type of production logging tool was called for. Starting in mid-1994, BP and Schlumberger joined forces to find a solution. Within 18 months, a totally new production tool prototype had been built. It was a monster, combining some traditional production logging sensors with a new sensor to visualize stratified flow and a nuclear measurement designed to measure the proportions of oil, gas and water in the borehole. It was run for the first time in November 1995 in a Wytch Farm well and caused a sensation. The log, and others made later, revealed for the first time how different was flow in horizontal wells. Since then, the tool has been engineered, shortened and had more sensors added.[41]

The second issue posed by horizontal wells was getting the tool to the end, or toe, of the well. In vertical and deviated wells up to 60 degrees, the wireline engineer could lower a production logging tool on an electric wireline, but at higher inclinations this proved difficult or impossible. Pushing the tools with coiled tubing was one solution, but even then there was a limit, particularly when well lengths reached many thousands of feet. What worked in the end was a downhole tractor, developed in the early 1990s and operated on the end of a conventional electric wireline. The downhole tractor has arms to engage with the walls of the well and pulls itself along, pushing in front of it whatever logging tool is required. Its main limitation is the large amount of power required for operation.[42]

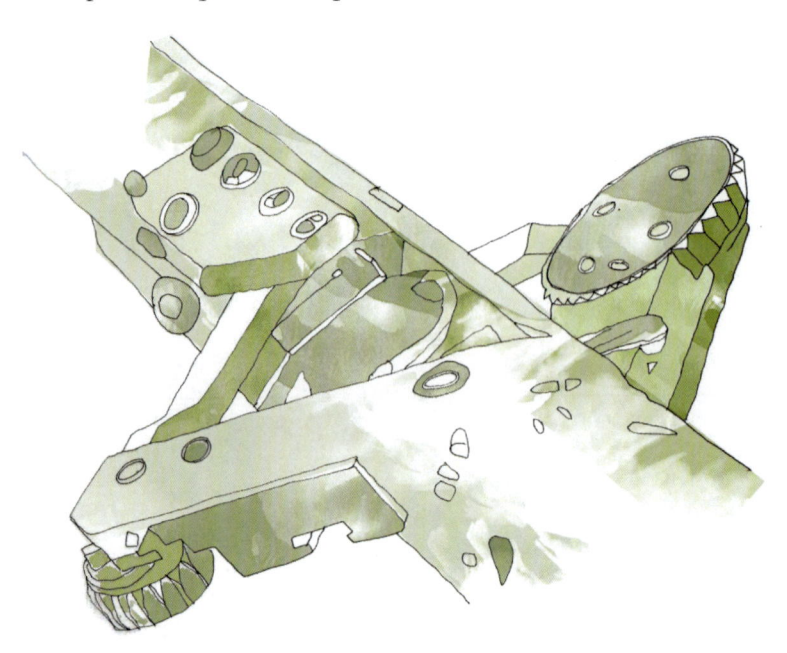

A downhole tractor. Cog wheels on the arms engage with the well wall to propel the tractor forward, pushing tools being deployed along the horizontal well.

An early tractor, called the Mule, was made in England by Robert Hunt-Grubbe's Sondex Company in the mid-1990s. Its motor was built around a novel and very powerful motor designed for sheep shearing by CSIRO, Australia's national science agency. "Its efficiency was so high that there was almost no lost energy and it remained cool in the user's hand," recalled Hunt-Grubbe. The Sondex publicity department put out a video showing a length of casing laid out across a field in the English countryside and its tractor inside the casing attached to a cable pulling a motor car across the field. For remaining skeptics, the demonstration was then repeated with the car's handbrake on. However, the Sondex downhole tractor never become a bestseller, partly because Sondex was not well equipped to provide field support and expertise.[43]

The Danish company Welltec was more successful. In 1987, Welltec CEO Jørgen Hallundbæk, who was based at the Technical University of Denmark, envisioned the idea of a downhole tractor and embarked on several years of development. In 1994, he founded Welltec, and its well tractor was introduced to the market two years later. It was capable of pulling more than 25,000 feet of wireline or coiled tubing into a highly deviated well. The tool was quickly adopted by operators, in particular in the North Sea by Statoil, who quickly spotted its potential. Downhole tractors had come of age and were adopted by all the big players.[44]

Inflow control devices

One result of production logging in horizontal wells was to rethink horizontal well completions. In the late 1980s, Norsk Hydro started drilling horizontal wells to tap a thin, high-permeability formation in its Troll field in the Norwegian sector of the North Sea. Production logs in the first well indicated that 75% of the production was coming from the half of the horizontal section nearest the heel of the well. For various reasons, there was more drawdown and more production near the heel than the toe.[45]

Norsk Hydro engineers proposed three options to distribute the production evenly along the length of the well. One was using a stinger assembly with multiple packing units, a second was to reduce the number of perforations near the heel, and the third, the one they tried, was to deploy an innovative piece of completion kit they called an inflow control device (ICD). The ICD basically divides the well into packed-off sections, with a choke in each section individually tailored to evenly distribute the drawdown along the well. Norsk Hydro built the first ICD in 1992.[46]

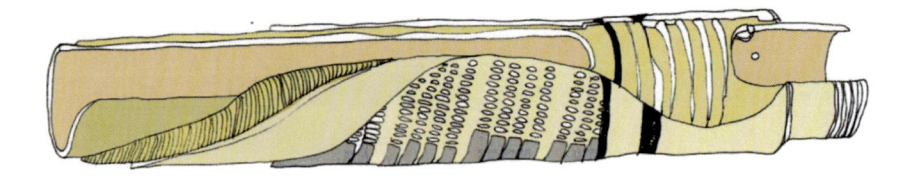

Cutaway view of an inflow control device (ICD). Formation fluid flows through multiple screen layers mounted on an inner jacket, and along the annulus between the solid basepipe and the screens. It then enters the production tubing through a tortuous pathway, and finally through an adjustable choke, the small hole on the right.

Service companies leapt on the ICD idea, with Baker Hughes initially leading the way, while operators such as Saudi Aramco were quick to see the benefits. When Nansen Saleri of Saudi Aramco saw the first results after Baker Hughes deployed ICDs in the Shaybah field in 2003, he went straight to the Baker Hughes representative to find out how many ICDs were available in Saudi Arabia. Saleri vividly recalled the following conversation, "They said only one! And I asked how quickly could you bring 10? They said we don't even have 10, this is just under development. This was a problem because I knew this was a game-changing technology."

Over the next few years, Saudi Aramco became the primary user of ICDs. Other service companies, including Halliburton and Schlumberger, soon got into the business, marketing their own ICD brands. Today, inflow control devices are a big market and an invaluable tool for production optimization in horizontal wells. More sophisticated versions now have chokes that react to varying downhole flow conditions.[47]

Chapter 21: A Seismic Revolution

DURING THE 1980s, a technology emerged that would change seismics and the industry forever, transforming how explorationists imaged the subsurface and upping the success rate in wildcat drilling. This breakthrough was none other than three-dimensional, or 3D, seismics.

The decisive move to 3D seismics

The principle of 3D seismics is simply to shoot data over a whole area rather than just along a few single lines. Processing then takes all the data and forms an image that is a subsurface cube rather than vertical slices through the earth. Guus Berkhout, professor of acoustical and geophysical imaging at the TU Delft University of Technology recalled the impact of 3D seismic, "You saw not only better quality data because you had more measurements, but you had directly for the first time a 3D view."[1]

As early as 1963, geoscientists at Esso Production Research led by Hugh Hardy, a geologic engineer who later pursued a military career and became a Marine Corps general, took the first steps. By 1964, Esso performed a first experimental 3D seismic survey over its Friendswood Field on the outskirts of Houston. However, they lacked the ability to migrate the data, to get everything into its right position, so the data proved difficult to use or interpret.[2]

About the same time, Donald Rockwell at GSI had been developing 3D migration, and by 1972, GSI was ready for its own experiment. William Schneider, who in 1970 had become GSI project manager in charge of the 3D technology, recalled: "I was involved in a brainstorming session about 3D seismic with my boss, Milo Backus, the director of GSI Research, Bob Graebner and M.E. 'Shorty' Trostle, both executives of GSI. We spent a lot of time speculating about what 3D could do for seismic exploration, but the fact remained, nobody was going to buy just an idea. We had to acquire some actual data."

Yet the cost of an experimental 3D survey was more than GSI could afford. So they persuaded a consortium of oil companies—including Chevron, Amoco, Texaco, Mobil, Phillips and Unocal—to share the cost. The chosen site was the Bell Lake Field in southeastern

New Mexico and data acquisition started in late 1972. The following year, Manuel Forero, a GSI geophysicist and interpreter, managed to migrate the data and produce time maps for several horizons. Interpretation of these maps proved consistent with nine wells that were producing and three that were dry. It also pointed to several new drilling locations, one of which was drilled and tested positive for oil.[3]

The Bell Lake 3D project was a defining event in seismic prospecting and paved the way for the first marine 3D survey conducted in 1975 by Sun Oil Company and GSI in the North Sea. By this time, the service company had combined a number of technologies into a functional 3D system. "GSI was using Texas Instruments computers specially adapted for seismic," recalled Milo Backus. "They were called the Advanced Seismic Computer, and at the time were some of the fastest computers around, matching even the famous Cray computer." The acquisition side also needed to be reconfigured. Not only was the air-gun reinforced, but compasses had to be placed in the marine streamer to transmit real-time information about its location as it was towed behind the seismic vessel.[4]

This first 3D offshore seismic survey was a mixed success. Norman Hempstead, then GSI manager in the UK, recalled, "The first 3D survey in the North Sea was performed by GSI in October 1975 as a means of demonstrating the principle and opening up the North Sea market. It was performed on the cheap—that is to say, with a spacing of 200 meters between adjacent lines. This sampling was not adequate for optimum resolution of the geology, but the results nevertheless provided clear indications of the potential of 3D." The oil company memories are slightly less bullish. As BP's Jim Hornabrook recalled, "GSI ran around entertaining presidents and chief executive of oil companies trying to persuade them to run 3D with decidedly mixed success."[5]

In these early days, the high cost of 3D seismic data acquisition made the technique much easier to sell in a reservoir development setting than for exploration since the coverage for the former was so much smaller than for the latter. One cause of marine 3D being so expensive was that the surveys were made with a single streamer. Covering the survey area therefore required endless zigzagging across the ocean, all the time making sure that the vessel and streamer kept to their correct trajectories, no mean feat given the vagaries of deep ocean currents. The obvious solution was to deploy several streamers at a time. This had the potential of cutting survey time and allowing closer spacings between lines, but the control of the various streamers then became even more challenging. The first multistreamer job, with two streamers, was performed in 1984 by the Geophysical Company of Norway, known as Geco, for Conoco.[6]

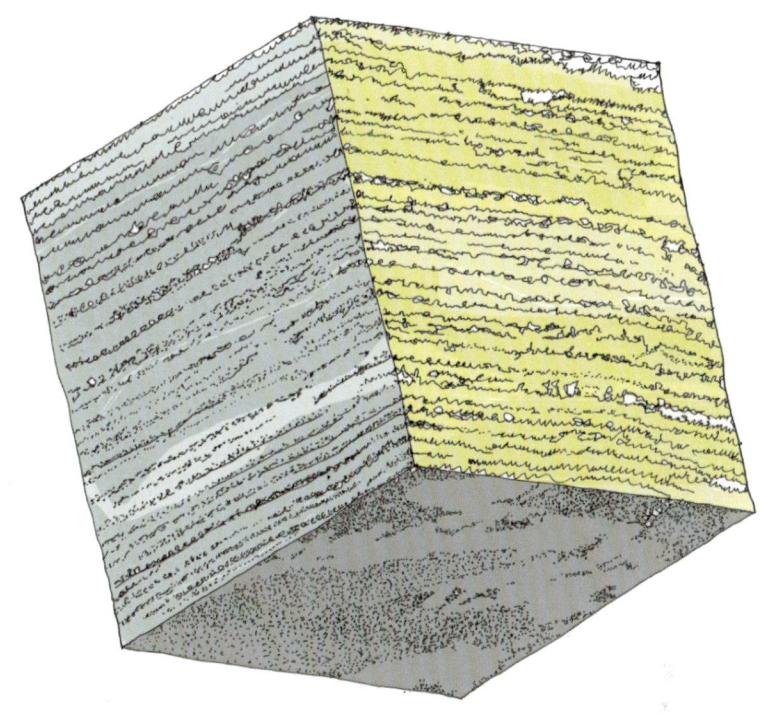

A three-dimensional seismic cube. This depiction is taken from the cover of the Oilfield Review in 1991, at a time when 3D seismic surveys started influencing exploration activities worldwide.

Another hurdle for 3D was interpreting the large datasets. GSI tried any technique it could think of. Bob Peebler, an engineer working for Schlumberger in the 1980s and later president of Landmark Graphics Corporation, an interpretation software company, remembers the type of hands-on solutions he saw at GSI at that time: "They were using plastic models to try to replicate what the data were telling them. They were actually building machines that would build a plastic model from the data!"[7]

One GSI interpreter looking at the North Sea survey was Alistair Brown. Brown could slice the 3D data vertically, producing lines and cross-lines in vertical sections. And he could slice it in horizontal sections, called time slices. But for Brown, these data slices yielded only a structural interpretation. The stratigraphy remained hidden, as he recalled: "We couldn't get the most out of the data. We didn't have the tools." The breakthrough idea came to Brown when he was returning to Dallas from a Southeastern Geophysical Society meeting in New Orleans with his boss Bob Graebner and chief geophysicist of Texas Pacific Oil Company Cornelius Dahm, aboard the oil company's private jet. The idea was simple: interpret as usual for structure but then subtract the structure from the data leaving only the stratigraphy. The horizon slice was born. It revealed stratigraphic features, such as river channels, that had never been seen before and became a key tool for the stratigrapher.[8]

The world of workstations

What the interpretation of 3D seismics needed was more powerful computers, to manipulate the datasets digitally at will. The promise, materializing with the advent of workstations in the 1980s, was that the interpreter would be able to scan huge data volumes at a speed and flexibility impossible with printed seismic sections.

Like the rest of the industry, the seismic community gained its first experience of computers with the huge, unwieldy machines of the 1960s and 1970s. These monsters had to be shared among users, making it impossible to imagine that one day something equivalent, but inexpensive and compact enough to sit on a desk, would become available for the individual. That dream became reality from an explosion of innovation in California's nascent Silicon Valley, and, in particular, from an Austrian immigrant named Andy Bechtolsheim.

Bechtolsheim grew up within sight of the Alps in an isolated farm without television and wiled away his teenage years designing digital electronics, earning enough in royalties to pay for his university education. By 1977, he had moved to the US and was studying for a PhD at Stanford University. For a young entrepeneur, already experienced in digital electronics, it was the place to be. In 1982, Bechtolsheim, together with Vinod Khosla, Scott McNealy and Bill Joy founded Sun Microsystems, and, thereafter, Sun's compact, desktop workstations became the workhorse of choice for the upstream industry, ever hungry for flexible and cheap computing power.[9]

These early workstations demanded software specifically designed to handle and interpret the massive 3D datasets, and this provided an opportunity siezed by two young companies originating in Houston: Landmark founded in 1982 by Royce Nelson, Bob Limbaugh, Andy Hildebrand and John Moulton, and GeoQuest Systems founded in 1984 by Rex Ross. The first challenge was handling, processing and visualizing 3D seismic data, and in this arena the two companies found they had a third competitor, the Geco seismic company, soon to be acquired by Schlumberger.

Geophysicist at a seismic workstation, with multiple screens for visualizing, processing and interpreting 3D seismic data sets.

Viewing 3D data in all perspectives proved a quantum leap for oil company geologists as Walter Ziegler recalled, "This allowed you to map the whole data set cohesively. Suddenly you could see depositional patterns, you could understand facies and rock type better—you could start to recognize hydrocarbons in the rocks. Suddenly you had a geologic tool you simply couldn't work without." The bottom line for oil companies was less dry holes.[10]

Amplitude variation with offset

Another seismic novelty that emerged in the 1980s was the amplitude variation with offset (AVO) technique, a method of directly detecting hydrocarbons even more efficiently than the bright spot method had done in the 1970s. The latter technique had proved quite successful, particularly in gas exploration, but operators were also experiencing false bright spots, amplitude anomalies caused by something else. From 1974 to 1976, relying on bright spots, Exxon and Champlin Oil drilled nothing but 15 dry holes on the extensive Destin dome off the coast of Alabama and Florida. In the fall of 1974, Chevron drilled a well on a very high amplitude seismic event in the Fallon Basin of Nevada, but it turned out to be a high-velocity basalt layer, an igneous rock not known for its hydrocarbon-bearing potential.[11]

This spurred Chevron to conduct further research into direct hydrocarbon detection. The leading figure was Bill Ostrander, a young geophysicist and skilled seismic interpreter who worked in the company's West Coast Division during the 1970s. Based on experiments at the Chevron's research center at La Habra near Los Angeles and theoretical work published by Shell in the mid-1950s, Ostrander figured that seismic reflections from gas sands might show a characteristic increase in amplitude as the angle of incidence increased at the reflector of interest. This was easily accomplished by simply increasing the distance, or offset, between seismic source and receiver.

Ostrander tested the idea in the Putah Sink gas field in the Sacramento Valley discovered by Shell in 1972, one of the best known examples of a gas-related bright spot. Ostrander shot and recorded traces at various offsets and found, as expected, greater amplitude at greater offsets by a factor of two or more. He looked at other gas fields in the area and also along the Gulf Coast and came up with similar results. The only exception was the Fallon Basin, Nevada, where no such amplitude build-up was observed—point proved. Thus, the AVO technique was born.[12]

In 1976, Chevron drilled a well in the Sacramento Valley based purely on AVO data and found gas. AVO remained proprietary to Chevron for many years, and only in 1982 did the company allow Bill Ostrander to present the technique and results to the geophysical world at large at the annual SEG meeting in Dallas. Since that time, AVO has been used extensively and successfully in many of the world's gas provinces. Detection of gas-bearing sandstones is by far the most promising application of AVO analysis, but there have also been notable successes in oil detection, for example offshore in the Niger Delta.[13]

More proprietary work was done in the Tenneco Oil Company by two geophysicists, Steven Rutherford and Robert Williams. They proposed a classification of AVO signatures that clarified the interpretation but were not allowed to publish until Tenneco was put up for sale in 1988. Their classification remains current and is still used.[14]

Factoring in anisotropy

The AVO technique required seismic surveys with large offsets between seismic source and receiver. This created reflection angles much greater than before and the need for deeper understanding of wave propagation in the subsurface. One aspect of wave propagation in particular that had been known about theoretically for years now moved center stage.

Because sediments are generally layered horizontally, sound tends to propagate slower in a vertical direction than it does horizontally. This is the phenomenon of anisotropy that until 1980s geophysicists safely neglected, because in almost all seismic surveys made up to that time, propagation was close to vertical. "Whenever you measured anisotropy it was a few percentages," explained Leon Thomsen who joined Amoco in 1980, "So people thought, why should we worry about such a small effect."[15]

But as wells got deeper and more expensive, and AVO surveys with increasing offset became the norm, these few percentages became too large to ignore. Amoco and Exxon started looking into anisotropy in the early 1980s. At Amoco, Leon Thomsen got in the midst of it, whereas Exxon chose to hire consultant Stuart Crampin, principal scientific officer at the British Geological Survey in Edinburgh. Thomsen and Crampin, developing theories much in parallel during the mid-1980s, discovered the key concepts for understanding and accounting for the phenomenon. Leon Thomsen's 1986 paper in *Geophysics* became the most frequently cited paper in the journal's history, and Stuart Crampin joined the University of Edinburgh in 1992 as professor of seismic anisotropy.[16]

By the early 1990s, most oil and service companies were attempting to master anisotropy. "Suddenly it was the flavor of the month," recalled Chris Chapman, a professor in physics from the University of Toronto who joined Schlumberger in 1991, "Everyone wanted to jump on the anisotropy bandwagon."[17]

Better images with full waveform inversion

As anisotropy became better understood, another related development called inversion was beginning to take shape. Like AVO, inversion held out the promise of converting seismic data into physical properties of the subsurface that would say something directly about oil and gas. In its purest form, inversion is best thought of as the opposite of something called forward modeling. Forward modeling describes a mathematical or theoretical process that assumes a certain physical model of the earth—for example a layered model of the subsurface with an assumed density and velocity for each layer—and then simulates the act of acquiring seismic data on the model. Inversion does the reverse: It starts with measured data, applies an operation that steps backward through the physical experiment and delivers an earth model that would theoretically produce the actual data.[18]

In the early 1970s, inversion started as something much simpler. Roy Lindseth, then at his new company Teknica Resource Development in Calgary, realized that by analyzing the low and high frequency components of a common-depth gather he could extract each sedimentary layer's acoustic impedance. This was an important new piece of information for the explorationist. In the endless varieties of inversion that would follow, Lindseth was actually performing 1D poststack inversion, trace by trace. The resulting sections created yet more opportunity for stratigraphic interpretation. Recalled Roy Lindseth, "These inversion sections were pretty phenomenal at the time as a stratigraphic tool—lots of wells drilled from them did produce." John Rees, a geologist and later president of Dragon Technology in Dubai, worked for Teknica in the late 1980s as an interpretation manager and seismic inversion expert. He recalled Lindseth's impact on the oil industry: "Lindseth's seismic inversion married geophysics and geology, and enabled us to truly explore the unknown."[19]

It was only a matter of time before inversion acquired several dimensions of complexity and effectively any number of methodologies and variants. First, it went 2D and then 3D. Second, it aspired to invert seismic traces before they were stacked, called prestack inversion, but at that point, it had to take into account the full complexity of how waves behaved when they hit interfaces at oblique angles of incidence. The issue is that two types of waves are present in seismic acquisition, compressional and shear, and both produce transmitted and reflected offsprings at each interface. Handling all this required some new theory, a clever way of adjusting the subsurface model to minimize the difference between observed and simulated data and, most important, a level of computer power that only began to be available in the 21st century. Thus, progress was slow and in steps.

The theory was established in the early 1980s by Spanish-born physicist Albert Tarantola, who was a professor at the Institut de Physique du Globe de Paris, and French geophysicist Patrick Lailly from IFP, but they were frustrated by a lack of computing power. In the 1990s, physicist Gerhard Pratt came on the scene. Pratt had worked for Schlumberger in the 1980s, joined Imperial College London in 1992 and six years later became a professor of geophysics at Queen's University, Kingston, Ontario. He found a new way to simulate actual data, by first matching low frequency content and then moving to higher frequencies. In 2004, BP challenged the entire industry and academia to a blind test, circulating a set of simulated data and inviting correspondents to invert it. Pratt won hands down, his inversion matching the BP model better than any other.[20]

Today, full waveform inversion remains the tool of choice for obtaining high-resolution reconstructions of geologic structure and properties. But ultimate goal, prestack 3D inversion, still seeks greater processing power, as Schlumberger's Chris Chapman noted: "The industry is trying to do full pre-stack waveform inversion in 3D, but the key word is trying . . . it's not done routinely, still. Sometimes it's working, sometimes not."[21]

Marine seismics matures

Until the late 1980s, marine seismic surveys were fairly simple, using a single vessel to tow a seismic source array and two streamers. But the acceptance of 3D and the need to explore in ever deeper waters in the Gulf of Mexico, Brazil and West Africa, where wells today can cost in excess of US$ 200 million, provided a catalyst to increase dramatically the acquisition coverage. Top of the list was increasing further the number of streamers, next was deploying increasingly longer streamers, and finally was the need to track the exact position of the resulting thousands of hydrophones in the streamers.

In the 1980s, conventional radio positioning systems were complemented by transit satellite positioning—an early satellite navigational system developed by the US Navy—and since the early 1990s by the global positioning system (GPS), also developed by the US Navy. By the late 1990s, seismic vessels were capable of towing 12 streamers, each extending 10 kilometers and packed with thousands of hydrophones. In 2001, the Norwegian company Petroleum Geo-Services (PGS) attempted a 16-streamer tow, but this was discontinued. Today, 12 streamers remains the standard.[22]

As the number and length of streamers increased, a given survey area could be covered with fewer transits, a plus for efficiency. At the same time, these new streamer configurations opened the window for increased offsets between source and receiver, the prerequisite for wide-azimuth prospecting that was proving essential for illuminating geometrically complex targets. Even better was to shoot marine seismic in several different directions, creating multiple wide-azimuth opportunities. In 2005, BP and PGS used six directions to image below the complex Messinian geology beneath the Egyptian Nile Delta. The main instigator behind this was Jim Keggin, head of BP seismic acquisition and a colleague of Ian Jack, who had been a BP marine seismic innovator since the 1980s.[23]

Yet more wide-azimuth possibilities came on the market in 2006 when separate vessels sailing some distance apart began to be used, each towing source arrays and streamers. Wide-azimuth surveys acquired a massive amount of data, so the technique was crucially dependent on computer power. But the other key ingredient was improving visualization, and for this the industry capitalized on advances in image processing that since the 1990s had been driven by consumer demand in the games industry. As these powerful wide-azimuth surveys became possible, geophysicists were at last able to resolve the most complex structures such as the subsalt reservoirs being explored, for example, offshore Brazil.[24]

But the best was yet to come and the main player was WesternGeco which had just become a full subsidiary of Schlumberger.

WesternGeco, the product of a merger in 2000 between Western Geophysical and Geco-Prakla, bears remnants of DNA from a roll call of the original seismic companies, including Geco, Mintrop's Seismos which became Prakla-Seismos in 1972, Seismograph Service Corporation and finally GSI. GSI had once ruled the world of exploration geophysics but it stumbled during the 1980s oil price collapse and was sold to Halliburton, then passed to Baker Hughes and merged with Western Geophysical, and finally ended up with Schlumberger.[25]

A coil-shooting seismic survey. The circles show the path of the vessel pulling airguns and multiple hydrophone streamers several kilometers long.

WesternGeco's innovation was to use a single vessel and have it shoot in a circular trajectory that progressed through the survey area, creating a sort of coil pattern on the ocean surface. This provided the same coverage as conventional shooting but created a complete range of azimuths in a remarkably efficient manner. The prerequisites for the new technique were complete control of the streamer trajectories and processing software that could cope with the new and unusual acquisition geometry. The WesternGeco solution for controlling streamer trajectory was to build into the streamer a series of so-called birds with controllable wings and fins that could hold the streamer in position and at the right depth, an idea first proposed by Conoco in the

1960s. This, in turn, required a very dense network of acoustic transponders to ensure that the position of each streamer was known at all times. The complete steering system came on the market in 1999 and ultimately made possible the first coil survey, performed for Eni offshore Indonesia during August and September 2008. Since then, coil surveys have become even more sophisticated, using separate vessels for shooting and recording, circling the likely area of interest in a sort of slow dance.[26]

Ocean-bottom seismics and shear waves

The marine seismic industry is never short of challenges. In the 1970s in the Gulf of Mexico, there had been so much field development and so many production platforms installed, large and small, that it was becoming difficult or impossible to navigate a seismic vessel with its long hydrophone streamers through the hardware cluttering up the ocean. One solution was to weight the cables and lay them on the ocean floor, rather than try to tow them. Thus began the era of ocean-bottom seismics.

There was an unexpected benefit in this new deployment. Because the hydrophones were deeper in the ocean and in direct contact with the ocean floor, data quality generally improved. By the 1980s, a geophone was added to the sensor package and data quality took another leap forward. By now it seemed logical to go one step further and somehow embed the sensors in the ocean floor, rather than deploy them in a cable laid somewhat haphazardly on the sea bottom.

The breakthrough came from Norway. Starting in 1988, a Statoil team led by Eivind Berg developed a subsea seismic (SUMIC) system at the company's research center in Trondheim. SUMIC boasted individual sensors, called nodes, comprising a three-component geophone that recorded waves in three mutually perpendicular directions—inline, crossline and vertical—and a hydrophone measuring pressure. The nodes were connected by cables and individually planted on the seabed using a remotely-operated vehicle (ROV).[27]

This brought the expected improvement in data quality but also introduced the new world of using shear as well as compressional data for seismic processing and interpretation. Of the two wave types available in seismic exploration, only the compressional wave is present in conventional marine exploration because shear waves don't propagate through fluids, for example seawater. In the early 1980s, Conoco and Shell had sought ways to record shear offshore but got nowhere. Now with SUMIC, the way seemed open to employing both wave types, and combining the two promised big benefits. Plenty of subsurface scenarios remained ambiguous when only compressional data were available but could be successfully resolved by having both waves recorded.[28]

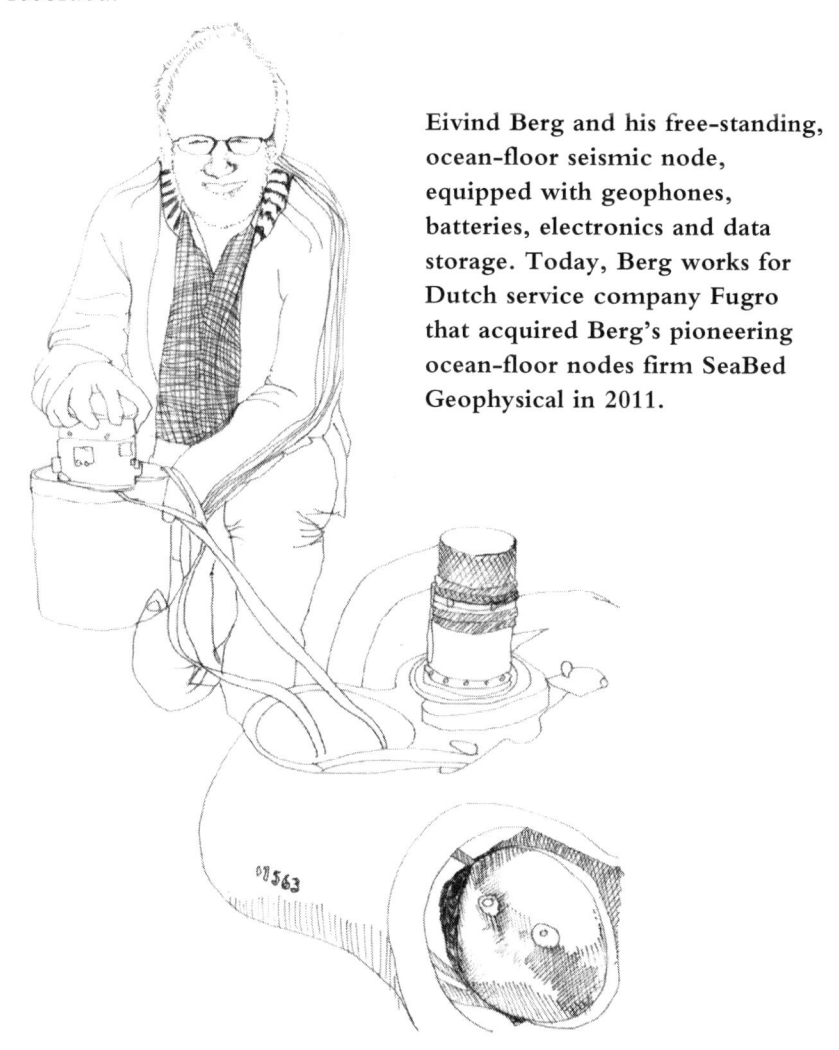

Eivind Berg and his free-standing, ocean-floor seismic node, equipped with geophones, batteries, electronics and data storage. Today, Berg works for Dutch service company Fugro that acquired Berg's pioneering ocean-floor nodes firm SeaBed Geophysical in 2011.

This was exactly the case in an early Statoil survey using the SUMIC system in its Tommeliten Gamma reservoir in the North Sea. Statoil was having difficulties imaging the field because, over geologic time, gas had leaked into the overburden creating a "gas chimney." In the first conventional survey, the gas chimney attenuated the compressional waves making the field more or less invisible. Surmising that shear waves would be unaffected by gas, Statoil decided to survey again, this time with a SUMIC system. Although the survey was only done in 2D, the shear data clearly outlined the reservoir.[29]

In 1994, Statoil licensed the SUMIC patent rights to Schlumberger, which has fielded it ever since. Meanwhile Eivind Berg remained active. In 1996, he and a few colleagues left Statoil and established SubseaCo with financial backing from CGG and started building their own ocean-floor, node-based seismic acquisition system. Early tests were made in the Exxon Balder field and the Shell Guillemot field, both in the North Sea. Soon Eivind Berg was moving on again. CGG wanted full control over SubseaCo, so Berg and his fellow entrepreneurs quit and set up another company, SeaBed Geophysical, to develop a system consisting of free-standing nodes, each equipped with battery, electronics and data storage, and placed and retrieved with an ROV. In 2002, a successful test was carried out in Statoil's North Sea Volve field.[30] More recently, a similar system has been developed by Fairfield Industries in Houston and deployed in BP's deepwater Atlantis field in the Gulf of Mexico.

Chapter 22: Engineering the Oceans

CONSTRUCTING AND operating offshore production facilities is a challenge quite unlike anything else in the oil and gas industry. From the moment offshore exploration and drilling went out of sight of land in the late 1940s, production platforms had to cope with an increasingly harsh environment. Winds offshore are stronger than onshore, ocean currents can turn treacherous and waves may reach tens of meters high.

Offshore production gets going

In May 1946, Magnolia Petroleum Company took the first steps when the company started constructing a combined drilling and production platform ten miles off the coast of Louisiana, farther offshore than anything similar at that time. Hundreds of wooden piles were used to support the structure in 18 feet of water, supplemented by 50 or more 15.5-inch steel piles. Magnolia built most of the 173-foot by 77-foot platform on land and barged it in large sections to the location. Their project marked the first use of steel to build an offshore platform. A year later, Houston-based Superior Oil introduced all-steel platforms that were entirely prefabricated onshore, making for easy installation and low costs.[1]

Superior Oil contracted the engineering to J. Ray McDermott Inc., a company that constructed drilling rigs, established in 1923 in Luling, Texas, by Ralph T. McDermott, J. Ray's son. In 1932, the company relocated to Houston and in 1938 began manufacturing drilling equipment for the Texas and Louisiana marshlands. After World War II, it wasn't a jump too far to go offshore. By this time, McDermott had procured a World War II barge equipped with a harbor crane able to lift 35 metric tons.

McDermott built the platform's steel tubular frame on land, transported it to the offshore site and then set it in position with its barge crane. The platform legs extended from the sea bottom to above the water's surface. Steel piles were then driven through the hollow legs to pin the structure to the bottom; bracings between the legs helped transmit the wave loading down to the seabed. The jacket, as this type of construction came to be known, also served as a steel cage to protect the wells.[2]

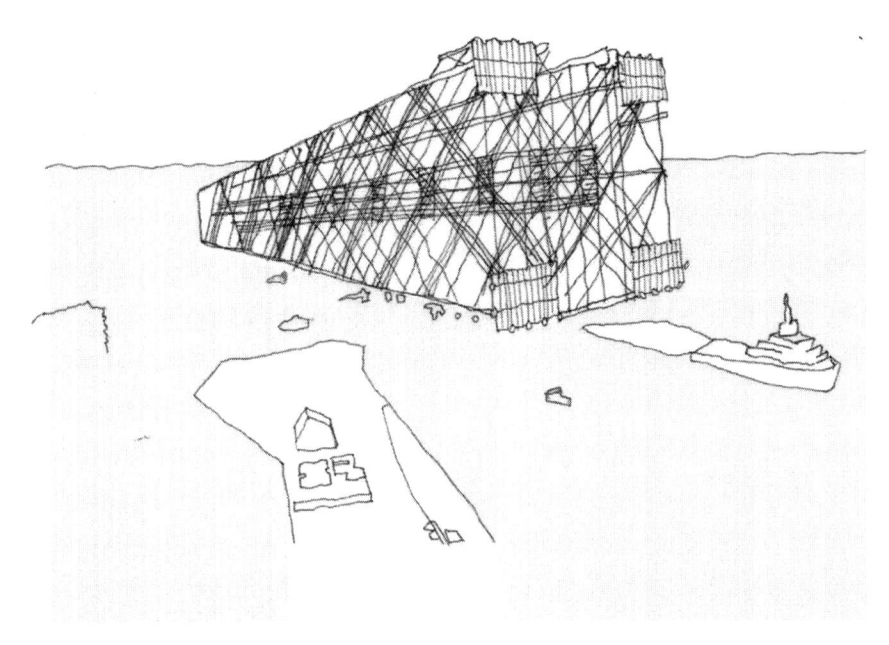

Shell's Bullwinkle steel jacket leaving Corpus Christi to be installed in the Gulf of Mexico in 1988. The 49,375-ton jacket would stand in waters 1,353 feet deep with a base measuring 400 by 480 feet.

During the 1950s and 1960s, McDermott's piled jacket became standard for offshore production in all parts of the world. The ultimate in pile-supported fixed production platforms was installed by Shell in the Gulf of Mexico in 1988. Their massive *Bullwinkle* platform was built in one piece by Gulf Marine Fabricators in Corpus Christi, Texas, and measured 1,365 feet high; for comparison the Eiffel tower in Paris is 984 feet high. "It was really pushing things as far as they could go," recalled Ken Arnold who was a Shell facilities engineer from 1964 to 1980. Twenty-eight steel piles held the *Bullwinkle* platform in place; each pile was seven feet in diameter with 2-inch thick walls and was driven 400 feet into the soft, muddy clay on the Gulf's sea-floor.[3]

As drilling went to even deeper waters, the steel jacket construction reached its limit. In 1976, Exxon proposed an alternative steel construction called the guyed tower for water depths up to 2,000 feet. This platform rested on a spud, or bearing, foundation and was held upright by multiple guy lines tied to anchors on the ocean floor. The tower was a lighter construction than a fixed jacket. Subsequently, the guys were dispensed with, replaced by a sturdy base with a smaller seabed footprint, a design that came to be known as a compliant tower. In 1983, Exxon installed a compliant tower in 1,000 feet of water in the Gulf of Mexico. The record, however, was set in 2000 when the *Petronius* compliant tower was built by J. Ray McDermott for Texaco. *Petronius* rests in 1,754 feet of water in the Gulf of Mexico and with its total height of 2,001 feet was the tallest freestanding structure in the world.[4]

Ken Arnold, an expert in surface production operations, first with Shell from 1964 to 1980, then with Paragon Engineering Services, and now as a consultant. He developed crucial recommended practices for offshore design and safety management.

The growing importance of crane vessels

As J. Ray McDermott, Inc. discovered in 1947, constructing offshore production platforms depended crucially on having a high-capacity floating crane. In 1949, McDermott commissioned the first crane barge built specifically for offshore use, the *Derrick Barge Four*, with the unheard of lifting capacity of 150 metric tons. This gradually ended the piecemeal construction offshore. Instead, both decks and production modules could be built on skids and transported offshore to be set in place by the high-capacity crane. By 1953, McDermott had a second unit with a 250-metric ton capacity.[5]

The other contractor that built and operated such crane vessels was Brown & Root, founded in 1919 as a road building company by Hermann Brown and his brother-in-law Dan Root in Belton in central Texas. Ever adaptive, the brothers soon diversified into dam building, underwater pipe laying, navy ship construction and finally, after World War II, offshore platform construction. In 1955, Brown & Root fielded a floating crane that could lift up to 300 metric tons. By the late 1950s these two main players, Brown & Root and J. Ray McDermott, could be said to dominate the design, fabrication and installation of offshore platforms throughout the Gulf of Mexico.[6]

When the industry moved into the North Sea and oil companies requested steel platforms for a much harsher environment, the offshore construction industry hit a wall. The key player in offshore cranes in the sector was Heerema Marine Contractor, established in 1962 by Pieter Heerema in The Hague. However, Heerema's cranes were mounted on old oil tankers and ore carriers and because of their instability in rough seas could only work half the year, particularly in the northern sector of the North Sea. Heerema therefore developed a semisubmersible carrier, since semisubmersibles had proved robust enough for drilling in the North Sea.[7]

The Heerema semisubmersible crane barge answered the demand for huge lifting capacity. The vessel could accommodate two cranes, which were able to operate independently on separate tasks or work together to lift a single, very heavy object. But even the semisubmersible concept had its limitations. "We needed some kind of system to compensate for eccentric loading and possible tilting, to make sure that the ship and the crane remained within pretty tight tolerances," remembered Jan Meek—former professor of offshore engineering at TU Delft and director of Heerema, "People weren't sure it could be done."[8]

Pieter Heerema, never one to admit defeat, sought help from Alexandre Horowitz, a Belgian-born Dutch engineer and professor emeritus of product design and mechanical construction at the Eindhoven University of Technology; he was known for his invention of the Philishave electric shaver. Horowitz solved the stability problem with a dynamic ballast system, which could best be compared with a flush toilet. When the cranes were about to hoist, the semisubmersible vessel remained stable by rapidly discharging or taking in water ballast. With this ballasting system, Heerema equipped two new semisubmersibles, named *Balder* and *Hermod*, with two large cranes each, initially with a lifting capacity of 2,000 and 3,000 metric tons respectively. Both were built at the Mitsui shipyard in Japan in 1978. A year later, the *Hermod* and *Balder* semisubmersibles crane vessels (SSCVs) did their first jobs in the North Sea, for Texaco's Tartan platform and Conoco's Murchison platform respectively. *Balder* and *Hermod* were the best in the business, not only in lifting capacity but also with their two cranes in versatility. Pile driving, for instance, could be done in half the usual time; modules could be bigger and installed more quickly. Heerema's oil company clients were happy, as Jan Meek recalled, "The semisubmersible crane vessels were critical in reducing costs for offshore facilities."

Eventually other companies caught up with Heerema. In 1985, J. Ray McDermott equipped an SSCV with a combined dual-crane lift capacity of 12,000 metric tons, and Italian salvage and offshore construction company Micoperi launched an SSCV hoisting up to 14,000 metric tons on two cranes. The McDermott SSCV was acquired by Heerema in 1997 and three years later was upgraded to become the world's largest heavy-lifting offshore vessel, *Thialf*, with a dual crane capacity of 14,200 metric tons.

Marine contractor Heerema's crane vessel Thialf. In March 2009, the Thialf crane lowered the topsides on to Shell's Perdido Spar production platform in the Gulf of Mexico, installed in a record water depth of 7,874 feet.

Surpassing the 10,000 metric tons milestone made possible the onshore assembly of complete production facilities rather than constructing multiple, prefabricated modules. This reduced topside installation from a whole season to a few weeks. In March 2014,

Heerema announced that it was commissioning a SSCV featuring two cranes that together will hoist 20,000 metric tons. But this was soon overtaken. In November 2014, Swiss company Allseas launched a platform removal and installation vessel almost 400 meters long and 150 meters wide with a lifting capacity of 48,000 metric tons.[9]

Massive concrete constructions

Alternative methods of constructing platforms had less need for massive cranes. Concrete constructions, for instance, barely need any offshore lifting because the topside modules are fabricated and installed onshore. Steel-reinforced concrete offshore platforms rest on the seafloor by their sheer weight, a handy solution for many parts of the Norwegian sector of the North Sea where a hard, firm clay seabed made it difficult to drive pilings.[10]

The first concrete structure in the North Sea, dating from 1973, was a huge storage tank for the Phillips Petroleum Ekofisk field in Norwegian waters that could hold a million barrels of oil. With a diameter of almost 100 meters and a height of 90 meters, it was built in the protected waters of a fjord by Ingeniør Thor Furuholmen A/S, Ingeniør F. Selmer A/S and Høyer-Ellefsen A/S—three Oslo-based engineering companies who jointly formed Norwegian Contractors—and then pulled out to sea and positioned in 75 meters of water. Høyer-Ellefsen's engineer Olav Mo christened this new concrete monster a "Condeep," short for concrete deepwater structure. The next step was to construct an entire concrete production platform.[11]

In July 1973, just a month after the installation of the Ekofisk tank, Norwegian Contractors received an order from Mobil to build and install the first Condeep production platform, for the Beryl field in the UK North Sea. Frank Manning, the Mobil manager in charge of the development recalled, "An offshore construction of this type and size was totally new technology, but given the advantages of the Condeep idea, Mobil's management decided the risk was worth taking." On July 4, 1975, the first Condeep platform, the *Mobil Beryl A*, was towed out to sea from a yard in Stavanger.[12]

In 1988, Norwegian Contractors was acquired by Aker Solutions of Oslo, and soon after, construction was started on the largest Condeep platform ever, the *Troll A*, for the Shell Troll gas field in the Norwegian North Sea sector. Installed in 1995, *Troll A* sits on a base with diameter 150 meters, reaches 430 meters high and stands in 303 meters of water. The entire concrete construction weighs a million tons. A Dutch newspaper reported: "The builders of the Norske Shell platform reverently call it the eighth Wonder of the World."[13]

A steel-reinforced concrete production platform, of the type that started to appear in the North Sea in the 1970s. This structure was pioneered by Norwegian contractors and featured a cellular base that could be used for oil storage.

But there was the occasional failure. In August 1991, Aker had almost completed a Condeep platform construction for Mobil that was towed out from the yard in Stavanger and into the 220 meter deep Gandsfjord where engineers started lowering it in the water in a controlled ballasting operation at a rate of one meter every 20 minutes. About halfway through the operation, rumbling noises were heard followed by the sound of water pouring into the unit—a cell wall had failed. Seawater poured in faster than the deballasting pumps could deal with, and within a few minutes the hull began to sink. When it struck the bottom of the fjord, it created reverberations with a 3.0 magnitude on the Richter scale at a local seismograph station. The entire construction was a write-off.[14]

Aker learned from this experience and by 1993 had built a successful replacement platform for Mobil. By that time, Aker was already a year into the construction of another spectacular concrete platform in St. John's, Newfoundland. This was the *Hibernia*, the first iceberg-resistant drilling and production platform that now sits in 80 meters of water, 315 kilometers off the coast of Newfoundland. The platform is operated by ExxonMobil and designed to resist a six-million-ton drifting iceberg. The *Hibernia* platform has a massive concrete base resembling a huge vertical gear wheel with sixteen teeth that attenuate the impact of any wandering iceberg.[15]

Drilling-inspired offshore production

Not all offshore production came from fixed platforms. The industry also piggybacked on established offshore drilling hardware. For example, a submersible rig could be equipped with oil and gas processing equipment, storage for a certain quantity of processed crude and the ability to offload the crude to a tanker or pipeline. The first such submersible was built by The California Oil Company, now Chevron, in 1960 for action in the Gulf of Mexico. Another early example, completed in 1961, was the purpose-built submersible by ODECO, now Diamond Offshore Drilling, called the OBM for the initials of the interested parties ODECO, Burma Western (later part of British National Oil Corporation and BP) and Murphy Oil.[16]

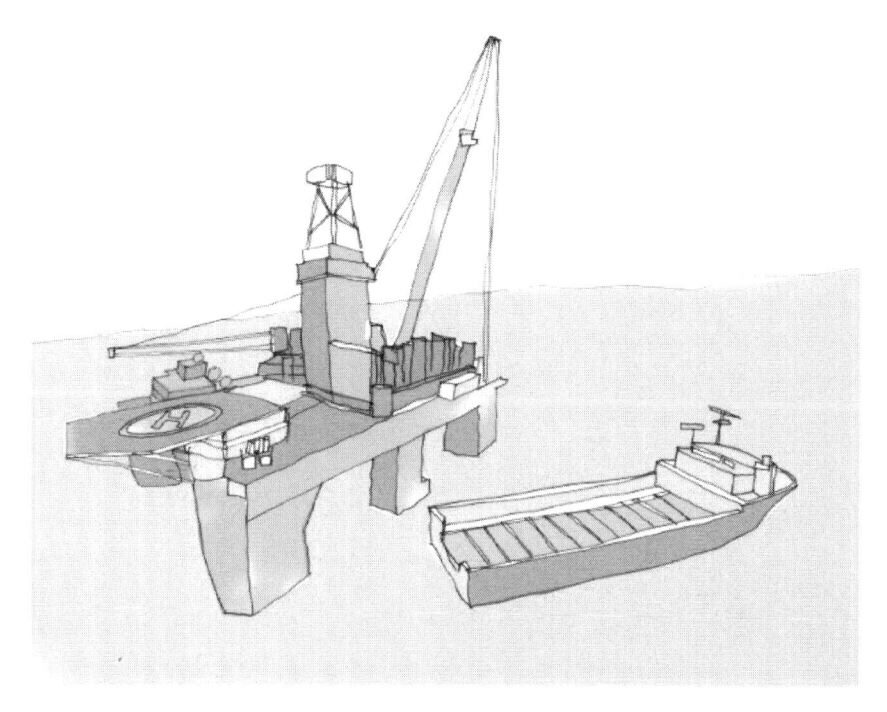

A modern supply vessel delivering cargo to a semisubmersible. In 1954 Alden J. "Doc" Laborde developed this unique design using a forward wheelhouse and a long flat afterdeck that became the standard for offshore supply vessels.

The next steps were production jackups and semisubmersibles. The first production jackup was installed in the Ekofisk field in Norway using the *Gulftide* drilling jackup to accommodate an early production system. Production began in 1971, 18 months after discovery. The first semisubmersible production system was installed in 1975, in the North Sea Argyll field at a depth of 86 meters, by Hamilton Brothers Oil and Gas Ltd. The semisubmersible was a converted first-generation drilling unit, the *Transworld 58*, and served until 1984, followed by another four years on the adjacent Innes field.

Semisubmersible production units had the advantage of having a rather quiet motion but the disadvantage of no storage capacity for produced crude. The units always required a permanently moored storage tanker or a pipeline to shore. However, semisubmersible production systems could boast many followers, in particular in Brazil, where offshore drilling had begun in 1968. The country's first semisubmersible production unit mobilized quickly to the the Brazilian national oil company Petrobras's Enchova field when the field began producing in 1977. In Brazil, semisubmersible production has been used in waters up to 1,830 meters deep, and today Brazil's semi-submersible production fleet accounts for almost 50% of such units worldwide.[17]

A personnel basket, used to transfer workers between offshore rig and supply vessel. Invented in 1955 by Billy Pugh from Corpus Christi, Texas, the basket featured hefty floatation rings at the bottom to absorb any impact when landing on a heaving vessel.

In the Gulf of Mexico, semisubmersible production was tried for the first time in 1988 when Placid Oil deployed a floating system at the Green Canyon field in 1,534 feet of water. When Placid experienced a downhole equipment failure and production problems, operators hesitated and semisubmersible production was not adopted again until the 2000s. The world record for semisubmersible production is the Anadarko-operated *Independence Hub* that has produced from a depth of 7,000 feet in the Gulf of Mexico since 2007.[18]

Tension-Leg and Spar platforms

By the year 2000, the heyday of giant fixed platforms for deep water was coming to an end. Offshore exploration had now reached water depths too deep for concrete or steel structures. Some kind of floating production system was the only option.

For many in the industry, semisubmersibles were too prone to movement in rough seas and a potential safety and environmental risk in ultradeep water. Some other solution was required, and it formed in the mind of an inventive petroleum engineer named Ed Horton. In the 1950s, Horton had started working on new platform ideas during his college years at the University of Southern California, but for him it was "more curiosity than anything else." He later joined Standard Oil of California, where there was a chance to test some ideas, as he recalled, "I filled an enclosed box with some water. I then submerged the box and tied it down. The box was amazingly stable."

Soon after, Horton left Standard Oil of California but continued developing his idea with Randy Pauling, the head of naval architecture at the University of California, Berkeley. Pauling and Horton constructed a 5-foot scale-model and tested it in San Francisco Bay. "From that moment," recalled Horton, "we got enough interest from the oil companies to build a large-scale mock-up." In 1973, Horton's newly established Deep Oil Technology Company formed a consortium of twelve oil companies and began constructing a one-third scale model, the *Deep Oil X1*. Testing was conducted offshore Catalina Island, California, and completed in 1975. By this time, Horton's idea had a name—tension leg platform (TLP).[19]

Ed Horton, inventor of the Spar and tension leg platforms (TLP). Today, he heads Houston-based Horton Wison Deepwater Inc., an offshore technology development company Horton founded in the early 1980s.

The TLP relies on vertical steel trusses that anchor the platform to the seabed ensuring up and down stability. This allows wellheads to be placed on the platform floor resulting in production processes similar to fixed platform production. Like the semisubmersible platform, however, the TLP lacks storage capacity, so it requires a separate storage tanker, pipeline or shuttle tanker for export.

A tension leg platform (TLP). A TLP is tethered to the seabed by a number of tendons kept in tension. The tendons are secured to a foundation template piled into the seabed. TLPs represent robust and relatively stable platforms in harsh seas. The TLP concept has since been adopted for offshore wind turbines.

Conoco, under the direction of L. B. "Buck" Curtis and N.D. "Scotty" Birrell, was the first adopter. They decided to test the concept in the relatively shallow yet harsh North Sea, installing the first TLP for its North Sea Hutton field situated in 486 feet of water in 1984. From then on there was no stopping. In 1989, the company installed a TLP, the *Jolliet*, in the Gulf of Mexico in 1,760 feet of water. And in 2003, again in the Gulf, the Conoco *Magnolia* TLP was set in water 4,674 feet deep. Another big player during the 1990s was Shell, which ordered five TLPs that reached water depths from 2,860 to 3,950 feet.[20]

Ed Horton and Randy Pauling, meanwhile, had moved on, developing what became known as the Spar platform, the term spar deriving from the upended wooden or metal pole that sailors used to mark a spot. A Spar platform is a floating system comprising a deep-draft cylindrical hull, or caisson, the bottom of which contains ballast material to provide stability and the top of which supports the topsides structure. The Spar is moored using a system of conventional mooring lines anchored to the seabed. Crucially, the Spar has the capacity to store produced oil. For their first prototype, Horton and Pauling turned to the Scripps Institution of Oceanography in San Diego, California, because Scripps already had built a similar, experimental platform for oceanographic studies, calling it a floating instrument platform. Horton tuned this construction for offshore oil production and soon gained the industry's attention. "He connected to other companies very well," remembered Gordon Sterling, a former Shell offshore engineer.[21]

Another joint industry program led by Shell helped commercialize the Spar concept. In 1976, Shell installed the first Spar on the Brent field in the North Sea in a water depth of 140 meters for crude storage and tanker loading. It was 147 meters high and 29 meters in diameter with a storage capacity of 300,000 barrels of oil. Since then, Spars have blossomed into many different guises, with Conoco, Shell, Exxon, Kerr-McGee and Technip particularly active. In 1996, Aker built a full drilling and production Spar for Oryx Petroleum of Calgary, Canada, for installation in 1,936 feet of water in the Gulf of Mexico. In 2003, London-based Dominion Energy plc installed a production Spar in the Gulf of Mexico in 5,610 feet of water. The deepest Spar currently is Shell's *Perdido* in almost 8,000 feet of water in the Gulf of Mexico. Today, engineers see no depth limit for Spar technology.[22]

Mark Mau and Henry Edmundson

Enter FPSOs

With the rise in oil prices after 1973, offshore oil operators increasingly looked to produce smaller oil fields they had previously disregarded. Producing smaller fields required relatively cheap production units, because the period of exploitation would be shorter. What were needed were mobile, floating production units that could be relocated without much effort. The solution was a ship-like floating production, storage and offloading system, in short, an FPSO.

The earliest experiments date back to 1977 when Shell and Single Buoy Moorings Inc. of Schiedam, in the Netherlands, deployed an FPSO vessel with a storage capacity of 350,000 barrels of oil in 117 meters of water in the Mediterranean Sea, offshore Castellon, Spain. During the following years, the FPSO idea slowly took off. In 1979, Petrobras employed an FPSO for the Garoupa offshore oil field in 122 meters of water off the coast of Brazil, and in 1982 the Italian oil company Eni started developing its Nilde oil field offshore Sicily using an FPSO.

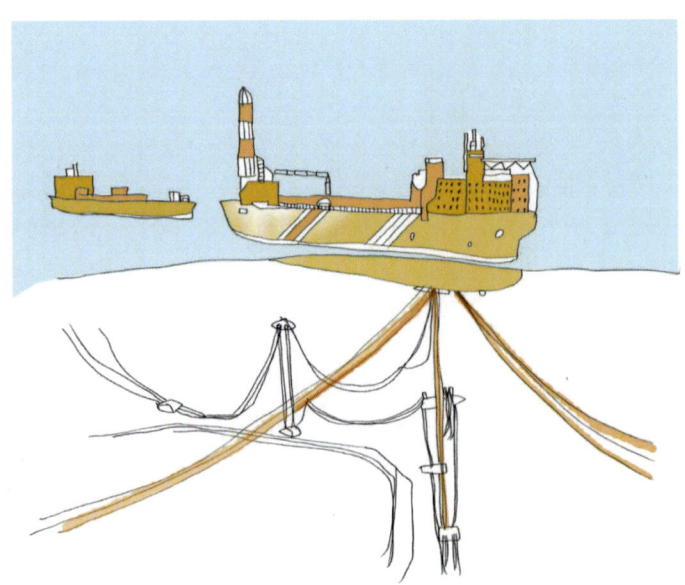

The floating production, storage and offloading (FPSO) vessel Aoka Mizu, designed, built and operated by Bluewater Energy Services B.V. from the Netherlands. The Aoka Mizu has been deployed at the Ettrick and Blackbird fields in the North Sea since 2009. This cutaway view shows how the vessel is linked to a number of wells via flexible pipelines.

Until 1986, FPSOs were converted secondhand oil tankers that were inexpensive and seemed fit for purpose. The necessary upgrading of structure usually more than outweighed the higher cost of a new build. Production equipment placed on the deck of the vessel produced the crude that was loaded into tanks in the ship. Every so often, a shuttle tanker moored up to the FPSO to take off the crude. The understanding at the time was that FPSO systems were fine for moderate to benign seas but had no practical value in rough waters like the North Sea.[23]

In 1986, Norwegian Gola Nor Offshore A/S from Trondheim, now Teekay Corporation, designed the first purpose-built FPSO, the *Petrojarl I*, and had it manufactured by the Japanese shipyard Nippon Kokan Koji Corporation. The vessel had onboard separating facilities and was dedicated to producing marginal oil wells in the rough waters of the North Sea. Reputedly, it could stay connected to production wells in North Sea 100-year worst-case conditions of 30-meter-high waves and 140 kilometers-per-hour winds. The trick was an advanced station-keeping system. Mooring winches located on the ship could adjust the tension in each mooring line, and a turret system allowed the ship to rotate in case of bad weather. Norsk Hydro was the first operator of *Petrojarl I*, using it in the Oseberg field until 1988. The development of FPSOs continued around the world, including offshore Australia and in the South China Sea, further proving the concept's applicability in difficult waters.[24]

Over the past 20 years, FPSO capability and size have increased rapidly. Today, more than 200 FPSOs are in operation, accounting for two thirds of all floating production system installations. The FPSO concept is increasingly accepted for quickly establishing production in both remote and developed areas.[25]

The birth of subsea technology

The growth of the offshore industry and its move into deeper waters in the 1950s and 1960s caused oil companies to question every engineering practice. In 1961, Shell took the momentous step of placing the wellhead on the ocean floor, heralding the beginning of subsea production. This was in the West Cameron field in the Gulf of Mexico in a water depth of 55 feet. Cameron Iron Works delivered the wellhead and Christmas tree, Hydril Corporation built the BOP, and

British Insulated Callenders made the flexible flow lines that tied production back to a fixed platform on the surface. This experiment was well in reach of divers who at the time could operate down to 100 feet.[26]

Deeper water was a problem for divers. In the early 1960s, however, the US Navy pioneered what came to be known as "saturation diving." Navy researchers discovered that replacing air with a mix of oxygen and helium could extend divers's depth limit to well beyond 100 feet and diving duration to a week or more. The divers were lowered in a pressurized vessel near the place of work and then, with the task completed, pulled to the surface again. But saturation diving came at a price—divers required long periods of decompression after each dive. In the summer of 1967, Taylor Diving & Salvage, founded ten years earlier in New Orleans by former US Navy divers Edward Lee Taylor and Mark Banjavich, performed the first saturation diving operation for the oil industry to install risers for the Shell Marlin subsea system at 320 feet in the West Delta field in the Gulf of Mexico.[27]

Diver at work at a Placid Oil subsea well in the Gulf of Mexico around 1970 when divers routinely installed and maintained subsea hardware. On top, a set of "fire-safe" underwater safety valves that were made compulsory after a platform fire at Vermilion Bay off the coast of Louisiana in 1967 and the Santa Barbara oil spill in 1969.

The offshore oil industry clearly needed some kind of nonhuman intervention and around 1960 began experimenting with robots. Shell was a frontrunner and teamed up with Hughes Aircraft Company, a major US aerospace and defense contractor owned by Howard Hughes, the son of the drilling-tools inventor. By 1962, Hughes delivered the first ROV, christened the Mobot, that was attached to the subsea installation and equipped with a socket wrench and a TV camera for monitoring operations. Shell used the Mobot in the Molino gas field off the California coast in 240 feet of water and successfully completed six subsea wells without the assistance of divers.[28]

Exxon had also been busy. In 1968, a small group of Exxon engineers led by Joe Burkhardt moved to Los Angeles to design, build and test a prototype of a submerged production system (SPS) with similar robot attachments. Following extensive tests on land, Exxon chose a field in the Gulf of Mexico off Louisiana in relatively shallow waters of 170 feet for its first offshore trials. The Exxon SPS featured three wellheads and a valve manifold mounted on a subsea template structure. The oil flowed through a pipeline to a nearby production platform. The experimental SPS was installed in 1974.

An evaluation of the pilot test concluded that subsea production in deepwater was practicable and that robots were competent to assemble and repair the complex systems. But another decade passed before subsea production gained industrywide acceptance. The fact remained that the early Exxon SPS project wasn't yet economic. "SPS cost Exxon three dollars for every barrel of oil," remembered Brian Skeels, a former Exxon subsea engineer, "At the oil price then, it was money-losing; they knew the technology needed to be improved and get cheaper."[29]

By this time, several commercial subsea ROVs had come on the market. One of the first was the RCV 225, conceived and produced by Hydro Products Inc. of San Diego, California. It looked like a "flying eye-ball" with propellers and boasted a highly sensitive camera still on the list of restricted military technology. The first major offshore installation to use the RCV 225 was the Shell *Cognac* platform in 1030 feet of water in the Gulf of Mexico in 1977.[30]

But soon, Exxon and many other companies were hammered by the oil price crash of 1986. As a result, service companies such as FMC Technologies and Cameron Iron Works took over much of the subsea technology development. Two national oil companies, Brazil's Petrobras and Norway's Statoil, however, remained active. In 1974, Petrobras had discovered the Garoupa field in the Campos basin offshore Brazil and ten years later the Albacora field, proving the existence of giant fields in great water depths. Until then, Petrobras had converted old drilling rigs to production platforms. Now they were challenged by water depths of more than 400 meters. The Brazilian oil company had no choice but to pursue its own research and development. Brian Skeels remembered: "Petrobras was basically doing deepwater at a time when nobody had even thought about doing it."[31]

As the offshore industry passed the 500-meter water depth in the 1980s, divers reached their natural limit, even using saturation diving. The move into deeper water resulted in a boom for ROV manufacturers and operators. As they became more sophisticated, ROVs even displaced divers in shallow waters. "It was the safety factor," recalled Drew Michel, a subsea expert and former owner of ROV Technologies, an engineering consultancy, "The ROV was a no-brainer."[32]

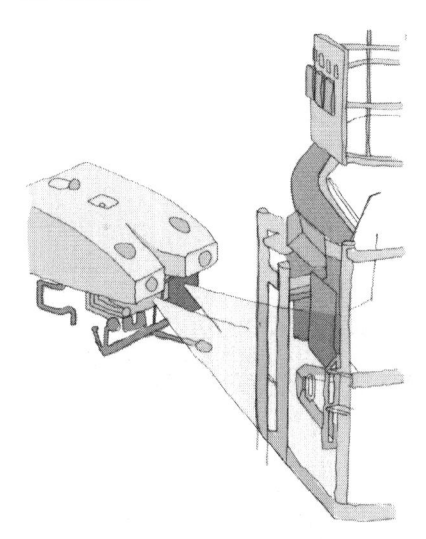

A remotely operated vehicle (ROV), able to operate at several thousand meters' water depth. The ROV has powerful headlights and robotic manipulator arms to perform maintenance work on a subsea completion.

One man who pushed the boundaries of ROV technology was Tyler Schilling from Davis, California, a dropout from engineering school but a natural tinkerer and inventor. At the Space and Naval Warfare Systems Center in San Diego, California, working on the US Navy's underwater research vehicles, Schilling learned all about robots and then returned home in Davis to start his own business. His mother mortgaged the family home, and in 1985, Tyler Schilling founded Schilling Robotics, operating for the first few years out of a rented garage.

Drew Michel's ROV Technologies was one of Schilling's first clients, and Michel recalled an early visit, "Tyler rang me to come over and look at this manipulator arm. They were all nervous because I was potentially a big customer. Years after, Tyler told me that the thing that scared him the most was that I was going to ask for the bathroom, because at that time we didn't have one!" Schilling Robotics soon outgrew the garage and became a well-respected manufacturer of robotic arms for ROVs, capturing 90% of the market. This success had much to do with Tyler Schilling's way of tackling technical challenges, as Skeels explained: "He is a very analytical guy; he takes a very abstract view of some problem and then comes up with amazing, practical solutions." In the 2000s, Schilling Robotics successfully ventured into the construction of ROVs, then in 2008 was acquired by FMC Technologies.[33]

Another key ingredient for subsea production was the invention of flexible pipe. Movement of the sea and the span connecting the subsea well to the floating production unit made solid pipe difficult or impossible to use. The solution was a flexible pipe of steel with thermoplastic sheath, first introduced by IFP in 1971 and later commercialized through the French subsea contractor Coflexip. "Flexible pipe was a big innovation for very deep water," recalled Murray Burns, a retired engineer who worked for Technip, another French firm that acquired Coflexip in 2001. In 1973, Coflexip's first commercial installation of flexible pipe was in the Elf Emeraude Field, offshore Republic of the Congo.[34]

Modern subsea systems operate at astonishing depths and at great distances from the production platform or facility, so the flexible flow lines, or tie-backs as they are called, can be long, typically up to several kilometers. The Ormen Lange subsea installation in the Norwegian sector of the North Sea, constructed in 2007 by FMC Technologies and operated by Statoil, holds the record for the longest tie-back at 120 kilometers to an onshore facility.[35]

In 2012, Statoil introduced the concept of the "Subsea Factory." The Norwegian oil company aims to place all subsea production and processing elements on the ocean floor, minimizing all surface installations. This brings together earlier developments working with FMC Technologies installing the first complete subsea system for separating and injecting water for its EOR project in the Tordis oil field and for injecting raw seawater in the Tyrihans field, both in the North Sea. The next step is a full-scale subsea gas compression system that will be installed by Aker Solutions in early 2015 to boost falling pressure in the Åsgard field, North Sea. Statoil's ambitious objective is to have a complete Subsea Factory available by 2020.[36]

Chapter 23: Reservoir Engineering Comes of Age

ALTHOUGH THE principles of reservoir engineering were well understood by the 1960s, oil companies were hampered by lack of computing capability. The computer revolution that was to come eventually allowed scientists and engineers to create complex models at both microscopic and macroscopic scales, and to develop a better understanding of how oil and gas flows through rock and out of the reservoir.

Research into porous media

It was in the 1970s that researchers began to seriously ponder how fluids behaved at the pore scale. The foundation for this work was sound laboratory technique and measurements on core samples. Permeability is one example. In an early method of measuring permeability, a cleaned and dried core plug was placed in a core holder, and gas, usually air, was forced through the rock. The measured pressure drop and knowledge of the air's viscosity yielded permeability. One popular version was patented in 1944 by Gerald Hassler of Shell from Berkeley, California, and later licensed to Core Laboratories Inc. in Dallas.[1] But operating these early permeameters was time-consuming and too costly to be used repeatedly for detailed rock studies. So a quicker alternative for single-phase fluids, called the mini-permeameter, was developed by Shell geologist Koen Weber in 1971.

Weber recalled how he developed the idea: "My father was a mechanical engineer from Delft. One day at a conference he learned about a device called a rotameter that measured air flow in small vertical pipes." Weber was taken with the rotameter, bought a few and started experimenting. Slowly his mini-permeater took shape. He pressed the end of a rotameter onto the surface of a cleaned core, pumped air through it and then measured the resulting back pressure as the air permeated into the rock. The rotameter did its job measuring the flow rate, and that, combined with the back pressure, yielded permeability. It was a quick if slightly less accurate measurement than that provided by a permeameter. But as Weber recalled, the mini-permeameter was a breakthrough for formations such as the Rotliegend formation in the

North Sea, where extreme heterogeneity would otherwise have demanded extensive and tedious permeability measurements.[2]

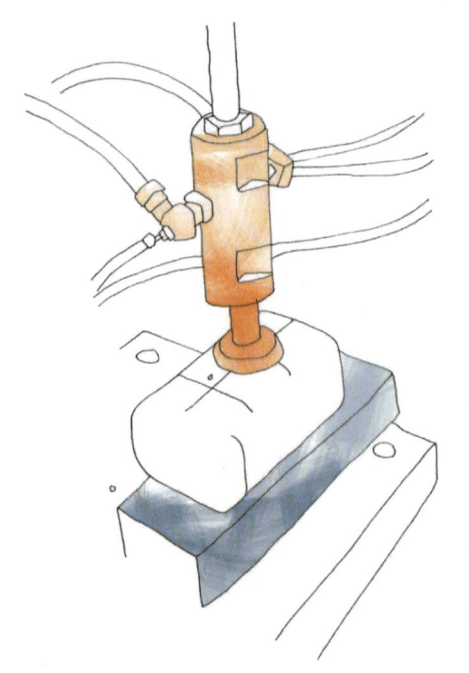

A mini-permeameter used for quick lab measurements of permeability. With an injection nozzle, red, pushed against the surface of the rock, pressurized air is pumped into the sample resulting in a specific flow and pressure drop from which permeability is derived.

Another example is capillary pressure, which is the pressure required to squeeze oil through a pore space against the opposing force created by oil-water interfacial tension, harder to do as the pore gets smaller. In the early 1940s, Humble Oil's Miles Leverett showed how to estimate a rock's capillary pressure from its pore size distribution. Leverett had examined the rock's pore space with a binocular microscope, an indispensable piece of kit in any core lab. In 1949, Bob Purcell, at Shell's Bellaire Research Center in Houston, devised a laboratory technique to measure capillary pressure directly from core samples by injecting mercury into them. How much mercury entered the sample at a given injection pressure revealed the pore size distribution and, by inference, permeability.[3]

Then in the early 1970s, the scanning electron microscope was invented. The first scanning electron microscopes had variable magnification from 20 to 20,000 and approximately 1,000 times the depth of focus of conventional light microscopes. The new microscopes provided stereoscopic images of pores and flow channels with a level of detail Leverett and Purcell would have killed for. This got theorists interested. One was Alkiviades Payatakes in the Department of Chemical Engineering in the University of Houston. Payatakes developed what came to be known as a pore network model, in which the pore space is represented by a network of nodes and tubes linking the nodes. Network models had some success in predicting flow properties, and work along similar lines engaged scientists at other institutions, such as IFP and Imperial College London.[4]

In 1978, Schlumberger created an entire department in their Ridgefield research lab to study the basic physics of rock. This included the development of statistical models to describe the microscopic characteristics of pore space and theories to better understand the physical response of their logging tools to rock at that scale, particularly new measurements such as the dielectric constant and nuclear magnetic resonance. Core samples from around the world were dissected and characterized, and in an experimental counterpoint to the theorizing, scientists created an artificial porous rock nicknamed Ridgefield sandstone by fusing glass beads together. Various porosities were obtained by varying the amount of fusing. Ridgefield sandstone was analyzed exhaustively in numerous experiments and cut into a sequence of thin sections to understand its pore structure.[5]

Difficulty in visualizing 3D pore structure in rock was a weakness in much of this early work. If that could be done, then it might be possible to calculate the rock's expected permeability and compare it with a laboratory measurement. That was the dream of Pierre Adler, a theoretician working at the Centre National de la Recherche Scientifique in France. In the early 1980s, Adler was invited by Howard Brenner, emeritus professor of chemical engineering at MIT, to work with him on the theory of flow through porous media, but neither at that time had access to real data about rocks at the pore scale. Then in 1983 during a conference, Adler met Christian Jacquin from IFP who had a large dataset on Fontainebleau sandstone, famed for being clean and homogeneous, including thin sections and measurement of permeability. With real data provided by Jacquin, Adler created a flow model to cope with the complex realities of the Fontainebleau pore space using finite difference methods and powerful computers to make the calculations. The computation of permeability matched the experimental results reasonably well.[6]

A short while later, Adler became acquainted with work at Brookhaven National Laboratory in Upton, New York, where X-ray tomography—which uses the same principles as medical CAT scanning—was being used to investigate the 3D structure of materials at a microscopic scale. This seemed a far easier method of mapping pore space than the labor-intensive cutting and digitizing of thin sections. Fontainebleau rock samples were soon being dispatched to Brookhaven and placed in its synchrotron particle accelerator. Permeability and electrical resistivity calculations using the new data again matched experimental values.[7]

By now, the industry was catching up. In 2004, Pål-Eric Øren, a project team leader at Statoil and acquaintance of Adler, saw an opportunity, and with his parent company's help, founded Numerical Rocks, based in Trondheim, Norway, to provide flow computations for porous media. Meanwhile a company called Digitalcore, which provided digital core imaging, took root in Canberra, Australia, as an offshoot of the Australian National University and the University of New South Wales. And in 2013, Numerical Rocks and Digitalcore merged to become Lithicon, based in Trondheim, effectively providing a soup-to-nuts service for rock pore characterization and flow modeling. A similar capability was developed in the US in 2007, when Stanford professor Amos Nur and colleague Henrique Tono founded Ingrain.[8]

More recently, new microscopy techniques allow zooming into the nanoscale. One novelty is atomic force microscopy (AFM) that uses a mechanical probe to magnify surface features up to 100 million times. Another breakthrough is the focused ion-beam scanning electron microscopy that promises to slice a rock sample physically at the nanometer scale and produce even better images than tomography.[9]

Well testing widens its scope

However good the science becomes, it is no substitute for measuring reservoir permeability in the field. The early drillstem tests carried out by Simmons's and Johnston's tools in the 1920s and 1930s were used only to measure reservoir pressure and test flow of hydrocarbons to the surface, but in the 1940s, oil company researchers realized that well tests could offer more than that.

The Atlantic Refining Company set the ball rolling. In 1950, using techniques from groundwater hydraulics, Charles Miller, A.B. Dyes and C.A. Hutchinson proposed a method for estimating permeability from noting how quickly bottomhole pressure increased after a producing well was shut in—the more rapid the increase, the higher could be assumed was the permeability in the formation surrounding the wellbore. This became known as a buildup test. The reverse process, the drawdown test, monitored the pressure when the well was opened and could be interpreted in a similar manner. A year later, D.R. Horner, working for a Shell subsidiary in Indonesia, established an alternative method for interpreting buildup and drawdown tests that not only established permeability but also the distance to a sealing fault and the static reservoir pressure. The Horner method is used to this day.[10]

Then in 1955, a team of Schlumberger engineers led by Maurice Lebourg introduced the first testing device that was lowered on wireline rather than drillpipe. The wireline tool applied a pad against the borehole wall to exclude borehole mud and took fluid samples from the formation through a tube into a sample chamber, all the time recording pressure with a slightly crude mechanical gauge. In 1962, Schlumberger presented a follow-up tool, the formation interval tester, which could fit into smaller diameter wells and even operate in cased holes thanks to a couple of perforating bullets that when fired, if you were lucky, provided a conduit to the formation through casing and cement.[11]

One disadvantage of these early wireline testing tools was lack of repeat capability—they had to be pulled to the surface after each test and rebuilt. Schlumberger's next move therefore was to develop the repeat formation tester (RFT), introduced in 1975. This could measure formation pressure along the wellbore at many different locations on one trip into the well and collect multiple samples. Thus for the first time, reservoir engineers could obtain a pressure-depth profile. That was a breakthrough because it allowed them to deduce fluid density and identify contacts between different fluids, such as oil and water. In the late 1970s and early 1980s, the tool was equipped with new, highly accurate quartz-crystal pressure gauges.[12]

These developments spurred a new interest in interpreting pressure data, whether from a drillstem test or the latest wireline tool. Until the early 1980s, most of the research took place in university departments, for example at Stanford University under the leadership of legendary reservoir engineer Hank Ramey. But increasingly Schlumberger took the lead. In 1971, the company acquired Flopetrol, a French company specializing in testing and oil well production, and in 1978, Flopetrol hired Alain Gringarten, who gained his PhD at Stanford under Ramey, "To create a well test interpretation service," as he recalled.[13] Another member of the Flopetrol team was Dominique Bourdet.

Alain Gringarten, professor of petroleum engineering at Imperial College in London and industry expert in well test analysis.

In 1983, Bourdet came up with the bold idea of using the derivative, or rate of change, of pressure rather than pressure itself. This sharpened the picture considerably and was "a milestone in terms of methodology," as Leif Larsen, a former Statoil reservoir engineer and early well testing researcher, recalled. "Also, the technique was increasingly viable as computer power increased." From then on, pressure derivative analysis was developed extensively to determine all manner of reservoir characteristics including reservoir size and heterogeneity, the presence of impermeable barriers or a dual-porosity system, and a host of near-wellbore effects.[14]

Meanwhile more sophisticated wireline testing tools were being developed. Although the RFT could take several samples in one run into the well, sample quality varied and volumes were small. "Each sample was less than a glass of water, maybe half of that," explained

Larsen. The problem with quality was that the samples were just as likely to be mud filtrate as pure reservoir fluids, and no way existed to check whether this was the case before pulling the tool to the surface. The industry needed clear and valid samples. "A new wireline testing tool able to take clean samples promised to be a technology milestone," recalled Fikri Kuchuk, one of Schlumberger's well testing gurus.[15]

Starting in 1986, a team of Houston-based Schlumberger engineers headed by Tom Zimmerman started working on a solution. Zimmerman clearly recalled his boss, Volker Reichert, pushing him to "think outside-the-box." Eventually, Tom came up with the idea of adding a pump-through capability, so that flow through the sampling tube could be maintained and then monitored until it was sure that the sample chamber was filling with reservoir fluids rather than mud filtrate. Another novel feature of the tool was its modularity. Introduced in 1989, the modular dynamics formation tester (MDT) could be configured for the specific needs of any individual testing program. Modularity also enabled the MDT to integrate new measurement techniques still to be invented. One example was an optical fluid analyzer tool added in 1993, the first of an increasingly sophisticated series of downhole measurements that today includes hydrocarbon composition and gas/oil ratio.[16]

A new milestone in well test interpretation was reached in 2001 when three researchers at Imperial College London—including Alain Gringarten, who in 1997 had joined the university as petroleum engineering department head—found an effective algorithm for the deconvolution of pressure and flow rate data. In well test analysis, deconvolution is a process that converts pressure data at a variable rate into a single drawdown at a constant rate, thus making more data available for interpretation than in the original dataset. One benefit is seeing reservoir boundaries more clearly, a considerable advantage compared with conventional analysis in which boundaries are often not seen and must be inferred. This had significant impact on the ability to certify reserves.[17]

Simulating the reservoir

Although everyone from the geologist to the reservoir engineer fully appreciated the incredible complexities of any oil or gas reservoir, both in terms of its geometry and key formation properties such as porosity and permeability, all parties had plenty of incentive to build a model of the reservoir, even if it was doomed to be highly simplistic. The dream was that if a model could be constructed to simulate production up to the present, which in most cases was known, then the model could be used for plan production in the future. The more down-to-earth use of simulation was simply understanding better the nature, shape and size of the reservoir.

The question was where to begin. "When I started in 1961," recalled ARCO's Fred Stalkup, "the standard reservoir description was the so-called pancake model. If you had a well and you knew the porosity and permeability in each layer, then that was your model and you made your calculation. But of course it was far too simplistic. Reservoirs are nothing like that." Alain Gringarten recalled a characteristic presentation from 1982 when the first horizontal wells were drilled at Rospo Mare by Elf Aquitaine: "The Elf engineers were showing a reservoir that was highly heterogeneous, sometimes there, sometimes not—actually it looked like a mousse. It was clear that this was not a layer-cake situation."[18]

The first reservoir models were physical mock-ups, one example dating from 1920 attributed to Robert van A. Mills at the US Bureau of Mines, Bartlesville, Oklahoma. A typical experiment featured layers of sand, saturated with oil on top and water below, and a small hole drilled through the oil sand to simulate a well. The water was then pressured to force the oil up the well. Experiments like this soon demonstrated reservoir effects that are nowadays well known, such as the coning of water above the oil/water contact, causing water entry earlier than might be expected. Another type of model used the analogy between electric current flow and fluid flow. In one example, an East Texas oil field was simulated with an electrolyte solution covering a plastic sheet contoured to represent the reservoir's thickness times permeability. The objective was to measure the potential distribution in the model for various configurations of current flow.

In the end, the mathematicians stepped in. The first computational treatment of reservoir flow can be attributed to Morris Muskat of Gulf Oil Research and Development Corporation in his seminal 1937 book *The Flow of Homogeneous Fluids through Porous Media*; William Hurst at Humble Oil and Antonius van Everdingen at Shell also contributed treatises. But this early work had two serious limitations. The models tended to be too simple, and in the precomputer age solutions had to be analytic, in other words expressible by some concise mathematical formulation, and even when they were, computing answers by hand was impracticable.[19]

Reservoir simulation with computers began in the early 1950s at Humble Oil Production Research in Houston with the hiring of two chemical engineers, Donald Peaceman and Henry Rachford. "When I came to work in 1951, we didn't have any digital computers," recalled Peaceman, "only occasional access to some accounting machines that the accounting department would let us use at night only." The first simulators could simulate oil and water or oil and gas but not all three, and the oil was "black-oil," meaning that any change in its composition with pressure and temperature was ignored. Within these limits, the simulator was flexible enough to model most simple reservoir geometries. Humble Oil was soon exporting its knowhow to its parent Standard Oil Company of New Jersey's most important joint venture, Aramco in Saudi Arabia. In later years, the nationalized Saudi Aramco would develop simulators to a scale that has never been matched.

Meanwhile, Peaceman and Rachford abandoned analytic methods for finite difference techniques that could model nonuniform formation properties and arbitrary reservoir geometries. But there were obstacles. The first was reformulating the differential equations of classical mathematics into a finite difference framework. The key was the discovery of a technique called Alternating Direction Implicit (ADI) on December 31, 1953. "We were celebrating New Year's Eve at our house," recalled Donald Peaceman. "Naturally we were very excited about the ADI discovery and could hardly talk about anything else. This shop talk was very distressing for the hostess, my wife. I think she finally forgave us a few years later." The second problem was lack of computing power, but help was on its way. In 1954, the first tests of the ADI method were performed on an IBM card-programmed calculator, and in 1956, Humble Oil gained access to an IBM 704, IBM's first mass-produced computer. Peaceman and Rachford were finally able to model real reservoir problems. One of their earliest attempts was modeling a field with injection water drive. The reservoir was represented with a rectangular grid containing approximately 1,400 grid points, and each iteration of the model took the IBM 704 three hours to compute.[20]

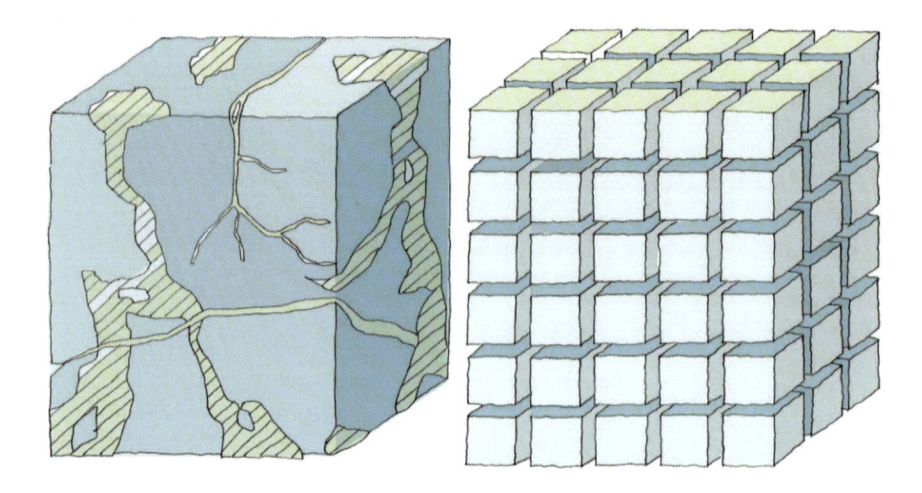

Idealized model, right, of a naturally fractured reservoir with both matrix and fracture porosity, first proposed in 1959. So-called dual-porosity reservoirs are characterized by a complex rock matrix, blue in left figure, surrounded by an irregular system of vugs and natural fractures, in green.

Reservoir engineers in the Soviet Union had different ideas, particularly for carbonate reservoirs which have a notoriously complex pore structure. In 1959, Grigory Barenblatt, Yuri Zheltov and Irina Kochina from the Institute of Petroleum at the Academy of Sciences in Moscow introduced the dual-porosity concept, defining one porosity relationship for part of the rock that is fractured and a different one for matrix blocks that represent rock with conventional intergranular porosity. Separate equations dealt with the two types simultaneously. "That was a very bright idea from the Russian school at the end of the 1950s," recalled Pierre Adler. But it wasn't until 1963 when the idea was taken up in the US by Joseph Warren and Paul Root of Gulf Research Development Company that the double-porosity model was introduced to the oil industry worldwide. In 1976, a group of Marathon Oil researchers in Houston led by petroleum engineer Hossein Kazemi were the first to incorporate the dual-porosity ideas into a numerical model. Today, the dual-porosity model remains a workhorse for the industry.[21]

During the 1970s, other companies got into the game, such as the Computer Modelling Group (CMG), the brainchild of University of

Calgary professor Khalid Aziz, now a professor emeritus at Stanford University. Aziz was able to convince the Alberta Government to fund research into modeling the complex process of extracting oil from the Alberta tar sands. Initially CMG was not for profit but later evolved into a major supplier of commercial modeling software. Two other players were John Nolen of Nolen & Associates in Houston, who built a simulator that was eventually purchased by Landmark, and Al Breitenbach who founded Scientific Software in Denver in 1968.

Grigory Barenblatt, currently professor emeritus of mathematics at the University of California at Berkeley. Barenblatt made important contributions to solid and fluid mechanics at the USSR Academy of Sciences in Moscow and in 1992 emigrated to the West when he was offered the G.I. Taylor Chair in Fluid Mechanics at Cambridge University as its inaugural holder.

Then, an added complication appeared on the simulation scene. A rise in oil price coupled with government trends toward deregulation and partial funding of field pilot projects led to a proliferation of enhanced recovery projects. Simulation had to be extended beyond conventional pressure maintenance and depletion to miscible flooding, chemical flooding, carbon-dioxide injection, steam or hot water stimulation, not to mention in-situ combustion. Each recovery method required a bespoke simulator. In 1967, chemical engineer George Hirasaki joined Shell in Houston to lead the company's reservoir simulation effort, and as Hirasaki recalled, "The first example problem I was given was polymer flooding." In 1968, Keith Coats, a graduate of the University of Michigan, cofounded Intercomp Resources Development and Engineering Inc. and marketed the only simulators available at the time for EOR and thermal applications. "Coats made simulation technology available to people other than the major oil companies who built simulators exclusively for their own use," recalled Aziz.[22]

Around the same time in England, a novel type of simulator was about to emerge. In the mid-1970s, a priority for the UK Department of Energy (UK DOE) was estimating future production in the North Sea. The UK DOE needed a reservoir simulator and for guidance turned to the Atomic Energy Research Establishment in Harwell, near Oxford, where Ian Cheshire was a theoretical physicist. Cheshire's expertise seemed to fit, so in 1977 he was put in charge of the project. Over the next four years, he and his team developed a software called PORES: Program for Oil Reservoir Simulation. PORES boasted many new features: It operated in three dimensions, it could handle oil, gas and water simultaneously, and its "fully implicit" computational technique made it more stable over larger time steps.[23]

Its appearance was just what a small London-based consultancy, Exploration Consultants Limited (ECL) established in 1972 by Ted Daniels, had been waiting for. ECL's bread and butter revolved around seismic acquisition and interpretation, but by 1980 the company was trying to move into the new field of reservoir simulation. Daniels initially tried to license the PORES simulator but in the end recruited the team that made it. "In 1981 he invited Ian Cheshire on board of his yacht and wouldn't let him off until Ian agreed to work for him," recalled Jonathan Holmes, who was eventually hired together with Cheshire and fellow team member John Appleyard. "Ian was a great character," recalled Tommy Miller, who worked with him at ECL during the 1980s and 1990s. "He was very down to earth and never gave the impression of being an intellectual, but of course he was."[24]

Ian Cheshire (1936-2013), a theoretical physicist and developer of reservoir simulators at ECL and, later, Schlumberger.

The simulator that Cheshire's team developed at ECL borrowed all the good features of PORES but also introduced an idea that Cheshire called "corner-point geometry." Rather than defining each reservoir cell by the four parameters traditionally used in finite-difference modeling—thickness, two widths and a midpoint depth—the cell could be defined simply by the location of its corners. This allowed cells to be flexible in shape and size, making it easier to model the chaotic geometry of most reservoirs. In 1983, as the release date of the new ECL simulator approached, Cheshire, Holmes and Appleyard needed a name for their new simulator. "We were in a bar in Copenhagen," recalled Holmes, "wondering what on earth we could call it and started thinking of names beginning with ECL. I suggested calling it ECL's Implicit Program for Simulating Everything, and, bingo, ECLIPSE was born!"

The first three ECLIPSE licenses were sold to three Australian companies—BHP, Santos and Woodside. Statoil was the next, but the company asked for a simulator that could handle deviated and horizontal wells. ECL buckled down and added this feature to the ECLIPSE functionality, and before long ECLIPSE became firmly established as the de facto reservoir simulator for anyone operating in the North Sea. A novelty in its day was that ECL let universities have free licenses for teaching and research. This helped spread the acceptance of ECLIPSE worldwide as the industry's benchmark simulator.

The dramatic oil price drop in 1986 caught almost everyone by surprise, including ECL boss Ted Daniels. When ECL started experiencing serious cash flow problems, the company was soon taken over by Intera Information Technologies Corporation, a Canadian geosciences and engineering consulting firm based in Calgary. Most of the ECL staff stayed, including Cheshire's team, and ECLIPSE continued to thrive under its new owner. By 1995, Intera itself got into financial trouble, and in November of that year its reservoir engineering business including ECLIPSE was bought by Schlumberger. More recently, Schlumberger entered into a joint venture with Chevron to create the successor to ECLIPSE.[25]

The most recent contribution to reservoir simulation has come courtesy of geomechanics. It was never in doubt that, as a reservoir produced, pressure would decrease and change the stress regime in the reservoir rock, sometimes significantly. The most famous example of reservoir depressurization is the North Sea Ekofisk field, where the seafloor has subsided at least 6 meters, requiring a US$ 3 billion spend to raise the production platform in 1987. Equally, an increase of reservoir pressure from water injection may cause stress changes in the reservoir rock. Two questions needed answering: how much does rock stress change during the life of the reservoir and how does that change affect the flow behavior in the reservoir.

Greek-born Nick Koutsabeloulis earned a PhD in numerical modeling from the University of Manchester and then worked for BP for eight years developing stress models, not for oil extraction but for slope stability in open-cast mining, in which BP were active at the time. Koutsabeloulis was attracted more by applying his ideas to oil and gas and in 1993 founded Vector International Processing System (VIPS) to explore that opportunity. Initial funding came from Statoil, Mobil and BP. The ensuing years saw the development of a sophisticated finite element model that could simulate exactly how reservoir stress and behavior changes as pressure increases or decreases. Then, in 2007, Schlumberger acquired VIPS with the express interest of linking its stress model to the ECLIPSE reservoir simulator. The simulator predicts changes in pressure as oil or gas is produced, then it's the turn of Koutsabeloulis's stress model to calculate for that pressure change how the reservoir porosity and permeability change. Then it's back to the reservoir simulator with a modified map of porosity and permeability.[26]

Reservoir monitoring and 4D seismics

However sophisticated simulators become, a simulation is still only as good as the data that are put into it. The data come in two varieties: static data that describes the fixed physical state of the reservoir and dynamic data that provide a glimpse of how the fluids are moving through the reservoir and up the wells, or down the wells in the case of injection.

Typically, petroleum engineers monitor the dynamic state of the reservoir by noting the production and injection rates at the surface and measuring pressures at the surface or at reservoir depth in the wells. Permanent gauges in wells providing real-time pressure data had become a reality in the early 1970s and for subsea wells in 1978. Logs are vital as well. Neutron spectroscopy logging tools measure change in fluid saturation behind casing, and production logs monitor fluid entries into the wellbore. Chemical tracers are also part of the reservoir engineer's arsenal. Injected into one well, their emergence at a nearby producing well tells of a direct path between the two. A tracer technique to measure residual oil saturation in a single well was developed at Esso Production Research in 1968.[27]

In 2010, the Norwegian company Resman A/S, based in Trondheim, pioneered the use of tracers embedded in polymer strips that are inserted into the completion hardware at any number of discrete depths. The tracer is only released when the polymer strip comes in contact with a fluid, and there are any number of distinct tracers for detecting both oil and water. Analyzing samples taken at the surface reveals the DNA of the tracers and shows where the well is producing and which fluid is being produced. The Resman technology is virtually the only way to understand that level of production detail in subsea completions where physical access is more or less impossible.[28]

For a bird's-eye view of the dynamic reservoir, however, nothing matches 4D seismic time-lapse monitoring. Time is the fourth dimension in 4D seismics, so 4D time-lapse monitoring simply means repeatedly surveying the reservoir with 3D seismics and noting changes from one 3D view of the reservoir to the next. Why these differences might reveal fluid movement in the reservoir was the subject of research conducted beginning in the early 1980s. The first discovery was that seismic propagation in oil saturated rock was significantly affected by temperature. The effects were first documented in 1982 by Amos Nur at Stanford University, using heavy-oil sand samples from Venezuela. "By sheer accident, we discovered that seismic velocities in rocks saturated with oil are extremely sensitive to temperature," recalled Nur.[29]

Nur's ideas were controversial. Up to that time, velocity measurements on rocks had been done only on rocks saturated with water, and no temperature effect had been noted. A different result for oil was rather unconceivable. When Nur showed some preliminary results to his research program's industry sponsors, one US company decided to drop out. "Their representative said that this is a sloppy work. He had been working on velocities for the last 10 years and hadn't seen anything like it," recalled Nur. Most assumed that his findings were a laboratory artifact.[30]

Some didn't though. Norm Pullin and Larry Matthews, two Amoco geophysicists, were convinced that Nur's work could be the answer for mapping the movement of heavy oil in a steamflood. For two years, Pullin and Matthews promoted the idea to management and in 1984 had the chance to try it in a steamdrive pilot in the Athabasca tar sands, south of Fort McMurray, Alberta. In April 1985, Western Geophysical carried out a baseline 3D seismic survey, and once the steamflood got underway, three more 3D surveys were carried out. Comparing one survey to the next, it was possible to identify hot regions of oil and follow how the steamflood was progressing.[31]

Applying the 4D idea to conventional reservoirs would take more investigation. During the late 1980s and early 1990s, Amos Nur, with associate Zhijing Wang, continued laboratory measurements of oil- and water-saturated rocks. Mike Batzle at ARCO was also involved in these experiments. Meanwhile, Al Breitenbach, by now head of the merged Scientific Software-Intercomp, was figuring how to use 4D seismic monitoring data to fine-tune the simulation of waterfloods. The initial premise of 4D monitoring was that rock doesn't change, so any variation between surveys could be attributed to fluid movement, and fluid distributions at any time could be interpreted by analyzing acoustic impedance and other seismic attributes. But then it had to be admitted that rock could change—the collapsing Ekofisk field in the North Sea was known to the whole industry. Changes in the rock would clearly complicate any interpretation.[32]

In spite of these difficulties, Statoil and Norsk Hydro moved forward during the 1990s with large-scale 4D projects over the Gullfaks and Troll West fields in the North Sea respectively. These offshore fields had few wells and interwell information was important. Their 3D surveys, repeated several times, clearly showed water sweep through the reservoir and revealed undrained compartments of oil. Since then, 4D seismic monitoring has seen increasing use.[33]

In recent years, oil companies have considered the next logical step, which is laying seismic sensors permanently on the seafloor and hooking them up to a recording vessel each time a new 3D survey is required. In 2012, Petrobras requested PGS to install such a system in their Jubarte field in the North Campos basin, but for most operators, the technology remains unproven and risky. Right now, the cost of manufacturing a system of subsea sensors and cables that is able to withstand a decade in deep water seems to outweigh any commercial benefits.[34]

Crosswell monitoring

As a sort of halfway house between monitoring the reservoir in individual wells with well logs and monitoring the whole reservoir with 4D seismic time-lapse surveys, one more monitoring opportunity remains—surveying the reservoir space between adjacent wells.

In 1978, the Los Alamos National Laboratory was building a deep-geothermal energy system by drilling two closely-spaced wells into the hot granitic basement rock of the Jemez Mountains, New Mexico. After forming a circulation loop by fracturing the rock between the wells, the idea was to pump water down one well and extract energy from the hot water emerging from the other. For this to work, it was essential that the fractures actually connected the two wells. In short, the researchers needed to map somehow the fracture network between the wells.

The Los Alamos National Laboratory turned to Dresser Atlas, where a team, led by Wendel Engle and John R. Smith thought through the options. One option was doing vertical seismic profiling surveys, placing an acoustic source at the surface and recording the acoustic wavefield in the wells. This was rejected because the near-surface, weathered layer would attenuate the high frequencies necessary to obtain good spatial resolution. The Dresser engineers therefore attempted a crosswell measurement, placing acoustic logging tools in both wells, using one as a transmitter and the other as a receiver. Experiments carried out at the New Mexico site during 1977 and 1978 proved the feasability of the measurement, but the tools struggled with the downhole temperature.[35]

Dresser Atlas and the Los Alamos Laboratory nevertheless continued developing the crosswell technology, and during 1983 and 1984 were asked by the US Department of Energy (US DOE) to deploy their system as part of their Multi-Well Experiment (MWX) in western Colorado, a government sponsored field experiment for finding ways to unlock the tight gas sands of the Mesaverde formation in southwest Colorado. The MWX site featured three wells spaced 100 to 200 feet apart, ideally suited for a crosswell survey. Although the MXW crosswell surveys helped in developing the Mesaverde tight gas resources, crosswell seismic systems remained at the level of a science experiment for many years. For the oil industry, the economic value looked rather limited, mainly because borehole acoustic sources were not powerful enough to reach between wells much more than 300 feet apart.[36]

An alternative was to use electromagnetics. The Lawrence Livermore Laboratory in California had some experience from the early 1970s when it carried out some megahertz experiments for tunnel detection in hard rock. For the oil field, however, high frequencies would never work, as Michael Wilt, leader of the Lawrence Livermore team, highlighted in a 1995 article: "In a soft rock oil field environment, megahertz signals are rapidly attenuated and don't propagate for more than a few meters."

By this time, Wilt was leaving Lawrence Livermore for ElectroMagnetics Instruments Inc. (EMI), based in Richmond, California. With Wilt in charge, EMI took up the challenge and built a system using frequencies as low as a few hertz up to a kilohertz. A first prototype was tested in October 1990 at the BP test facility in Devine, Texas, in two wells 300 feet apart. Three years of further development resulted in a successful test at the Mobil Oil Lost Hills field in California where the wells were 2,000 feet apart. The EMI system was commercialized in 1995.[37]

One of the first clients for the new service was Sinopec's Shengli Oilfield Co. Ltd. in China's Shandong province. The company's chief petrophysicist Zeng Wengchong tried the new EMI technology in the Gudao field because circumstances seemed ideal for the new measurement. The wells were close together, and the reservoir was a stratigraphically complex series of channel sands that after years of waterflooding had yielded only 25% of the oil in place. Wengchong figured another 15% could be extracted if only they knew where the trapped oil remained. Surveys were made in three pairs of wells, and the resulting clarification of oil distribution enabled the company to locate new wells in bypassed zones.[38]

Having followed these developments for several years, Schlumberger acquired EMI in 2001 and launched the service worldwide. Improvements continued to be made, including being able to place the receiver tool in cased wells. Up to that point, the service had been available only in open hole. Several surveys were made in the

Middle East, including for Saudi Aramco, but the service could never claim to be mainstream. Meanwhile, crosswell seismics reappeared, driven by new startups such as Z-Seis Corporation, established in Houston in 2003 by Bruce Marion. In 2009, Schlumberger acquired this company as well, looking to combine both electromagnetic and seismic crosswell opportunities.[39]

Juggling the data

From the 1980s onwards, everyone concerned with finding and producing oil and gas reservoirs risked drowning in data. The geophysicist struggled with huge datasets generated by 3D seismic surveys, the log interpreter found that well logs now featured vast data streams, undreamt of in the early days of analog recording, and for simulation, the reservoir engineer increasingly juggled production data from all wells in a field rather than individual wells. Data management became a challenge in itself, and vast amounts of data acquired throughout the life of a reservoir needed to be addressed. It was a perfect storm that could only be mitigated by an increased focus on software development.

Two of the three established workstation companies from the 1980s, Landmark and GeoQuest, tried to match this new demand by developing proprietary software and acquiring a multitude of small, niche software providers and integrating their new applications into the company offerings. At the same time, they had to contend with the march of computer technology, as increasingly powerful PCs looked capable of displacing the Sun workstation on the upstream technologist's desktop, which they eventually did. Both companies were snapped up by the big players: Schlumberger acquired GeoQuest in 1992, and Halliburton acquired Landmark in 1996.

By the mid-1990s, there was a clear need to link or somehow integrate applications relevant to each phase of the exploration and production cycle, so the geophysicist or engineer could transition smoothly from one phase to the next, without spending a disproportionate amount of time dealing with data management or, indeed, losing track of where they were in the process. The first person to recognize and then address this need was Jan Grimnes, who founded Technoguide A/S in Oslo in 1996, acquired by Schlumberger in 2002. Grimnes's software, called Petrel, not only linked better the usual exploration and production applications but had the additional feature of tracking and, when required, reproducing a sequence of computations. The idea of an application workflow was born, and linking software applications in this way became standard industry practice.[40]

Chapter 24: The Boom in Unconventionals

AS THE price of oil fluctuates between its highs and lows, so does the industry's interest in unconventional resources, which always cost more to extract than conventional oil and gas. The industry waxes enthusiastic when the price is high, only to cool off when the price dips.

Steam-assisted gravity drainage

During the late 1970s and early 1980s, it seemed a good time for Canadian operators to develop Alberta's Athabasca tar sands, 80% of which are too deep for mining. Imperial Oil, then the largest oil and gas operator in Canada, accelerated its heavy oil research and in 1975 appointed British-born chemical engineer Roger Butler to lead the charge. The key problem was that the Athabascan heavy oil was so viscous that conventional steamflooding didn't work.[1]

Butler found inspiration in some Russian literature that described heavy oil extraction in the Yarega oil field in the northwestern Urals using a variant of steamflooding. In the 1960s, working from underground tunnels, the Russians drilled horizontal wells through the reservoir and injected steam down vertical and inclined wells drilled from the surface. They then started drilling the injection wells from the tunnel, horizontally and positioned higher than the producers. As reservoir temperature increased, the hot oil drained under the force of gravity toward the production wells. During the 1970s, production increased rapidly and surpassed production from nonstimulated surface-drilled wells 15 times over.[2]

In 1978, Butler decided to conduct a pilot following similar lines. At its heavy-oil Cold Lake field near Edmonton, Alberta, Imperial drilled a vertical steam-injection well just above a 150 meter long horizontal production well near the base of the heavy oil reservoir. The results were encouraging. In a 2006 interview, Butler recalled, "The oil came at about the right rate—I felt pretty darn good!" Over the next 10 years, the pilot accumulated 50,000 cubic meters of oil production. But Butler had a better idea. Why not, like the Russians, drill both wells horizontally, with the steam injector just meters above the producer; drilling horizontal wells was becoming good enough that this promised to be achievable. The heavy oil should heat up, become less viscous and drain into the producing well. He called this technique steam-assisted gravity drainage (SAGD) and validated the concept in the laboratory during 1979 and 1980 using glass-sided boxes packed with sand saturated with Athabasca crude. But an actual field test never materialized.[3]

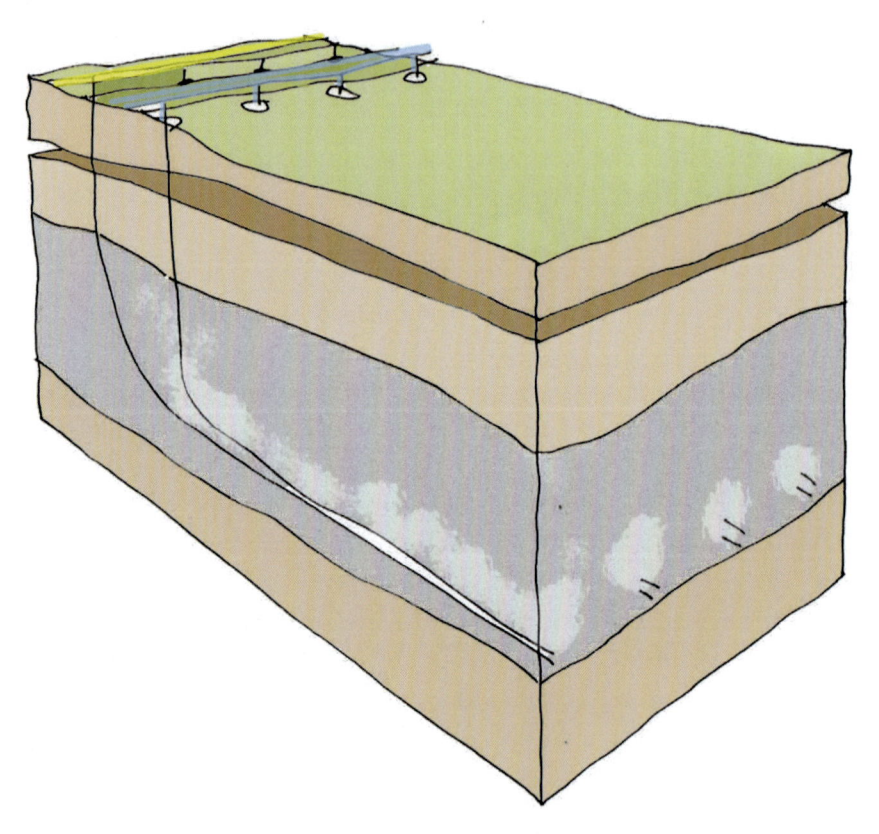

Steam assisted gravity drainage (SAGD), a technique developed in the mid-1990s to extract heavy oil in Alberta, Canada. Hot steam is injected through the upper well of a pair of closely spaced horizontal wells, heating the formation and encouraging the oil to move by gravity to the lower production well.

In 1983, Butler left Imperial Oil to join the Alberta Oil Sands Technology and Research Authority (AOSTRA) as director of technical programs. AOSTRA's management supported the SAGD concept and agreed to construct a purpose-built testing facility, the

Underground Test Facility in Fort McMurray, Alberta. Experiments began in 1986. Three 600-meter long well pairs were drilled from an underground tunnel into the base of the oil sand pay zone. The wells in each pair were separated vertically by four meters, and the pairs were spaced horizontally from each other by 70 meters. Steam was injected into the top well of each pair. The experiment operated successfully for as long as 10 years and at one point produced a very respectable 2,000 barrels of oil per day. By the mid-1990s, horizontal drilling and completion technologies had matured to the point that a SAGD pair of horizontal wells, 650 meters in length with 6 meters vertical separation, could be drilled from the surface. Two well pairs were drilled and put on production. The results were as good as from the tunnel-drilled well pairs.[4]

In 1996, SAGD went commercial. The Alberta Energy Company Ltd. was the first to take the plunge in its Foster Creek field near Fort McMurray. By 1997 operators worldwide had completed some 28 SAGD well pairs, mostly in Canada but also in Venezuela and California. Much of the success of SAGD can be attributed to a new technology that enabled real-time monitoring of the steamflood. This was fiber-optic sensing, permanently deployed along the length of the wellbores and able to measure temperature anywhere in the well. Fiber-optic sensing had been developed in the 1990s by specialized companies such as Cidra Corporation from Wallingford, Connecticut, and Southampton UK-based Sensa, which was acquired by Schlumberger in 2001. The temperature profiles measured using these fiber-optic systems provide a minute by minute picture of where and when the oil is being heated, allowing the whole SAGD process to be fine-tuned.[5]

On the average, SAGD projects are estimated to increase recovery to as high as 50%. In addition, water usage in SAGD is far more economical than in mining the tar sands. Today in Canada, SAGD produces more than 300,000 barrels per day, which is 20% of the country's total tar sand production.[6]

The shale gas bonanza

Of all the unconventional ways of producing oil and gas, there is no doubting the significance of shale gas, a resource that to a large extent has reversed the US dependence on energy imports. Shale is often organic rich and in fact thought to be the hydrocarbon source for a large percentage of conventional reservoirs. But shale is virtually impermeable, so the received wisdom in the oil patch back in the 1950s was to let this particular sleeping dog lie.[7] Then along came George Mitchell, a stubborn petroleum engineer and entrepreneur from Galveston, Texas.

In 1949, Mitchell founded the wildcatting partnership Oil Drilling Inc. in Houston with his brother Johnny and H. Merlyn Christie, a well-connected oil business broker and family friend. "George Mitchell didn't have any money to drill wells," recalled Schlumberger's Robert Freedman," so he got some investors in Houston to back him. One of them was my uncle, Jake Oshman, a businessman from Houston. I remember after my uncle died, my aunt was still getting pretty big checks from George Mitchell."[8]

George Mitchell (1919-2013), the legendary shale gas pioneer. He was also an environmental advocate who directed much of his wealth towards green projects. Among his land holdings was a large tract of pine woods north of Houston where he developed both an enthusiasm for sustainable forestry and the development of an ecologically progressive new town, named The Woodlands.

In 1952, the partnership began drilling in north Texas and made a huge gas strike with its discovery of the Boonsville field in Wise County near Fort Worth. By the late 1970s, Mitchell had taken full control of the company, renaming it Mitchell Energy and Development Corporation, but the Boonsville was beginning to run dry. Mitchell recalled, "You could see it fading. In a few more years I knew we'd be in trouble." Then in 1980, he read a research paper that one of his veteran geologists, Jim Henry, had submitted to the Dallas Geological Society. The paper described a shale formation deep below the Mitchell Energy acreage in Wise County, called the Barnett. Mitchell Energy had drilled through the Barnett shale every once in a while on its way to deeper sedimentary rock formations, and often noticed gas shows but had not given it a second thought. In his paper, Henry analyzed the Barnett shale in some detail and conjectured that it might be chock-full of producible oil and gas. Mitchell got excited.[9]

Elsewhere in the US, shale gas was also getting noticed. In 1976, the US DOE, concerned about declining natural gas production, initiated two significant initiatives. One was the creation of the Gas Research Institute, which encouraged gas research and development, including subsidies to all segments of the natural gas industry. The other was the US DOE Eastern Gas Shales Project, which was evaluating the gas potential of the extensive Devonian and Mississippian organic-rich black shales within the Appalachian, Illinois and Michigan basins. Some wells were drilled, but the virtually zero gas produced confirmed what the industry had expected all along. Shale was a no-go.[10]

George Mitchell believed otherwise. If he could somehow tap the Barnett shale with its permeability of just 0.0001 millidarcies, it would reinvigorate his company's gas production. In 1981, Mitchell Energy drilled and completed the first Barnett well, a vertical well called Slay no. 1, and fracced it. "The well logs were quite similar to logs from the Devonian shales in the Appalachian Basin," recalled Dan Steward, Mitchell's chief geologist from 1981 to 2000. During the next eight years, another 62 vertical wells were drilled and fracced. The wells produced very little gas. "We couldn't figure out how to make it work," Mitchell recalled.[11]

By 1990, George Mitchell's quest for Barnett shale gas was becoming well-known, and as BP's Martin Rylance recalled, "Most people still didn't believe that he was ever going to get anywhere." But Mitchell was already considering the novel technique of combining horizontal drilling with fraccing. "In those days people didn't realize the implications of horizontal drilling for shales," recalled Exxon's Claude Cooke. In need of financial support, Mitchell convinced the US DOE to subsidize the next experiment, and in 1991 drilled the first horizontal well in the Barnett shale and fracced it. Some gas was produced, but still not enough.[12]

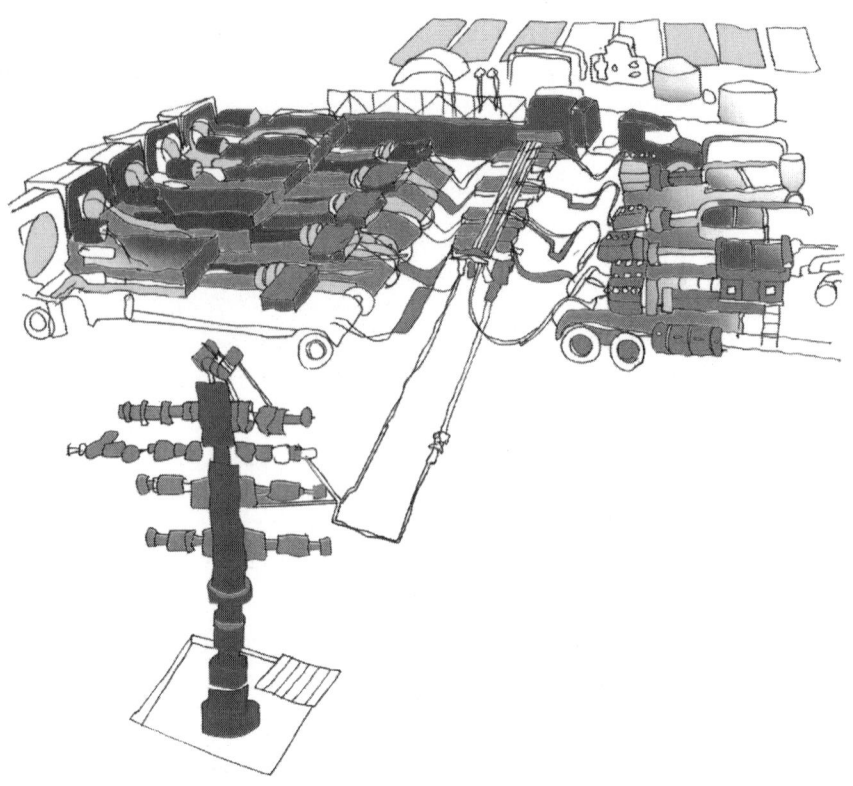

A typical hydraulic fracturing operation, resource-intensive and demanding a lot of horsepower. A small fleet of fraccing trucks combine to achieve the desired high pumping pressures. This depiction shows a Halliburton frac job in 2005 near Caldwell, Texas, where injection pressure reached 20,000 psi.

Meanwhile, his business took another hit. His longtime customer, the Natural Gas Pipeline Company of America, bought out its 20-year contract to buy his natural gas, two years before it was scheduled to expire. Now Mitchell no longer had a guaranteed price for his gas but was dependent on market prices, which were dropping, reaching an all-time low in 1995. Throughout the 1990s, however, Mitchell Energy continued drilling the Barnett, about 100 wells each year, some horizontal but mostly vertical, and continued fraccing them. "It was clear to him that Barnett held a lot of gas, and he wanted us to figure a way to get it out," recalled Dan Steward. "If we couldn't, then he would hire people who could."[13]

The eventual breakthrough would be delivered by Nick Steinsberger, a petroleum engineer who had been working for Mitchell Energy since 1987. While supervising a well in the Barnett shale in 1997, Steinsberger noticed that the fraccing gel wasn't mixing properly with the water, and that the well was being fracced by something more resembling a fluid than a viscous gel. The well produced better than usual, and Steinsberger began to wonder if a watery mix might be the answer to opening up the shale. A few weeks later, Steinsberger met with Mike Mayerhofer, an old friend working for rival Union Pacific Resources, and learned that he had had a similar experience. Steinsberger decided that the watery mix was worth a try. Fellow Mitchell colleagues, such as Steven McKetta, the company's fraccing guru, thought he was crazy: "You can't pour so much water on shale, shale has clay in it. It's almost like hardened mud. Adding more water will create an awful mess." But by now the idea had a name, "slickwater" fracturing.

With George Mitchell's backing, Steinsberger fracced three more Barnett wells with his slickwater mixture, and they all produced gas at a good level without the steep decline in production that usually occurred in gel-fracced wells. The water-base liquid seemed to go in every direction in the rock, creating complex mini-networks of cracks enabling gas to flow to the surface. By September 1998, Mitchell engineers had dropped gel fracs entirely. Mitchell Energy began refraccing old wells with slickwater and by mid-2000 had doubled their production. As natural gas prices started to rise in the early 2000s, so did the shares of Mitchell Energy. In 2002, George Mitchell, now 83 years old and up to his ears in real-estate in Houston, decided to sell his company to Devon Energy of Oklahoma City. The agreed price was US$ 3.5 billion.[14]

That year, Devon drilled seven horizontal wells in the Barnett shale and 55 wells the following year and used the novel technique of multistage-fraccing in all of them. Using a variety of downhole mechanical devices, multistage fraccing directs the fraccing fluid to sections of the well one at a time, rather than to the whole well. Multistage fraccing had been developed by Gulf Oil Corporation in 1965 and used to stimulate vertical wells. It was famously applied in horizontal wells by Maersk Oil in the Dan field in Denmark in 1989. "If you drill a long, horizontal well and fracture it at multiple locations," explained Exxon's Cooke, "You get a lot more production than just fraccing the whole well." Natural gas prices meanwhile climbed, and by 2004 others were joining the race, such as Chesapeake Energy, also from Oklahoma City. The shale boom had begun.[15]

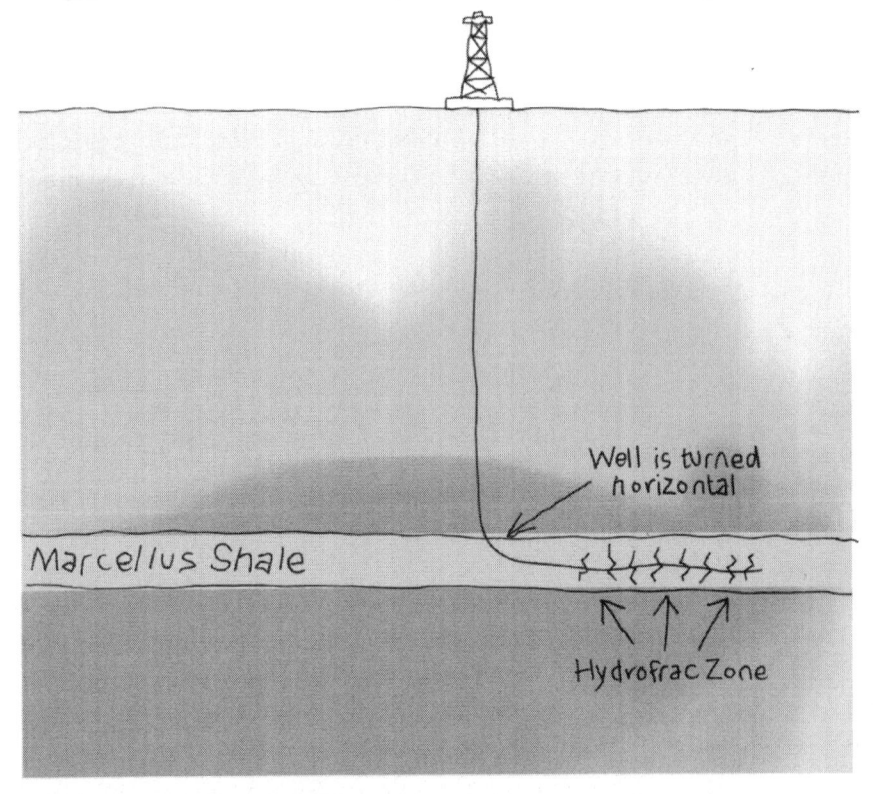

A shale gas well in the Marcellus Shale in the Appalachian Basin, in which multistage hydraulic fracturing has created fracs along the horizontal portion of the well.

As shale gas took off, one question kept resurfacing: Where did the fracs actually go? The answer came from the experience of Phillips Petroleum as the company sought to understand the subsidence of their Ekofisk field in the North Sea in 1997. By placing geophones in boreholes, it was possible to record microseismic events as the reservoir compressed, even triangulate on individual events by recording in several wells simultaneously. Beginning in 2000, Mitchell Energy had used the technique to triangulate on fracs as they propagated since each incremental opening of a fracture produces a small sound that can be detected. Since then, the technique has seen widespread use for visualizing fractures and ultimately designing better frac jobs.[16]

In 2000, shale gas provided just 1% of the natural gas produced in the US. By 2012, it was 39%, as the boom spread to the Haynesville shale deposits in Louisiana and Texas and the Marcellus shale in Pennsylvania and West Virginia. The US was now a gas exporter and the world number one natural gas producing country. Around the world, the potential for shale gas production is considered huge, but progress is slow. No other country has the US's favorable mineral rights law that encourages landowners to engage, and environmentalists worldwide are putting up a fight, highlighting water contamination, surface disruption and earthquake risks among other fears.[17]

Shale oil in the Bakken

The second shale revolution was oil. Interest in shale oil goes back to the mid-1970s, when US oil imports were rising and US President Jimmy Carter announced a plan to supplement the country's existing production with more than two million extra barrels per day from shale and liquefied coal. At the time, mining was the only known way to extract oil from shale, and several companies, including Occidental, Unocal and Exxon, were active. Mining shale for oil remained niche and not very economic.[18]

The dream of course was to drill for shale oil. The stage was set in 1978 with a groundbreaking paper by Fred Meissner, an exploration geologist working out of his home state of Colorado. Meissner proposed looking for oil, not in the usual compartments where it had migrated, but in "the hydrocarbon kitchens" where it was being formed. He was referring in particular to the Bakken shales, a formation sprawling across 200,000 square miles of North and South Dakota and Montana, and beyond into Saskatchewan and Manitoba in Canada. But it would be almost two decades before his idea would be tested, and the most successful exploiter of the Bakken oil shales would be Harold Hamm.

Hamm, an Oklahoma oilman, had founded Shelly Dean Oil Company in his hometown Enid in northern Oklahoma in 1967 at the tender age of 21. Harold Hamm was the quintessential Oklahoman oil man, but after many years operating locally, he dreamed of an historic discovery that would put him on the map and win him fame and fortune.[19]

In 1995, Hamm ventured into North Dakota where he discovered the Cedar Hills field, which by 1999 produced 7,000 barrels of oil per day and rated the seventh-largest onshore field in the Lower 48. As oil prices were rebounding, Hamm looked for new opportunities and started thinking about the Bakken oil shale. If George Mitchell could tap shale for gas, why couldn't he do the same for oil? Porosities in the Bakken averaged about 5% and permeability was very low, averaging 0.04 millidarcies. Hamm's company, now called Continental Resources, had already drilled a few vertical wells into the Bakken producing on the average a meager five or ten barrels per day.[20]

In 2002, Continental Resources focused on eastern Montana where oil prospects in the shale initially seemed better than in North Dakota and began leasing land. Along with a few other operators such as Dallas-based Headington Oil and Lyco Energy Corporation, and at the suggestion of his senior geologist Jim Kochic, Hamm tried horizontal drilling combined with fraccing. The first well in August 2003 was a winner, producing 1,300 barrels per day. The Bakken was starting to reveal its secrets. Hamm meanwhile turned his attention back to North Dakota, leasing 300,000 acres there in 2003. But the first wells quickly turned into a trickle. It turned out that the Bakken in North Dakota was very different from Montana. When Hamm's drilling teams tried putting a wellbore in the top layer of the shale, the rock caved and collapsed. And the fraccing wasn't working either. Their liquid concoction went anywhere and everywhere. Sometimes it even exited the Bakken Formation, opening up fractures in a water-producing zone. "It was pretty frustrating," Hamm recalled.

One day, Hamm heard that Houston-based EOG Resources, a competitor, who in April 2006 started drilling east of the Continental North Dakota Bakken acreages, were using multistage fraccing and getting good results. The new method was quite expensive, twice the cost of conventional fraccing, but soaring oil prices that would reach an all-time high of US$ 147 per barrel in July 2008 meant that it was worth a try. In late 2006, Continental Resources engineers began multistage fraccing, and by early 2007, the company's North Dakota production reached 7,000 barrels per day with almost all of its expanding acreage still undeveloped.

When, in April 2008, the USGS estimated that the North Dakota and Montana Bakken shale held between 3 and 4.3 billion barrels of "undiscovered, technically recoverable oil," a 45-fold increase from their 1995 estimation, Hamm felt vindicated but was sure the USGS was still on the low side. He ordered his landmen to lease more acreage, and by March 2009 Continental Resources controlled more than 600,000 acres, making it the largest leaseholder in the area. Within just seven months, the oil price had tumbled downwards, and Harold Hamm was on the verge of moving elsewhere. But during the spring and summer of 2009, the US economy and oil prices recovered. Hamm still had enough cash to carry on. Finally, Continental Resources engineers figured out the last piece of the puzzle. Through trial and error, they realized that fraccing in about 30 stages helped extract huge amounts of crude: fewer than 30 stages led to too little oil, more than 30 was too costly and not worth it. Soon, Continental Resources wells were producing over 100,000 barrels a day, and by 2010, production reached 225,000 barrels per day.[21]

In a 2011 investor meeting, Hamm dubbed the Bakken, "America's Saudi Arabia." The following year, North Dakota became the second-largest oil producing US state after Texas. In 2014, production from the Bakken surpassed one million barrels a day, joining the top-10 list of the most productive oil fields of the world. Since 2009, the success at Bakken has spread to other shale oil plays in the US, notably the Eagle Ford shale in Texas. So far, the shale oil revolution has increased US domestic oil output by 30%, to almost match that of Russia.[22]

Chapter 25: Looking Ahead

PREDICTING WHICH new upstream technologies will succeed and which will pass forgotten into history remains a difficult call. A recurrent theme in this history is the risk-averse nature of the oilfield professional, a mindset born through performing a task that is potentially difficult and very dangerous. But inventive minds are as active now as in previous times, and the industry demands new solutions every day. Some examples of today's evolving technologies give a flavor of what might be mainstream tomorrow.

The search continues

While exploration geophysics has marched from one technical innovation to the next, the geologic side of the equation has progressed through a continual amassing of field data and the occasional breakthrough concept that does a better job at explaining the data than before. Collecting data, accessing data and integrating data is the key. Taking this to the limit has been the brainchild of three former BP geologists, Dave Casey, Roger Davies and Peter Sharland who founded Neftex, a UK-based geoscience consultancy in 2001.

Centerpiece of their approach is a global earth model that connects together multiple regional models. Through detailed evaluation of the geologic record of the world's major sedimentary basins using every piece of publicly available data they could lay their hands on and leveraging Peter Vail's sequence stratigraphy approach, Casey and his colleagues have identified, correlated and mapped more than 100 depositional sequences around the world ranging in age from Late Precambrian to Pliocene. The range of data runs the gamut of anything that's geologically relevant, including surface field studies, paleontology, geochemistry, logs and more. And by linking these modern-day correlations to a tectonic model of the major continents that can wind back 600 million years, the entire geodynamic relationship between the Earth's plate tectonic history and the occurrence of hydrocarbon deposits can be viewed and analyzed.[1] In 2014, Neftex was acquired by Halliburton.

Unconventionals provide another direction for geology research. Dietrich Welte, who now is an exploration advisor for Schlumberger, recently explained in an interview with the *AAPG Explorer* magazine: "The current activity in unconventional resource development has simply shifted the emphasis from the end of the petroleum systems line, existing accumulations in classical reservoirs, to the beginning of the petroleum systems line—the hydrocarbons that have been retained in source rocks, or their organic-lean immediate neighbourhood."[2]

In total numbers, conventional oil and gas still represent the majority of global oil and gas production and will do so until at least 2035, according to a forecast of the International Energy Agency (IEA). However, there is no doubt that unconventionals will provide the driver for innovation in the future. According to the IEA, the net increase in global oil production up to the year 2035 will be driven entirely by unconventional oil with unconventional gas accounting for nearly half of the increase in global gas production.[3]

Latest and greatest in seismics

Meanwhile seismics marches on. A recent innovation for marine seismics has been to record data separately from every single hydrophone in every streamer: literally thousands upon thousands of individual recordings. In a conventional setup, a number of adjacent hydrophones in a given streamer is grouped together and delivers an averaged, collective signal. This practice was inherited from the early days of land acquisition to cut background noise, and it conveniently limited the sheer volume of data that had to be recorded and processed. But geophysicists have known for a long time that recording every signal separately would produce much sharper subsurface images, mainly because it would systematically eliminate the effect of aliasing, a phenomenon causing confusion about the timing between seismic signals. Recording data from every hydrophone separately required managing much greater complexity in the field, being able to record an order of magnitude more data without running out of disk space, and having enough computer power to crunch all the data. The first such system, called Q-Marine, was launched in 2000 by Schlumberger's WesternGeco. The practice is now mainstream for that company.[4]

Another game changer has been to upgrade the conventional streamer by adding accelerometers that can decompose pressure variations into their spatial components, vertical and horizontal. Norwegian PGS managed to add a vertical accelerometer, but it was

WesternGeco that succeeded in miniaturizing a 3D accelerometer sensor and packaging it en masse into their streamer, calling it IsoMetrix. After 10 years of development, the first surveys included one for Statoil in 2012, and another for Thombo Petroleum Ltd. in February 2013 using 10 fully instrumented streamers and covering an area of 700 square kilometers off the west coast of South Africa. The expected bonus is an ever sharper image after migration.[5]

WesternGeco's seismic vessel WG Vespucci built in the Norwegian Ulstein shipyard in 2010. The vessel has a characteristic inverted bow form that improves handling in rough sea and lowers the fuel consumption by causing less hydrodynamic drag.

Meanwhile, the design of marine acquisition vessels has diversified. Whereas the WesternGeco boats look more or less conventional, relying on birds to maintain streamer separation, PGS opted for a delta-shaped vessel with an extraordinarily wide stern, 70 meters wide, to more efficiently deploy the streamers. Their *Ramform Titan* was launched in April 2013.[6]

Another stab at efficiency has different seismic sources firing simultaneously or at close time differences. Some detectors record all reflections, others only a few of them. One needs to know the exact locations of sources and receivers and which is doing what. The only way to make sense of it all is to label each source for later tracing the records to the right detectors. In this coding technique, the signals are recorded in a blended state, and then in the processing center "deblended."[7]

This idea actually goes back to the early days of Vibroseis in land seismics in the 1980s. In the Vibroseis technique, the long signal sweeps of the source required to obtain a satisfactory signal-to-noise ratio typically took 10 to 20 seconds, so it was reasonable to try doing two sweeps simultaneously and somehow coding them so the resulting signals could be deblended. The first method for simultaneous recording was patented in 1979 by Daniel Silverman, an electrical engineer and physicist who became director of geophysical research in Amoco and continued developing new ideas even after his retirement in 1970. The idea has seen moderate success, and recently the idea has migrated offshore, with two airguns firing in close sequence. The challenge here is that deblending must cope with the fact that airgun sources cannot shape the downward energy so nicely as Vibroseis. Nevertheless, it seems to work, and the first trials were conducted in 2012 by WesternGeco for Apache oil company offshore Western Australia.[8]

Other ideas are a long way from being commercial, the following example coming from the Delphi Consortium run by Guus Berkhout. Traditionally, seismic acquisition is carried out with a centralized architecture, with the number of detectors having continuously increased over the years. In 2013, some systems in use had up to 100,000 geophones or hydrophones. Oil companies would even like to go to one million detectors. But instead of creating ever more complex acquisition systems, Berkhout has proposed breaking this trend by abandoning the central control and using a networked architecture: "Instead of having a big central system, you get a swarm of self-organized small systems or units with sources and detectors that are communicating with each other and that move around. This system is not controlled by a number of people. It is controlling itself. Then of course, you are much more flexible."[9]

Another area of research is dealing with unwanted artefacts inherent to the seismic technique, such as multiple reflections between sedimentary layers. Since the arrival of super-fast computers from the late 2000s, the industry began to accept that multiples, rather than being a nuisance, were sending a message. "Don't throw away multiples," said Gerard Schuster, a professor of geophysics at the King Abdullah

University of Science and Technology in Thuwal, Saudi Arabia, who has done much research in this area. "The multiples provide new opportunities for reducing exploration risk. The use of coherent multiples gives second and third extra views of the subsurface and can sometimes improve the view of the subsurface dramatically."[10]

In interpretation, some new paradigms also deserve attention. One is the fact that hydrocarbons, particularly in deeper formations, may well create the opposite of a bright spot, in other words an amplitude low that has come to be called a "dim" spot. Dim spots are pretty hard to work with though. Sudden amplitude increases attract attention, but amplitude dips are easy to miss. And psychologically, convincing oil companies to drill a dim spot is a challenge. But, as interpretation expert Alistair Brown puts it, "There is a great deal of hydrocarbon actually "hiding" in dim spots in the subsurface. Dim spots are an opportunity of the future, and the emerging generation of geoscientists should accept the challenge."[11]

Electromagnetics as a complement to seismics

Today, seismics remains the dominant method for investigating the subsurface, but that hasn't stopped other techniques being tried out. Recently, marine electromagnetics has experienced some success. The key innovator was Charles Cox at the Scripps Institution of Oceanography in San Diego, California. Cox started working in this area in the 1970s with funding from the US Defense Advanced Research Projects Agency, which was interested in how the seafloor affected submarine communications. Cox measured the electrical conductivity of the shallow ocean-bottom layers using electromagnetic transmitters and receivers placed on the ocean floor. His first experiment was carried out on a midocean ridge in the Pacific in 1979, and the technique came to be called the Controlled Source Electromagnetic Method (CSEM).

Two features of CSEM then caught the eye of oil explorers. First was its ability to map subsurface structures below overburden layers such as salt, albeit with very poor resolution, and second was its high sensitivity to thin resistive layers, a characteristic of many oil-producing formations. The promise of CSEM was complementing a seismic interpretation, reducing the risk of expensive exploration decisions. Tuning CSEM for the oil industry started at Scripps in the early 1980s, but development was slow. First, the early equipment was quite cumbersome to deploy on the ocean floor. Second, there was too much coupling between transmitter and receiver through the atmosphere as typical exploration water depths had not yet passed 1,000 feet. Exxon was one of several companies that started looking into the technique.[12]

Meanwhile, Martin Sinha, a marine geophysics researcher at the University of Cambridge in the mid-1980s, introduced a neutrally buoyant antenna that allowed the deep-towed transmitter to be "flown" about 300 feet above the seafloor. This approach had solved the problem of surveying over the rough terrain of a deep ocean ridge for the US Navy but was also promising for hydrocarbon exploration. The first trials of the Cambridge system were carried out in 1987 and 1988, and were followed by collaborative Cambridge and Scripps experiments on midocean ridges of the East Pacific Rise, the Reykjanes Ridge in the North Atlantic and the Valu Fa Ridge in the southwest Pacific.

By the late 1980s, oil prices had dropped to a record low, and exploration was drastically curtailed as a result. Also, increasingly successful results from 3D seismics further discouraged investment in new unproven techniques. CSEM languished. But the rising oil prices in the mid-1990s, combined with the move to deepwater oil provinces, gave it a new lease on life. Exxon continued to pursue CSEM, as did Statoil.

Both Exxon and Statoil began modeling studies and field trials of CSEM in water 3,000 feet and deeper specifically for measuring the electrical resistivity of previously identified drilling targets in an effort to mitigate drilling risk. Trials offshore Angola from 2000 to 2002 by both companies showed that CSEM data could measure reservoir resistivity for targets several thousand meters deep. Both Exxon and Statoil leveraged instrumentation and expertise from Scripps's Seafloor Electromagnetic Methods Consortium, established in 1994, which counts 26 industry members today, including all major oil companies as well as service companies.[13]

In the 2000s, the success of the electromagnetic exploration method continued, with the technique also being deployed to assist in the delineation of basins. In 2011, marine CSEM was regarded as an established technique for hydrocarbon exploration, and some, such as Leonard Srnka, Exxon chief research geoscientist, and Scripps professor

Steven Constable, see electromagnetics as a key exploration tool of the future, writing: "The ability to determine the resistivity of deep drilling targets from the seafloor may well make marine CSEM the most important geophysical technique to emerge since 3D reflection seismology."[14]

Expandable casing

Drilling is not short of innovation either, expandable casing being an example. The classic well construction, using smaller casing sizes as the well is drilled, is certainly the safe and well-proven option. But as wells get deeper, production risks being increasingly restricted by the final casing size.

In the 1990s, Shell started looking at alternatives. The basic idea was to drill below a given casing string with an underreamer, a special bit that expands and drills at a larger diameter than the casing it passes through, then descend a casing that could be expanded to match the diameter of the casing string above it, and in fact seal into it. If successful, this strategy would result in a smaller diameter surface pipe, better isolation of problem zones and, eventually, improved production at depth. The first laboratory trials were performed in The Hague in 1993.[15]

The casing was constructed from special automotive steel, similar to that used in the crumple zone of an automobile. Once the casing was in place in the well, a cone-like plunger was pulled up the casing to mechanically deform the pipe to a larger diameter. The pipe in the trial was four inches in diameter, and it got expanded by about 22%.[16]

Later, the development continued jointly between Shell and Halliburton, and in late 1999, Halliburton set the first expandable-tubular system in a well in the Chevron West Cameron 17 field, just outside of Louisiana state waters in the Gulf of Mexico.[17] During the early 2000s, expandable casing gained increasing popularity for deepwater and ERD wells. In September 2006, Baker Hughes and BP used expandables to run a casing string of the same diameter all the way to the bottom of a 4,138-foot deep well in southeast Oklahoma. The liner at the very bottom was expanded from 8 to 8.625 inches, which helped to increase production.[18]

From underbalanced to managed pressure drilling

In the early days, any well drilled with a cable tool and that had no mud filling the hole was, by definition, underbalanced. The wellbore pressure, if not close to atmospheric, was certainly less than the formation pressure. Therefore, once the bit struck oil, there was a natural rush to the surface creating the gushers seen in photographs of that era.

However, since circulating drilling mud was introduced at the turn of the century and BOPs introduced in the 1920s, wells have been drilled with a wellbore pressure slightly higher than the pressure in the rock. This so-called overbalanced drilling certainly controls the well and circulates out the cuttings, but the mud invades and contaminates the formation, and risks compromising forever the formation's ability to produce.[19]

As wells were drilled deeper and longer, formation damage increased, and from the 1960s, researchers in both the US and the Soviet Union starting looking for a new approach. One way to reduce the mud pressure was to mix natural gas, nitrogen or air with the liquid phase of the drilling fluid, but the trick was not lowering the pressure so much that the well blew out.[20]

Gauging the right amount to lower the pressure required knowing the formation pressure at a given depth, and that information was initially unavailable. In 1969, Ben Eaton of Continental Oil published a seminal and widely used method for predicting both formation pressure and fracture pressure, the pressure at which drilling fluid causes the rock formation to split open. His method hinged on estimating the total weight of rock above any given point in the well using well logs. Eaton's method assumed typical US Gulf Coast conditions, but his method soon proved to be applicable elsewhere. Four years later, Stan Christman of Exxon extended the method for use offshore.[21]

Drilling with wellbore pressure less than formation pressure—underbalanced drilling—was the next logical step. A prerequisite was surface pressure equipment to control the upward flow of mud and cuttings, and even oil and gas shows, in the annulus between casing and drillpipe. Already in the 1930s, drilling equipment and BOP manufacturer Shaffer Tool Works had introduced a special device called a rotating control head that was added on top of the BOP to seal and control annular pressure while allowing the drillpipe to turn. These devices enabled the first underbalanced drilling trials in the US in the 1960s. For their own underbalanced drilling in the 1970s, Soviet drillers added hardware to separate produced gas from the mud.[22] In the late 1980s, several operators in the Austin Chalk formation of South Texas adopted underbalanced drilling to limit formation damage, but they also confirmed another benefit: underbalanced drilling resulted in higher rates of penetration. This had been noted 20 years earlier by D.I. Vidrine and E.J. Benit of Drilling Well Control, Lafayette, Louisiana.[23]

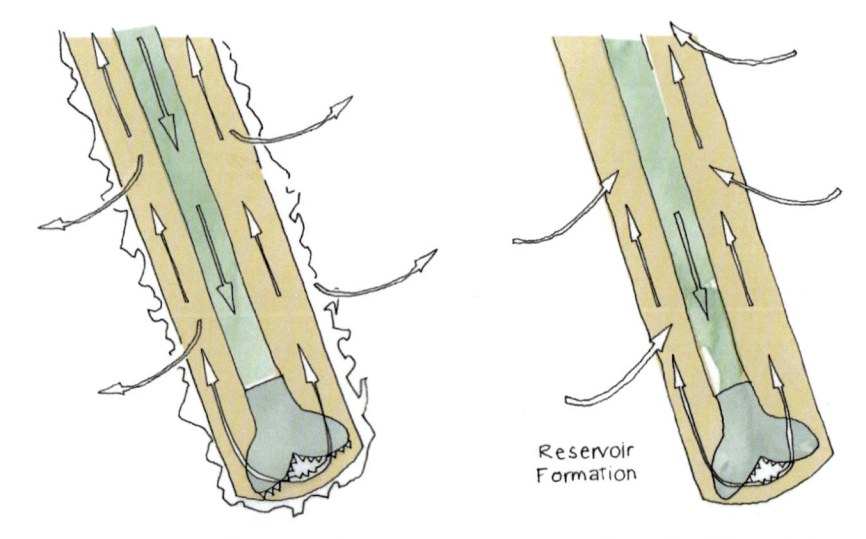

Underbalanced drilling. With conventional, overbalanced drilling, left, wellbore pressure is higher than in the formation causing formation damage and possibly lost circulation. In underbalanced drilling, right, formation pressure exceeds that wellbore pressure, avoiding these problems, but requires careful well control.

Underbalanced drilling technology went through tremendous growth in the 1990s, spurred by successes in the Austin Chalk. Operators increased drilling rates, reduced formation damage and improved productivity. Worldwide, underbalanced drilling started to take off. The technique reemerged in Russia when Lukoil started using it in the late 1990s. Shell introduced underbalanced drilling in the offshore environment in 1997. By 2005, 75% of US land-drilling programs drilled at least one section with a closed and pressurized mud return system, up from 10% in 1995.[24]

Over the past 10 years, underbalanced drilling has spawned a related technique called managed pressure drilling, introduced by Weatherford in 2003. Managed pressure drilling reduces wellbore pressure just the right amount to secure a smooth and safe drilling process, with less well kicks, less lost circulation, higher penetration rates and, most important, a significant reduction of formation damage.[25]

Going superdeep offshore

Moving to ever deeper water depths posed continuous challenges. At depths greater than say 4,000 feet (1,200 meters), drillers began to worry about the increasing weight of the mud column, effectively the water depth plus well depth, and the effect that this had on pressures deep in the well. The margin between too little pressure, allowing the well to blow out, and too much pressure, causing the rock to fracture and suck in all the mud, was becoming too narrow for comfort. The only way to manage this margin was to place more and more casing strings in the well, which is not good for production. The alternative idea of decoupling the fluid in the riser from the subsurface mud column occurred to the CUSS group in the 1950s but was not realistically addressed until the 1960s.

At that time, Shell had the idea of simply removing the riser—the pipe between the drilling rig and the ocean floor that protects the drillpipe assembly and provides a mud return—a concept that became known as riserless drilling. A key player was Charles Peterman, later to become director of research for Hydril. The technology hurdles, however, were massive and the project was not pursued. Offshore wells had yet to reach water depths that really merited the approach. But by the 1990s, they had done.[26]

Around 1996, a five-year, US$ 50 million joint industry project was launched by Conoco, Chevron, Hydril and several other operators and service companies to establish a reliable and safe dual-gradient system, as the decoupling idea had become known. In their approach, the riser was preserved but filled with seawater, basically serving to connect the rig with the BOP on the ocean floor; incidentally, having it filled with light seawater rather than heavy mud dramatically reduced the stresses on the riser and its supporting cables. On top of the BOP, a new device collected the returning mud while still allowing the drillstring to continue rotating. On the seafloor, powerful pumps returned the mud to the rig. In August 2001, the new system was tested successfully in the field, albeit in only 1,000 feet of water. In May 2012, Chevron announced that the *Pacific Santa Ana*, the first drillship designed with the capacity to perform dual-gradient drilling, had arrived in the Gulf of Mexico.[27]

Meanwhile, another joint industry program was promoting a riserless approach to dual-gradient drilling. This technology was developed and commercialized by the Norwegian firm AGR Drilling Services and designed specifically for drilling the tophole portion of a wellbore. This riserless drilling system saw its first commercial test application for BP on the Azeri field in the Caspian Sea in 2003. In the following three years, the technique was used successfully on 15 additional wells. By 2008, the system could operate in water depths up to 460 meters. Since then, AGR Drilling Services and BP have joined forces with the Integrated Ocean Drilling Program, the most recent scientific research program that drills seafloor sediments and rocks in oceans worldwide, to modify and deploy the system in water depths up to 12,000 feet.[28]

Nevertheless, offshore drilling with a marine riser is still the norm and can achieve remarkable results. On April 11, 2011, Transocean announced that the ultra-deepwater drillship *Dhirubhai Deepwater KG2*, equipped with a 21-inch riser, set a world record drilling in 10,194 feet of ocean for Reliance Industries in India. The rig surpassed the prior record of 10,011 feet of water, set in 2003 by the Transocean *Discoverer Deep Seas*, working for Chevron in the US Gulf of Mexico.[29]

Automating the drilling rig

By the end of the 20th century, the drilling rig finally succumbed to a makeover. Over the past two decades, drilling rigs have become increasingly automated, increasing efficiency and making operations safer. A key driver was the new safety regime put in place for the North Sea following the Piper Alpha disaster, a production platform that burned and sank with the loss of 167 lives.

Already in the late 1980s, Norwegian regulators pushed for adapting the Boyadjieff's Iron Roughneck for use offshore. Simultaneously Varco International started several developments to automate various rig operations, for example working with Maersk to produce a pipe pickup and lay down system that eliminated human intervention.

Regulation steadily became tighter, and in 1994, Norway passed stringent laws concerning safety for all floating drilling units working the Norwegian sector of the North Sea. By the 1990s, small specialist companies such as Hitec and Maritime Hydraulics started to fill an increasing need for safer drilling and increased automation.[30]

A totally automated drilling rig in the Groningen gas field in the Netherlands. This rig breaks down into container-sized modules for rapid deployment. To the right are the mud tanks, to the left the rig with automatic pipe handling.

A key development in rig automation was the introduction of computer systems. Aside from some automated mud systems in the early 1990s, computers were absent on drilling rigs. The industry believed, probably correctly, that they were not robust enough to withstand the harsh drilling environment. Whatever automation existed, such as pipe handling, the control was purely hydraulic. Then in 1999, Varco International replaced a joystick-based pipe handling and racking system on a Maersk jackup with a computer-controlled system. It was a watershed moment. In 1991, Varco computerized its Iron Roughneck, but the real tipping point didn't arrive until 2003 when the company launched a complete computer-based roughneck system that could screw drillpipe together more safely and faster than ever before, reducing the connection time from two minutes to 19 seconds. Combining the new Iron Roughneck with automated pipe handling resulted in a drilling operation more or less completely automated.[31]

Meanwhile in 1997, George Dotson, the CEO of US drilling contractor Helmerich & Payne, invested massively in the development of a semiautomated land rig, reinventing it from the ground up. Helmerich & Payne not only introduced innovations such as touchscreens for primary drilling control and the unheard of luxury of air conditioning for drillers's cabins but also seemingly marginal elements such as round mud tanks that were safer and easier to clean and move than the traditional rectangular ones. In 2005, the final piece of the puzzle, the drawworks, was integrated into the Helmerich & Payne automation scheme. The new automated rig had immediate impact, drilling faster than any other rig, moving faster between locations and delivering better wells. In fact, drilling performance improved a dramatic 30%.[32]

Soon, other drilling contractors were playing the same game. An example is the Bavarian group Streicher. Their automated rig looks as different from a traditional rig as can be imagined. The automatic pipe handling system picks up a 30-foot length of pipe from a rack with a gigantic steel arm, lifts it to the vertical before screwing it into the string. No mud-spattered roughnecks risking life and limb, just a drilling technician in a soundproof cabin consulting multiple computer screens and periodically checking on the robot's work.[33]

Nanotechnology

For the reservoir engineer, nanotechnology has for years inspired a lot of dreaming. In broad terms, nanotechnology is the science of materials in a range from 1 to 100 nanometers (1 nanometer is 10^{-9} meters). It is the use of such very small pieces of material by themselves or by manipulation to create new larger-scale materials that has spawned a new, exciting world. Still in its infancy, nanotechnology is beginning to be explored in the upstream industry in several scenarios. In the words of Wade Adams, the director of the Richard E. Smalley Institute for Nanoscale Science and Technology at Rice University, as reported in 2010 to the *American Oil & Gas Reporter*, "Nanoscale technology has moved from science fiction into really hard engineering."[34]

One approach is injecting nanoparticles into reservoirs to "illuminate" the front edge of waterfloods or delineate the extent of hydraulic fracture networks. The idea is that nanoparticles could be designed to augment a rock's electromagnetic, acoustic or other properties, so that conventional imaging, either from the surface or borehole, could yield a sharper image of fluid distribution. Saudi Aramco is a company now engaged in this area, and research is ongoing at Texas A&M and The University of Texas at Austin.[35]

The idea of injecting nanoparticles into an oil reservoir as a contrast agent came from the medical industry that has pioneered treatments placing very small sensors in the human body's circulatory system. Using nanosensors in the oil industry is proving much harder. "The human body is a lot easier to work in than the earth, because the body works at atmospheric pressure and temperature," said Scott Tinker, director of the Bureau of Economic Geology at The University of Texas at Austin. "In medicine, if you don't know what happened, you can in the limit cut the body open and have a look. Well, the earth is at high pressure and temperature with hot, nasty chemicals, and you can't just cut it open. It's a much bigger challenge."[36]

One attempt to tackle the nanotechnology challenge has been the creation of the Advanced Energy Consortium (AEC) in 2007. Hosted and managed by the Bureau of Economic Geology in Austin, Texas, the consortium includes seven oil companies, including heavyweights Shell, BP and Total, plus Schlumberger. Today, the consortium funds 25 universities around the world to do fundamental research in micro- and nano-sensing for the subsurface. This research covers not only the use of nanoparticles as contrast agents but also the issue of getting nanoparticles to move through the reservoir, the development of nanosensors that change in response to environmental conditions and the development of nanoparticles that contain physical sensors, some electronics and data storage.[37]

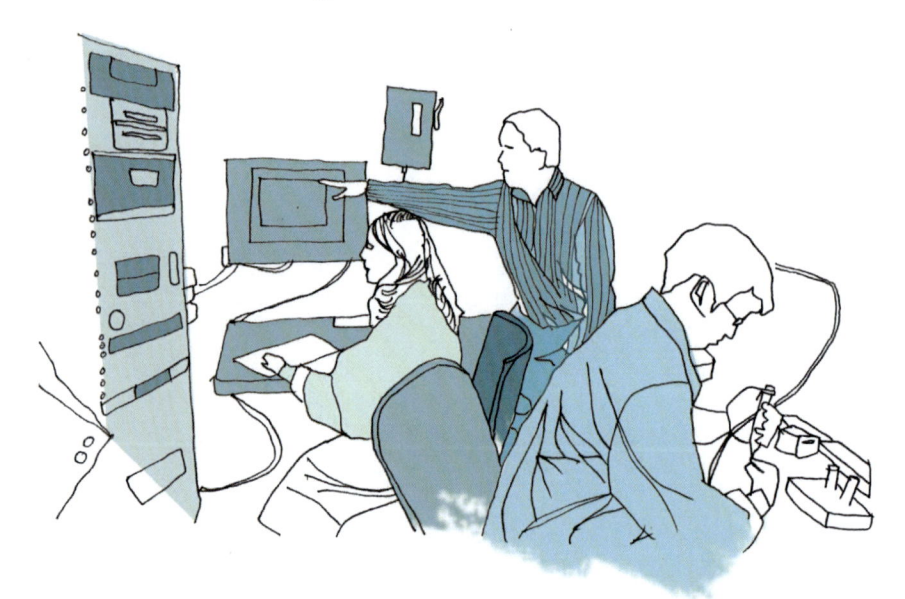

Scientists at the Advanced Energy Consortium (AEC) nanotechnology lab at the University of Texas in Austin, investigating magnetic nanoparticles for their application in illuminating waterfloods and fracture imaging.

Rice University in Houston is the main beneficiary and collaborator within the AEC. In April 2014, Rice scientists developed nanoscale detectors to detect hydrogen sulfide levels in crude oil—crude with an hydrogen sulfide concentration of one percent or more is highly toxic to workers and corrosive to wells, pipelines and transportation vessels. The Rice nanosensor is derived from earlier work at the University of Oregon where scientists developed nano-sized carbon black that was thermally stable yet underwent changes in its fluorescent properties in the presence of hydrogen sulfide. The plan is to inject the Rice nanosensors into the reservoir and then analyze them when they emerge up a production well. Analysis with a spectrometer will determine the hydrogen sulfide concentration. Laboratory testing in beds of sandstone and dolomite is helping the team perfect the size and formula for the nanosensors to ensure their survival during the trip through the reservoir. Field testing is some way off.[38]

The Advanced Energy Consortium is not the only upstream player in nanotechnology. Oceanit, a research and development company from Hawaii, is developing bespoke nanoparticles to add to the cement slurry during well cementing. Once the cement sets, the particles will allow the cement sheath to be interrogated electrically or acoustically to indicate cement pressure and well integrity. In this ongoing research program, Oceanit is partnering with the US Department of Energy and four major oil companies.[39]

Nanotechnology is exciting because it is dramatically different from established technologies, but commercialization remains elusive. The situation was nicely summarized by the Advance Energy Consortium on its own website: "We have developed an impressive base of understanding, but now the Advanced Energy Consortium needs to take the next step and integrate the research into solutions that meet the top business needs of our members."[40]

An endless story

The progress made since that first oil well in Azerbaijan in 1848 is truly astonishing. But it becomes doubly so if you look at some of the most amazing wells ever drilled.

Major Alexeev would probably appreciate that 167 years after he drilled the first oil well for Count Mikhail Semyonovitch Vorontsov in Baku, Azerbaijan, the deepest hole on Earth remains in Russia. As a Soviet relic of deep drilling research, the deepest hole ever drilled is the Kola borehole in the northwestern corner of Russia. The Soviet drillers reached a final depth of 12,262 meters (40,230 feet).[41]

By any standards, wells being drilled today are an extraordinary testament to the industry's continuous innovation since Major Alexeev's time, and records get broken every month. On January 28, 2011, a well drilled from an onshore location to tap the Exxon offshore Odoptu field in Sakhalin, Russia reached a length of 12,345 meters (40,502 feet), the true vertical depth at the end of the well being about 1,828 meters (6,000 feet).[42] This was a record for length then and may well be beaten today. A competition akin to building a skyscraper higher than your neighbor's excites operators and drilling contractors alike.

The Petrobras development of the giant presalt Tupi discovery, now renamed the Lula field for the former President of Brazil, broke many records—drilling in 2,000 meters (6,600 feet) of water and then navigating through 5,000 meters (16,000 feet) of the subsurface of which over 2,000 meters (6,000 feet) is exceedingly unstable salt. All this took place more than 250 kilometers off the Brazilian coast. There will be other records to beat. For developing their Stones field in the Gulf of Mexico, Shell will be drilling in 9,500 feet (2,900 meters) of water and tapping a reservoir 17,000 feet (5,200 meters) below the ocean floor, making the total depth from rig to reservoir of 26,500 feet (8,000 meters). Exxon in partnership with Statoil will face similar depths and length of drillstring in its nearby Julia field.[43]

These spectacular wells, drilled or planned, provide a useful yardstick for the health of the industry. Oil and gas continue to power our planet, and the exploration and production of these natural resources must move forward to new territories with new technology. This poses a challenge for the coming generation of geoscientists, petroleum engineers and drillers tasked with developing new concepts, techniques and industry practice. The question is where will the innovation come from? It is certain that all players will contribute: oil companies, service companies and academia. But it is the dynamism of these organizations, the knowledge transfer across the entire global upstream industry, the industry's market forces and, most of all, the inventive spark of the individual that will dictate the speed of success.

Epilogue

ANYONE JOINING the industry a generation ago would have ridiculed, let alone predicted, the notion of drilling horizontally, seeking reserves in water 10,000 feet deep or monitoring a reservoir with 3D seismics. It is the industry's good fortune to have been well served by innovators who believed so much in their particular vision of the future that they were willing to sacrifice almost anything to see their ideas get developed. We met many of them in this history.

In the early days, innovators such as Drake, DeGolyer, the Schlumberger brothers, Erle Halliburton and Howard Hughes Sr. were the garage entrepreneurs of their day. There was no government funding, no vast corporation backing them to the hilt. With limited capital, they flew by the seat of their pants. Their fledgling enterprises were fragile, but the owners inspired and tough. Schlumberger, for example, almost went broke in the 1930s and survived partly thanks to a series of contracts to provide services for the Soviets; both Conrad Schlumberger and his son-in-law Henri-Georges Doll made regular trips to the Chechnyan and Azerbaijan oil fields, no easy transit in those days, to ensure operations ran smoothly. Witness Erle Halliburton's tenacity in demolishing Schlumberger's well logging monopoly in the New Orleans court of appeal. He couldn't afford to be excluded from that lucrative business.

By the end of World War II, innovation started its long journey to becoming institutionalized. Many famous oil company labs took form: Shell's lab in Bellaire, Houston—where Gus Archie and King Hubbert reigned—Humble Oil's research lab also in Houston and the Amoco labs in Tulsa. This was a natural evolution because the large international oil companies (IOCs) dominated the industry and needed technology to evolve. The turn-of-the century innovator, previously free as a bird yet fighting for survival, now had to exist inside an organization with its budgets and business priorities.

The IOCs saw two benefits from investing in research and development. One was to do things that had never been done before. If the geologists said follow those turbidites and go into deeper water, they had to find a way. If tapping vertical fractures was the only way to produce the Rospo Mare field, Elf Aquitaine was obliged to accept

Jacques Bosio's suggestion of horizontal drilling. But technology was also harnessed to reduce the cost of what oil companies could already do and do it better. The total cost of finding and producing conventional oil in the Gulf of Mexico is estimated to have dropped from US$ 15 per barrel in 1986 to US$ 5 per barrel in 1996 thanks mainly to the introduction of 3D seismics and new drilling technology.

As we travel through the latter half of the 20th century, a shift occurred in the relationship between the IOCs and the companies providing oilfield services. Oil companies that were integrated vertically—meaning they actually owned and operated seismic crews and drilling rigs—increasingly outsourced these activities to the service industry. As a consequence, the service companies proliferated and grew, and so did their need to innovate technically. This required labs, more scientists and engineers and bigger budgets.

Then in 1986, the action sped up. The Saudis, frustrated by OPEC production quotas, opened their taps and flooded the market with oil. The price plummeted to US$ 10 per barrel, causing all players to cut their workforce by at least a third, from top to bottom, including research and development. Recruitment of new talent ground to a halt. Squeezed for resources, the IOCs reinforced their outsourcing strategy with a mantra that dictated "buy rather than build."

This provided fresh impetus to the service companies to invest further in research and development, of course with the proviso that they could extract commensurate reward from their oil company clients. By the late 1990s, the service industry was spending more on research and development than the major oil companies, an extraordinary turnaround that would have been unimaginable a decade before. If patents are any measure of technical innovation, the service industry is still racking up at least twice as many as the IOCs in the upstream arena.[1]

With a continuing low and volatile oil price during the 1990s, the large oil companies then went into the mixer, and four supermajors emerged joining Shell in that league. John Browne at BP led the charge acquiring Amoco and ARCO, Exxon absorbed Mobil, Chevron took over Texaco, Conoco took Phillips, and Total absorbed Elf and Petrofina. Every acquisition had to boast cost savings, and research and development provided an easy target. Amoco's world-famous labs, for example, were virtually wiped out.

At the same time, these big players were in the process of decentralizing decision making and budget control to their field assets, in whichever remote part of the globe they operated. This caused the technical expertise to become dispersed and made less capital available in any given asset to risk trying new technology. It soon reached the point where assets began to lack the necessary expertise to even assess, let alone purchase new technology coming out of the service sector. By 2000, it was time for a corrective, and two other factors came into play.

One was the IOCs' relationship with the emerging and increasingly influential national oil companies (NOCs).[2] The NOCs own 85% of the world's oil reserves and produce about 50% of global supply, the lower number reflecting their typically conservative production strategy. A key business opportunity for the IOCs was therefore partnering with NOCs, and one of their key selling points to the NOCs was technology. So it didn't look good if they were running down their research and development activity. The second factor was the IOC's pursuit of reserves in deep water and unconventional reservoirs, both requiring a serious investment in new technology. IOC research and development began to pick up.

In the meantime, the NOCs started engaging in the technical marketplace, to the extent that today, companies such as PetroChina, Petrobras and Saudi Aramco can boast research and development budgets larger than those of most supermajors. Meanwhile, another group of technology players emerged: hundreds of small high-tech startups financed by venture capital groups, some managed directly by oil companies such as Shell, Chevron and Statoil. There is something about human nature that innovation happens more naturally in a small start-up than in a large mature organization, so there is a healthy appetite among the heavyweights to snap up the most promising of these small minnows. The large service companies are particularly active.

Whoever is plying upstream technology, whether an oil company, the service industry or a startup, battles must be fought that are unique to the oil field. By instinct, the industry shies away from novelty, and for good reason. The natural hazards of operating remotely at the bottom of deep oceans, releasing pressure deep in the earth and producing highly combustible substances to the surface warrant a highly conservative and safety-conscious attitude. In addition, incremental technology improvements often require a re-jig of a bigger engineering picture. In spite of the money being thrown at new ideas, progress is slow.[3]

Fortunately, research and development budgets and return-on-investment calculations provide only part of the story. All the money in the world is worthless without the right idea. And ideas come from people. Luckily for the industry and the world, there has been no shortage of exceptional individuals who have been determined to find a better way, whatever it takes. Some today are household names, others perfected their art in the back room. For all of them, it is the authors' hope that this book satisfactorily honors their inventiveness and contributions.

Timelines

Exploration

1793 Belsazar de la Motte Hacquet attributes the origin of petroleum to marine animal matter

1795 James Hutton publishes his "Theory of the Earth: or an Investigation of the Laws Observable in the Composition, Dissolution, and Restoration of Land upon the Globe"

1798 Based on laboratory studies of bitumen extracted from seeps in Trinidad, Charles Hatchett suggests that these heavy oils consist of decomposed plant and animal matter

1815 William Smith makes the first geologic map of England; his time separation of different rock units and color code for different rock units are still used by the UK Geological Survey

1820-50 Numerous studies of strata and fossils throughout Europe gradually produce the sequence of geological ages and periods still in current usage

1830-33 Charles Lyell publishes his "Principles of Geology: Being an Attempt to Explain the Former Changes in the Earth's Surface, by References to Causes now in Operation" and pens the memorable line "The present is the key to the past"

1841 John Phillips, William Smith's nephew, locks down a geological time scale using fossil records in each era and standardizes the use of terms like Paleozoic and Mesozoic

1861 Ebenezer Baldwin Andrews reports that in Western Virginia the productive wells are closely associated with the axial trend of anticlines

1870 Swedish engineers Thalen and Tiberg invent the magnetometer to explore for iron ore deposits

1880s John Franklin Carll articulates how stratigraphy can trap oil

1886 Baron Roland von Eötvös announces his invention of a torsion balance to interpret subsurface geological structure

1894 David W. Brunton patents the so-called Brunton Compass, a device that can measure both dip and direction of the strata and that is introduced into the oil field in the early 20th century

1900-01 Captain Anthony Lucas strikes oil in the Spindletop salt dome

1901 Baron Eötvös takes his torsion balance on the frozen Lake Balaton in Hungary, mapping the irregular surface of the lake bottom; he also maps the subsurface extension of the Jura Mountains in France

1908	The first successful wildcat in the Middle East, following a number of dry holes, at Masjid-i-Sulaiman, Persia
1909	Malcolm J. Munn of the USGS surmises that oil might originate in shale and then be channeled over geologic time toward reservoir rock
1911	British geologist Arthur Holmes uses radioactivity to date rock samples
1912	Conrad Schlumberger conceives the idea of using surface electrical measurements for mapping subsurface rock formations
1913	Reginald Fessenden and his assistants succeed in detecting both refracted and reflected acoustic waves from the subsurface, in Framingham, Massachussetts
1914-18	During World War I, Ludger Mintrop uses a portable seismograph and highly sensitive carbon-grain microphone to locate Allied artillery firing positions
1915	Alfred Wegener proposes the idea of continental drift in his seminal work "The Origin of Continents and Oceans"
1919	August Udden, the director of The University of Texas Bureau of Economic Geology, draws attention to stratigraphic petroleum traps in a lecture at the annual AAPG convention in Dallas
1920	Union Oil geologists use aerial photography to define the Santa Fe Springs and Richfield prospects in California
1921	John Clarence Karcher of Geological Engineering conducts a reflection experiment in Oklahoma and obtains a clear reflection from the interface between two known strata, the Sylvan shale and the Viola limestone
	Karcher also performs migration, or imaging, of seismic data, using pencil and paper
1924	Everette Lee DeGolyer uses the torsion balance to survey a prospect at Nash in Brazoria county, Texas, and finds a salt dome
	A Seismos crew discovers the Orchard salt dome in Fort Bend County, Texas, using refraction seismic
1920s	The magnetometer is popular for mapping potential oil structures but declines in use from the early 1930s when other geophysical methods become available
	The mechanical seismograph is superseded by a variety of electrical geophones, the most popular being the moving-coil electrodynamic type
	Vacuum-tube amplifiers make it possible to amplify seismic recordings

Electronic filters eliminate extraneous seismic reflections

Several seismographs are recorded on the same strip of photographic paper

1926 Ludger Mintrop is granted a patent for seismic refraction called "Method for the determination of rock structures"

1929 Austrian-German engineer Fritz Pfleumer creates magnetic tape by coating a long strip of paper with ferric oxide powder

1930 The discovery of three reservoirs in the Seminole field, Oklahoma, using the seismic reflection technique, secures the technique as the most efficient and practical method for hydrocarbon exploration

1933 Experiments carried out by Vincent C. Illing of the Royal School of Mines in London corroborate the idea of petroleum migration

1934-8 Max Steineke, chief geologist for the California-Arabian Standard Oil Company (CASOC), the predecessor of Saudi Aramco, makes detailed geologic surveys of Saudi Arabia and the key discoveries at and around the Dammam Dome, near Dhahran

1930s Philip Kuenen, a geologist at Groningen University in the Netherlands, surmises that earthquakes could trigger undersea flows of muddy sediments, as seen in the 1929 Grand Banks earthquake

1936 Geologist Arville Irving Levorsen strongly promotes stratigraphy in his AAPG presidential address "Stratigraphic versus Structural Accumulation"

Frank Rieber of Continental Oil devises a sonograph capable of recording several geophone traces with various offsets simultaneously in variable density on a moving 35-mm photographic film, a technique borrowed from the movie industry

1938 A Shell crew performs a seismic marine trial four miles off the Louisiana coast in 65 feet of water

1940s Trimetrogon photography, developed during World War II, starts being used in reconnaissance aerial geology

The wide-angle stereoscopic camera, another breakthrough in aerial technology, is introduced for petroleum geology

1946 Kerr-McGee Oil Industries makes an offshore seismic survey out of sight of land

Air-borne magnetometers are available for use and get licensed to Aero Service Corporation and Fairchild Aero Surveys

1947 Roy Pasley, George Pavey and Pershing Wipff of Marine Instruments Company patents a marine seismic streamer, an oil-filled float containing hydrophones

Maurice Ewing of Columbia University finds sediment layers on the Atlantic sea floor to be far thinner and much younger than hitherto expected, supporting Wegener's continental drift theory

1950 API Project 51, a multidisciplinary study of sedimentation in the northern Gulf of Mexico, is an early example of the oil industry's growing interest for stratigraphically trapped oil that would result in establishing sedimentology as a distinct discipline in petroleum geology

Shell recruits its first sedimentologist with a chemical and biological training, James Lee Wilson

1952 Creation of the first organized industry-academia consortium, the Geophysical Analysis Group (GAG) at MIT

1953 Shell geologist Marion King Hubbert develops the hydrodynamic theory of oil in his AAPG paper "Entrapment of Petroleum under Hydrodynamic Conditions"

1954 Magnolia Petroleum Company develops a 13-channel magnetic recorder and playback system for their seismic crews

1955 Geophysical Services Incorporated (GSI) launches its strategy to exploit the new digital seismic acquisition and computer processing to map out complex stratigraphic petroleum traps

1950s Reconnaissance geology groups, the most famous being Exxon's "Rover Boys", help oil companies locate and describe basin areas around the world

Deconvolution, a processing algorithm that sharpens time-dependent data, is adapted for reflection seismology in the 1950s by both Enders Robinson of MIT and GSI's John Burg

GSI develops a system to record on magnetic disks

A proliferation of mechanical migration devices appears, including those invented by Colorado School of Mines student Albert Musgrave, Shell's Hagedoorn and GSI's Don Rockwell

1956 Harry Mayne is granted a patent for his common depth-point (CDP) data stacking method, a key signal-to-noise enhancing technique

Burton McCollum of the US National Bureau of Standards patents a so-called "thumper" truck for land seismic acquisition

1958 Shell adopts magnetic recording and develops the ability to plot cross sections from tape

Groundbreakers

1960 Carl Savit of Western Geophysical proposes the idea of direct detection of hydrocarbons in the subsurface, based on the principle that a rock filled with hydrocarbons can display markedly different acoustic properties from similar rocks filled with water

1961 Commercialization of Vibroseis, a seismic source technology for land seismic, patented seven years earlier by John Crawford and Bill Doty of Continental Oil

1962 Shell geologist Robert J. Dunham introduces a classification of carbonate rock in terms of depositional fabric and porosity

Sedimentologist Arnold Bouma, at Groningen University, describes the characteristic depositional sequence in underwater avalanches, the "Bouma" sequence

1962-70 Shell and Mobil Oil independently develop the so-called "bright spot technique" to detect hydrocarbons directly from the seismic record

1963 Frederick Vine and Drummond Matthews at the University of Cambridge study the magnetism patterns of the ocean floor and make a decisive contribution to the acceptance and modification of the continental drift theory, now known as plate tectonics

Lawrence Sloss, a professor at Northwestern University, recognizes six major stratigraphic sequences in North America controlled by eustatic sea-level changes

First transistorized digital field recording system, built by GSI

1964 Esso performs an experimental 3D seismic survey over their Friendswood Field on the outskirts of Houston

1965 Stephen "Bolt" Chelminski invents the airgun for marine seismic acquisition

1960s Jim Cole of Continental Oil invents the Condep Fin, a depth steering device for marine seismic streamers

1967 Chevron's John Sherwood performs seismic migration using an IBM accounting machine

1970-1 Jon Claerbout, a professor at Stanford University, publishes his imaging principle and a wave-equation approach for opening new possibilities in seismic migration

1971 Soviet geologist and geochemist Nikolai Lopatin develops a method of using both time and temperature to calculate the thermal maturity of organic matter in sediments, a method widely used for petroleum and basin modeling in the Soviet Union, as well as in the West

1972	The US space agency launches the first of several Landsat satellites capable of repeatedly photographing practically every spot on the globe
	Emiliano Mutti coins the term "turbidite" for describing the subsea avalanches previously noted by Kuenen and Bouma
	Amoco geologist Wallace Dow develops the concept of an oil system, later known as a petroleum system, that encompasses all the components necessary to create hydrocarbons—a migration pathway and a reservoir rock with its seal
1973	The Stanford Exploration Project (SEP), an industry-academia consortium, is initiated
	Manuel Forero, a GSI geophysicist and interpreter, migrates 3D seismic data
1974	William Dickinson, professor of geology at Stanford University, suggests five major basin types: oceanic basins, rifted continental margins, arc-trench systems, suture belts and intercontinental basins
	Vladimir Prorfir'ev, a Soviet geology professor, publishes in the AAPG Bulletin the concept of an inorganic origin of petroleum, a concept heavily debated in the Soviet Union in the 1950s and 1960s, symbolizing alternative pathways pursued by Soviet geologists
1975	First marine 3D survey, conducted by Sun Oil Company and GSI in the North Sea
1970s	At his company Teknica Resource Development in Calgary, Roy Lindseth performs 1D post-stack seismic inversion, trace by trace
1977	Pete Vail of Exxon links seismic section studies to Sloss' sequence stratigraphy ideas
1978	Dan McKenzie, at the University of Cambridge, presents the basic principles of sedimentary basin modeling
1979	Discovery of the sub-salt Tengiz field in Kazakhstan, using the gravity method
	GSI develops the horizon slice, a key tool for stratigraphic interpretation
	Charles Cox of the Scripps Institution of Oceanography in San Diego, California, carries out an experimental electromagnetic survey on a mid-ocean ridge in the Pacific, a technique that came to be known as the Controlled Source Electromagnetic Method (CSEM)
1980	Workstations begin to impact the work of geoscientists

1981	Zaitian Ma, working for the Chinese Bureau of Geophysical Prospecting, develops a migration technique to image highly dipping structures
1983-4	Albert Tarantola at the University of Paris and Patrick Lailly at the Institut Français du Pétrole (IFP) establish a theory of full seismic waveform inversion
1984	Chevron's Bill Ostrander publishes his AVO technique of direct hydrocarbon detection
	Geco-Prakla deploys multiple streamer in a marine survey for Conoco
1980s	Transit satellite positioning, an early satellite navigational system developed by the US Navy, gradually replaces conventional radio-positioning systems
	Mac Jervey and Henry Posamentier of Exxon refine Vail's stratigraphy to analyze the integrated effects of eustasy and tectonics
	Basin studies and petroleum system modeling become key for every major oil company
1986	Leon Thomsen of Amoco publishes his paper on anisotropy in *Geophysics*, which later becomes the most-frequently cited paper in the journal's history
1987-8	University of Cambridge's Martin Sinha carries out trials of a device for marine electromagnetic surveying, a neutrally buoyant antenna that can be deep-towed or "flown" about 100 meters above the seafloor
1990s	Global Positioning Systems (GPS) are introduced to offshore seismic operations
	Seismic vessels become capable of towing 12 streamers, each extending 10 kilometers with thousands of hydrophones attached
2000	WesternGeco launches Q-Marine, an advanced marine seismic system that includes single-sensor recording for noise reduction and improved imaging, and improved steering of the streamers with controllable wings and fins
2000-2	Exxon and Statoil perform controlled-source electromagnetic (CSEM) trials offshore Angola showing that CSEM data can measure reservoir resistivity for targets several thousand meters deep
2001	Petroleum Geo-Services (PGS) attempts a 16-streamer tow that proves impractical and is discontinued
	Three former BP geologists, Dave Casey, Roger Davies and Peter Sharland, found geoscience consultancy Neftex and begin developing

	models of the world's major sedimentary basins, using publicly available data and leveraging Vail's sequence stratigraphy approach
2005	BP and PGS perform a marine seismic wide-azimuth survey in six directions to image below the complex
	Messinian geology in the Egyptian Nile Delta
2006	More wide-azimuth possibilities come to the market, for example separate vessels sailing some distance apart, one towing the source arrays and the other the streamers
2008	WesternGeco performs a coil shooting job for Eni, offshore Indonesia
2010	WesternGeco performs a dual-coil survey in the Gulf of Mexico
2012	Trials of simultaneous source acquisition in marine seismics are conducted by WesternGeco for Apache oil company in Western Australia
	WesternGeco introduces Isometrix, a marine technology that measures the 3D seismic wavefield, using three-axis accelerometers in the streamer

Drilling

1848	Major Alexeev drills the world's first purposeful oil well in Azerbaijan, using the cable-tool drilling technique originating from ancient times in China
1859	Colonel Edwin Drake's well strikes oil in Titusville, Pennsylvania
	Drake uses wooden casing to shore up an oil well
1882	Rotary drilling is pioneered by the brothers M.C. and C.E. Baker, to drill water wells in the Dakota Territory of the Great Plains
	Christian Otto Mohr invents a graphical technique, called the Coulomb-Mohr circle, to predict rock failure
1892	The first submerged oil wells are drilled from platforms built on piles in the fresh waters of Grand Lake St. Marys in Ohio
1897	Henry Lafayette Williams drills oil wells offshore to develop the Summerland field of California, using a wooden drilling and production platform
1901	Spindletop becomes the first gusher to be drilled using the rotary technique; fishtail bits and water-base drilling mud are also introduced at Spindletop
1903	Frank Hill of Union Oil uses Portland cement to shut off water in an oil well at Lompoc, California
1907	Reuben C. Baker develops a casing shoe for cable-tool drilling, then for rotary drilling
	Drilling rigs are constructed, using a steel framework
1907–08	Shell and Standard Oil begin using rotary drilling
1908	Lee C. Moore introduces the standard derrick
1909	Hughes and Sharp patent the roller-cone drill bit featuring two steel cones
1911	Gulf Refining Company starts drilling operations in inland waters on Lake Caddo, Louisiana, using wooden platforms
1913	Myron M. Kinley kills a well fire with explosives at Taft, California
1915	Rotary table with a square kelly comes into use
	An all-steel sprocket chain is introduced for the rig's drawworks
	Rotary rigs begin replacing cable-tool rigs
1918	Victor York and Walter G. Black of Standard Oil Company of California receive a patent for driving the rotary table with a shaft

1919	Erle P. Halliburton establishes the New Method Oil Well Cementing Company
1920-24	National Lead introduces barite as an effective weighting agent for drilling fluids
1922	C.C. Scharpenburg of Standard Oil patents a multistage axial-flow turbodrill
1922-25	Matvei Alkuvich Kapelyushnikov, a Soviet engineer, builds and tests a prototype single-stage turbodrill
1923	Karl von Terzaghi develops a theory to account for stress and pressure in porous rocks
1924	The New Method Oil Well Cementing Company is renamed Halliburton Oil Well Cementing Company and is granted a patent for the Jet Mixer
	Harry S. Cameron and James Abercrombie build a ram-type BOP
1924-28	Lago Petroleum begins drilling wells in Lake Maracaibo, Venezuela, using wood and concrete piles to support the rig
1928	Louis Giliasso patents a submersible barge for rapidly moving a drilling unit between locations
1929	H. John Eastman patents a multi-shot instrument and whipstock, and introduces controlled directional drilling
	Hughes Tool Company applies tungsten carbide to the cutting elements of bits
	The Link-Belt Company introduces a shale shaker in the Kettleman Hills oil field of California
1932	Hu Harris of Humble Oil constructs a unitized drawworks, saving days between rig moves
1933	Hughes introduces the tricone roller-cone drill bit
1934	H. John Eastman drills a relief well to kill a blowout in Conroe, Texas
	The Texas Company, later known as Texaco, completes the construction of a submersible steel drilling barge and uses it for drilling on Lake Pelto, Louisiana
1934-35	Soviet engineers Rolen Ioannesyan and Mikhail Gusman construct a multistage turbodrill, featuring special seals that prevent mud from entering the bearings
1937	Experimental horizontal drilling takes place in Yarega, Soviet Union

	Lee C. Moore develops a jackknife cantilever-type drilling mast that could be erected as a single piece
	Pure Oil and its partner Superior Oil use a fixed platform to develop a field one mile offshore of Calcasieu Parish, Louisiana, in 14 feet of water
1938	Halliburton performs an offshore cementing job, Gulf of Mexico
1940	Halliburton introduces a new method called bulk cementing with the opening of a plant in Salem, Illinois
1941	Alexander Grigoryan, a young Soviet driller, directs horizontal well drilling in Azerbaijan
1942	Oil-base mud is commercialized by Oil Base Drilling Company, founded by George Miller
1943	John Eastman and John Zublin develop short-radius horizontal drilling tools
1945	Brown Oil Tool Works introduces the hydraulic power swivel
1946	Continental Oil Company, Union Oil Company of California, Superior Oil Company and Shell Oil Company form the CUSS consortium to develop drilling technology for deep waters
	Magnolia Petroleum drill at a site 18 miles off the coast of St. Mary Parish, Louisiana, erecting a platform in 18 feet of water
1947	Kerr-McGee Oil Industries make the world's first oil discovery drilled out of sight of land
	Halliburton markets the Super-Cementer truck
1948	Hughes Tool Company introduces jet bits
1949	Byron Jackson of Varco designs the 3-arm derrick-mounted pipe-racking system
	The first offshore mobile drilling platform, Hayward-Barnsdall's *Breton Rig 20,* works in up to 20 feet of water depth
1951	Hughes Tool Company introduce tungsten carbide insert bits
1952	The Hydril Corporation launches an annular-type BOP, based on a 1946 patent of Hydril engineer Granville Sloan Knox
1954	Colonel Leon B. Delong of Delong Corporation builds the world's first jackup drilling rig
1953-59	Pozzolan becomes a key cementing additive, enabling the drilling of deep and hot wells

1955	The first multilateral well is drilled in the Bashkiria field in the Soviet Union
1956	The CUSS Group develops the drill ship, *CUSS I*, as a technological test bed for the nascent offshore oil industry
1959	Hughes Tool Company introduces the self-lubricating sealed-bearing bit
	Dyna Drill Technologies patents a positive displacement motor
1961	Blue Water Drilling Company and Shell convert an existing submersible rig, the *Blue Water Rig No. 1*, into a semisubmersible drilling unit for operation in the Gulf of Mexico
1962	Red Adair kills the famous gas well fire in the Gassi Touil field in the Algerian desert, nicknamed "The Devil's Cigarette Lighter"
	Bill Rehm of Magcobar, a division of Dresser Industries, successfully introduces a well choke to restrict production rate
1963	Alden J. LaBorde builds a purpose-built semisubmersible drilling rig, the *Ocean Driller*
1967	George Savage and Robert Bradbury of The Offshore Company patent a "Marine Drilling Apparatus," describing the marine riser
1969	Hughes Tool Company introduces the steel-tooth bit with bush bearings, also called journal bearings
	Ben Eaton of Continental Oil publishes a method for predicting both formation pressure and fracture pressure
	Sid Green founds Terratek Inc. in Salt Lake City to provide rock mechanics services
1969-71	Southeast Drilling Company (SEDCO) and Cameron succeed in placing the blowout preventer on the seafloor
1970	Hughes Tool Company introduces the bush-bearing and tungsten-carbide bits
	Martin Chenevert of Humble Oil Production Research presents his paper "Shale Control with Balanced-Activity Oil-Continuous Muds," describing ways to control wellbore stability
1970-75	Polymers and oil-base muds start making an ever growing impact on the drilling fluids industry
1972	Christensen Diamond Tools make a prototype of the PDC drill bit
	Teleco Oilfield Services begin development of mud-pulse telemetry to convey data uphole about wellbore trajectory
	SEDCO 445 is the first dynamically positioned drill ship

1973	Leon Robinson of Exxon Production Research patents a mud-cleaning system that includes a hydrocyclone, working on the principle of a centrifuge
1975	Varco introduces the Iron Roughneck, nicknamed Big Foot
1977	*SEDCO 709* is the first dynamically positioned semisubmersible
	Newpark introduces a polymer PHPA drilling-fluid system
1977-79	The Soviet Union, as well as Halliburton, develop methods to tackle gas migration after cementing operations
1978	Teleco Oilfield Services introduce the first measurement while drilling (MWD) tool
1980-85	Weld-on connectors improve and quicken the process of running surface casing into the well
1980	Chevron begins manufacturing a logging cable featuring a central optical fiber
1982	Varco commercializes Top Drive
	Elf Aquitaine drills a horizontal well at Rospe Mare, in the Adriatic Sea, Italy
1984	Norten Christensen introduces a new steerable drilling motor
1986-89	Underbalanced drilling is used on a wide scale in the Austin Chalk formation in Texas
1989	On the Kola peninsula, Soviet drillers reach a depth of 40,230 feet (12,262 meters), still the deepest hole ever drilled
1990	Baroid launches a synthetic-based drilling fluid for the North Sea
1990-92	Baker Hughes introduces the anti-whirl PDC bit
1990s	Sandia National Laboratories construct a spooling system for running disposable optical fiber through drillpipe
1997	Both Baker Hughes and Camco introduce rotary steerable systems
1999	Varco replaces a joystick-based pipe-handling and racking system with a computer-controlled system
	Shell and Halliburton set an expandable-tubular system in a well in the Gulf of Mexico
2002	Saudi Aramco drills the first maximum reservoir contact (MRC) well
2003	Schlumberger introduces its rotary steerable system
	Varco presents a complete computer-based roughneck system

	Weatherford introduces managed pressure drilling
2005	Helmerich & Payne launches a semi-automated land drilling rig
2006	National Oilwell Varco commercializes a wired drillpipe system
	Petrobras discover the giant pre-salt Tupi field (now called Lula) drilling in 6,600 feet of water and through 16,000 feet of the subsurface, including 6,000 feet of salt
2011	Exxon Neftegas drills the world's longest extended reach (ERD) well in Sakhalin, Russia, with a length of 12,345 meters

Reservoir

1863   Rodolphe Leschot invents a coring tool for drilling blast holes, for tunneling in the Alps

1870s Leschot's simple device is replaced by a double-barreled tool, with inner and outer barrels separated by ball bearings

1880   John Franklin Carll takes cores from the Pennsylvania Venango Sands, the formation in which Colonel Drake had struck oil in 1859, and visually estimated their porosity

Carll raises the possibility that oil recovery might be increased by injecting water into the reservoir to push oil toward producing wells

1885   Frederick Haynes Newell presents a thesis at the Massachusetts Institute of Technology, describing how he forced water, kerosene and crude oil through small disks cut from cores, in an effort to understand the fluid-flow properties of the rock

1888   James Dinsmoor discovers the principle of gas flooding, in Pennsylvania

1892   Milan Bullock from Chicago patents the double-barreled coring tool

1890s Józef Grzybowski, of the University of Krakow, investigates mud returns from oil wells in the Carpathian mountains and reports that one group of marine microfossils, called foraminifera, was especially suited to providing a record of sediment age

1900s Early application of waterflooding in the Bradford field in Pennsylvania

1912   First temperature measurement in an oil well by John Johnston and Leason Heberling Adams, in Findlay, Ohio

1919   C.E. van Orstrand of the USGS begins experiments in Oklahoma on the effects of fluid entry on the geothermal gradient, and of subsurface structure on temperature

Jan Koster of the Holland Geological Survey invents a double-barreled core drill for loosely consolidated formations

1920   Robert van A. Mills at the US Bureau of Mines builds a reservoir model, a physical mock-up featuring layers of sand, saturated with oil on top and water below, and a small hole drilled through the oil sand to simulate a well

1921   Mining engineer John "Brick" Elliott improves the double-barreled coring tool, featuring reaming teeth to prevent the tool balling up in sticky formations

	Marcel Schlumberger makes resistivity measurements over a few feet at the bottom of a 760-meter hole in Molières-sur-Cèze in southern France, proving the feasibility of a downhole resistivity measurement
1923	Edson Wolcott of Los Angeles and Frank Howard of Standard Development Company are granted a patent for in-situ combustion, also known as fire-flooding
1925	Perley Gilman Nutting, a geophysicist at the USGS, introduces viscosity, the resistance of a fluid to flow, into Darcy's permeability equation
	C.E. Beecher and Ivan Parkhurst of the Cities Service Oil Company develop one of the earliest downhole pressure measurement techniques
1920s	Mud logging evolves into a structured industry activity
	Arles Melcher at the USGS and Charles Fettke at the Pennsylvania Geological Survey develop classic laboratory methods for determining reservoir properties from cores
	First commercial core laboratories are established
	Earliest attempts to do sidewall coring, using a boring device or knife arrangement, lowered on the drill string, that is pushed obliquely into the borehole wall to cut a sample out of the formation
1926	Mordica Johnston and his brother Edgar invent the drillstem tester
1927	Schlumberger conducts the first electrical logging operation in an oil well, near Pechelbronn, Alsace, enabling correlations of resistivity from one well to the next, revolutionizing the stratigraphic understanding of an oilfield
	Gulf Oil makes a bottomhole temperature survey
	Amerada Petroleum Corporation conducts a velocity survey by setting off explosive at the surface and recording the acoustic arrival time at various depths in the well
1929	Charles Millikan, Amerada's chief petroleum engineer, develops a downhole pressure gauge
1930	Schlumberger uses electrical logging to locate oil, in Indonesia
	Schlumberger starts measuring the spontaneous potential (SP), a natural occurring potential caused by electrochemical interactions between the borehole and adjacent sand and shale formations
1931	Amerada initiates subsurface measurements of both temperature and pressure, recording both on a rotating drum.

Groundbreakers

1932 Schlumberger develops the so-called teleinclinometer for measuring borehole deviation and direction

Schlumberger develops a logging tool to record temperature continuously versus depth

1933 Ralph Wyckoff, Morris Muskat, Holbrook Botset and Donald Reed of Gulf Research and Development Company describe a technique for measuring the permeability of porous media; they name the permeability unit a "darcy"

James Lewis, Kenneth Barne and George Fancher from the Pennsylvania State College publish the now classic article "Some Physical Characteristics of Oil Sands," providing extensive references to the various core analysis methods and techniques

1933-35 Soviet engineers conduct a successful in-situ combustion operation in a pressure-depleted reservoir in Azerbaijan

1935 Vladimir I. Kogan, of the Petroleum Institute of Azerbaijan, carries out experiments linking electrical resistivity to oil saturation in porous rock

1930s Sperry Sun Well Surveying Company builds a sidewall-coring tool lowered on a cable, a miniature drill that could be projected laterally, penetrate a few inches into the formation and take a sample

Schlumberger develops an improved sidewall-coring tool based on the idea of using an explosive charge to shoot a cup into the formation

Schlumberger lays the foundations for interpreting geothermal gradients

The technique of waterflooding to improve reservoir recovery is adopted throughout several oil provinces of the US, including Oklahoma, Kansas and Texas

1937 First mathematical treatment of reservoir flow by Morris Muskat of Gulf Research and Development Company

1938 William Green and Serge Scherbatskoy of Engineering Laboratories Inc. build a gamma-ray log and run it in a cased oil well in the Oklahoma City oil field

1940 Well Surveys perform a gamma-ray survey for Stanolind Oil and Gas Company at Spindletop, Texas

Shell engineer Folkert Brons patents a logging device that links neutron transport to rock properties

1941 Bruno Pontecorvo, Lane Wells, builds a neutron logging tool

Schlumberger commercializes a dipmeter with SP electrodes on each arm of a three-arm caliper and runs it in Louisiana

1942 Shell's Gus Archie publishes his now famous article "The Electrical Resistivity Log as an Aid in Determining Some Reservoir Characteristics," establishing the link between resistivity and porosity for clean sands

1944 Gerald Hassler of Shell invents a practical laboratory tool to measure permeability

1940s Miles Leverett of Humble Oil, Houston, shows how to estimate a rock's capillary pressure from its pore size distribution

1948 Schlumberger introduces the induction log that detects oil-bearing layers only a few feet thick

1949 Schlumberger initiates a nuclear logging program at its newly founded laboratory in Ridgefield, Connecticut

1950 In the *AAPG Bulletin*, Gus Archie coins the term "petrophysics" for a new discipline describing how the physical properties of rock relate to its solid components, its pore space and the pore fluid

Schlumberger builds a logging device, later called the "Microlog," that measures the resistivity of a small volume of rock just behind the borehole wall

Charles Miller, A.B. Dyes and C.A. Hutchinson of The Atlantic Refining Company propose a method for estimating permeability from noting how quickly bottomhole pressure increases after a producing well is shut in, later known as a "buildup test"

1954 Magnolia Oil successfully tests an experimental sonic tool

At Humble Oil in Houston, Donald Peaceman and Henry Rachford build a numerical reservoir simulator

1955 Maurice Lebourg of Schlumberger introduces a well testing device that is lowered on wireline rather than by drillpipe

1950s The Atlantic Refining Company leads the way in making miscible flooding a viable enhanced oil recovery (EOR) technique

1956 Schlumberger introduces a combined induction and electrical logging tool

The Wyllie time-average equation, established at Gulf Research Development Corporation, opens the door to using the new sonic log as a porosity measurement

1957 Schlumberger develops the density tool, a logging tool based on a chemical gamma-ray source

Schlumberger markets a new, improved sonic tool with two receivers

	Shell accidentally discovers cyclic steam stimulation in Venezuela
1958	Tulsa-based oilfield service company Birdwell builds a downhole camera
1959	Grigory Barenblatt, Yuri Zheltov and Irina Kochina from the Institute of Petroleum at the USSR Academy of Sciences in Moscow introduce the dual-porosity concept, helping model complex pore structures in carbonate rock
1961	Schlumberger adapts the sonic tool to create a cement bond log, superseding earlier attempts to find poor bond using temperature logging
1962	Schlumberger presents the formation interval tester for cased holes, operating with a couple of perforating bullets that could provide a conduit to the formation through casing and cement
1963	Joseph Warren and Paul Root of Gulf Research Development Company introduce the double-porosity model to the oil industry
1964	Shell develops the borehole compensated sonic tool and licenses it to Schlumberger
	Arthur Youmans of Lane-Wells builds a pulsed-neutron tool enabling the measurement of water saturation behind casing
1960s	Polymer flooding starts being used in the West, as well as in the Soviet Union
	Evsei Iosifovich Galperin, at the Institute of Physics of the Earth at the USSR Academy of Sciences in Moscow, creates the discipline of borehole geophysics, introducing the new method of vertical seismic profiling (VSP)
1967	Monroe Waxman and Lambert Smits of Shell extend Archie's law to shaly sands
	Joseph Zemanek at Mobil Research & Development Corporation constructs a borehole televiewer, the first borehole imaging tool to function in opaque liquids
1968	Esso Production Research develops a tracer technique to measure residual oil saturation in a single well
1970-73	Scanning electron microscopy technology begins being used in core laboratories
	Permanent gauges in wells provide real-time pressure data
1971	Schlumberger introduces the density-neutron crossplot for interpreting lithology from logs
	Shell geologist Koen Weber develops the mini-permeameter

1972	Chevron uses carbon dioxide in miscible flooding operations
1975	Gearhart-Owen Industries introduce a fully digital logging system, the direct digital logging (DDL) system
	Schlumberger develops the repeat formation tester (RFT) that can measure formation pressure and collect samples at many different locations during one trip in the well
1970s	Reservoir simulation is extended beyond conventional pressure maintenance and depletion to enhanced oil recovery projects
	Shell uses Stoneley waves that propagate along the borehole wall to estimate permeability
1975-80	Alkiviades Payatakes of the University of Houston develops pore network models
1977	Schlumberger's George Coates, Christian Clavier and Jean Dumanoir present their dual-water model as a user-friendly way to evaluate shaly sands
1977-78	Dresser Atlas Industries perform experimental crosswell measurements in New Mexico for the Los Alamos National Laboratories, using acoustic logging tools
1977-83	Quartz-crystal pressure gauges start being used in well testing operations
1978	Schlumberger creates a rock physics department at its research lab in Ridgefield, Connecticut
	Permanent gauges are installed in subsea wells for the first time
	Canadian Roger Butler, of Imperial Oil, pioneers the concept of steam-assisted gravity drainage (SAGD)
1982	Amos Nur, of Stanford University, discovers that seismic propagation in oil saturated rock is significantly affected by temperature, a first step towards 4D seismic time-lapse monitoring
1983	Sperry Sun Drilling Services commercializes an LWD tool developed by Rich Meador, comprising an electromagnetic wave resistivity measurement
	Dominique Bourdet at Schlumberger's Flopetrol introduces the use of derivatives in well testing analysis
1983	Exploration Consultants Limited (ECL) launches its new reservoir simulator, ECLIPSE
1983-89	Pierre Adler, at the University of Pierre and Marie Curie, and Christian Jacquin, at IFP, create a flow model for the Fontainebleau sandstone pore space

1980s	Vertical seismic profiling becomes standard practice in the West
1985	Amoco and Western Geophysical perform the first 4D seismic time-lapse monitoring operation in a steamflood project in Canada
1986	Schlumberger introduces the Formation MicroScanner for providing an electrical image of the borehole wall
1987	*Ekofisk* platform in the North Sea is raised six meters to counter ocean floor subsidence due to reservoir compaction
1989	Schlumberger introduces the modular dynamics formation tester (MDT)
1990-95	X-ray tomography, using similar principles as in medical CAT scanning, is introduced for mapping the pore space of rock samples
1991	Numar performs a nuclear magnetic resonance (NMR) logging job for Mobil in Oklahoma
1992	UK-based service company Sondex introduces a multi-finger caliper with 40 fingers for inspecting casing
1993	Nick Koutsabeloulis founds Vector International Processing System (VIPS), provides the industry with a 3D reservoir geomechanics modeling capability
1994	A Statoil team led by Eivind Berg develops the subsea seismic (SUMIC) technology for ocean-bottom seismics
1995	ElectroMagnetics Instruments Inc. (EMI) commercialize an electromagnetic crosswell logging system
1990s	Specialized companies such as Cidra Corporation from Wallingford, Connecticut, and Southampton UK-based Sensa develop fiber-optic temperature sensing for real-time monitoring
1995-99	Statoil and Norsk Hydro demonstrate the feasibility of 4D seismic time-lapse for the Gullfaks and Troll West fields in the North Sea respectively
1996	Steam-assisted gravity drive (SAGD) goes commercial, with the Alberta Energy Company Ltd using the technique in their Foster Creek field, near Fort McMurray
2001	Researchers at Imperial College find an effective algorithm for the deconvolution of pressure and flow rate data from well tests, allowing better reservoir definition and evaluation of reserves
2000s	Digital core imaging and flow computations for porous media are introduced by companies such as Norwegian Numerical Rocks from Trondheim, Norway, and Digitalcore, Canberra, Australia

	Atomic force microscopy and the focused ion-beam scanning electron microscopy start being used in the industry, allowing investigation at the nanoscale
2007	Creation of the Advanced Energy Consortium (AEC) for facilitating research in micro- and nanotechnology materials and sensors for improving oil recovery
2012	Norwegian company Resman based in Trondheim pioneers use of chemical tracers for monitoring production in subsea completions

Production

1848	Oil production from the world's first oil wells in Azerbaijan
1859	Edwin Drake starts oil production in the US in Pennsylvania, using an improvised iron water pump with extra pipe sections to reach the bottom of the well and rigging the pump handle to a steam-driven oscillating arm
1864	Thomas Gunning of New York patents an air-lift mechanism for an oil well, called a "blower"
1865	John Ross Cross from Chicago invents the seed bag for sealing off productive zones in an oil well
	First attempts at separating gas from oil and water, using a single-stage separator or "gas trap"
1880	Solomon Dresser of Michigan is awarded a patent for a rubber packer actuated by tubing
1892	Julius Pohlé of New Jersey patents an air-lift method of introducing air in controlled stages
1895	Hermann Frasch, at Solar Oil in Ohio, treats an oil well with hydrochloric acid, providing a more controlled stimulation than detonating nitroglycerine
1899	British geologist and engineer Arthur Beeby-Thompson, working for the European Petroleum Company, introduces air lift to the Baku oil fields, improving productivity significantly
1907-08	Andrew Lockett of New Orleans and Joseph McEvoy of Houston are granted patents for gas-lift valves that guarantee sufficient pressure to kick-start production
1910	John Swan of Marietta, Ohio, patents the first perforating method, a cutting tool with a rolling knife called a "ripper" or "splitter"
1911	Union Oil of California begins using natural gas produced from their wells rather than air to lift production
1914	William Barnickel, of St Louis, Missouri, patents a new separating method harnessing chemical demulsifiers
1920s	Gas lift technology spreads around the world
	Completion wellheads, called Christmas trees, become widely used
1926	The Atlantic Refining Company builds a manufacturing plant developing and building gas-lift valves

	Los Angeles oilman Sidney Mims comes up with the idea of shooting through the casing with hardened steel bullets
1928	Colonel Delamare Maze carries out experiments with a gun perforator for the Astra Română and Steaua Română companies in Romania
	Successful test of an electric submersible pump (ESP), built by Armais Arutunoff, who founded Reda
1932	Lane-Wells develop practical oil well perforating guns
	Jordan & Taylor Inc., of Los Angeles, conceives the idea of installing many gas-lift valves in the same tubing string, at intervals of a few hundred feet
	Pure Oil achieves promising results using acid stimulation in a well in Michigan
	Layne-Atlantic successfully applies the gravel-packing method, patented eight years earlier by Lester Uren from the University of California at Berkeley, in oil wells in the US Gulf Coast
1935	Jeddy Nixon of the Wilson Supply Company in Houston develops wireline-operated gas lift
1930s	Oil-gas separators become standard with several stages, each stage removing more gas and decreasing the pressure
1936	Jay Walker of Tulsa, Oklahoma, constructs a separator for emulsified oil
1947	Well Explosives Company completes development of a commercial shaped-charge perforator
	First successful test of hydraulic fracturing in Kansas, performed by Floyd Farris, a Stanolind engineer
1948	Stanolind invents a logging tool to measure flow in gas wells
1949	Halliburton conducts a commercial fracturing operation in an Oklahoma oil well
	McDermott commissions a crane barge built specifically for offshore use
1952	Humble Oil and Schlumberger pioneer through-tubing perforating
1953	Slick water of low viscosity used for hydraulic fracturing
1954	W.E. Gilbert, of the Shell subsidiary Asiatic Petroleum Corporation, proposes a systematic analysis of well production
1955	Shell's Marion King Hubbert and David Willis prove that in relatively deep wells hydraulic fractures will form vertically, not horizontally
	Humble Oil constructs the first downhole flowmeter for oil

1957	Schlumberger wireline engineer Simon Noik starts building a production logging tool locally, in Brunei, North Borneo
1959	Humble Oil's Jim Rike develops the key concept of "managed fatigue" to prove the feasibility and safety of repeatedly coiling and uncoiling steel pipe, a requirement for coiled-tubing operations
1960	The California Oil Company, now Chevron, builds a submersible production platform for the Gulf of Mexico
1961	ODECO, now Diamond Offshore Drilling, constructs a purpose-built submersible production platform
	Shell completes the first subsea well in open water, in the Gulf of Mexico
	Thomas Perkins and Loyd Kern of The Atlantic Refining Company improve the understanding and measurement of hydraulic fracture width
	Shell's Mike Prats finds a formula to find the optimum length-to-width ratio for optimizing hydraulic fractures
	Halliburton uses guar, a water-based fracturing fluid, on a job in West Texas for The Atlantic Refining Company
1962	Hughes Aircraft Company delivers the remotely operated vehicle (ROV) to Shell
	Brown Oil Tools, together with The California Oil Company, construct a functional coiled tubing unit
1963	The Venezuelan Ministry of Mines and Hydrocarbons and the Creole Petroleum Corporation combine hydraulic fracturing and sand control technique
1965	Gulf Oil Corporation develops multistage fraccing to stimulate vertical wells
1967	Taylor Diving & Salvage performs a saturation-diving operation to install risers for Shell's Marlin subsea wellheads, at 320-foot water depth in the Gulf of Mexico.
1970	Roy Vann performs a successful tubing-conveyed perforating (TCP) operation, for an independent operator in southeast New Mexico
1970-75	Ed Horton develops the concepts of the tension-leg platform (TLP) and the Spar platform
1971	The *Gulftide* jackup is installed in the Ekofisk Field in Norway to accommodate an early production system
	IFP invents flexible pipe for tying subsea completions to production

1973	Norwegian Contractors build the first concrete structure in the North Sea, a huge storage tank for Phillips Petroleum's Ekofisk field in Norwegian waters
	French subsea contractor Coflexip install flexible pipe for subsea production, in Elf's Emeraude Field, offshore the Republic of Congo
1974	Exxon installs a diverless, totally integrated submerged production system (SPS), featuring three wellheads and associated production equipment mounted on a subsea template structure
1975	Norwegian Contractors constructs a "Condeep" platform, a concrete deepwater structure, the *Mobil Beryl A*
	Hamilton Brothers Oil and Gas Ltd installs a semisubmersible production system, a converted first-generation drilling unit, in the Argyll field in the North Sea
1975–80	Progressive cavity (PC) pumps, also called Moineau pumps, begin to be used for artificial lift
1976	Shell installs a Spar platform in the Brent field in the North Sea
1977	Shell and Single Buoy Moorings Inc. convert a secondhand oil tanker into a floating production, storage and offloading system (FPSO) vessel and deploy it in the Mediterranean, offshore Castellon, Spain
	Exxon deploys sintered bauxite as propping material in deep, hot wells
1979	Heerema deploys semisubmersible crane vessels (SSCVs) with a dynamic ballast system, in the North Sea
	Schlumberger's Joe Mach and Eduardo Proano and University of Tulsa professor Kermit Brown introduce nodal analysis for analyzing well production
1983	Exxon installs a compliant tower in 1,000 feet of water in the Gulf of Mexico
1984	Conoco installs a TLP platform for the Hutton Field in the North Sea
1980s	Sohio's Bob Hannah develops frac-packing, a technique that impacts sand-control operations during the 1990s
	High-strength steel and bias welding enable reliable coiled tubing operations for workover
1986	Norwegian Gola Nor Offshore A/S, now Teekay Corporation, in Trondheim, Norway, designs a purpose-built FPSO, the *Petrojarl I*
1988	Shell installs the massive 1,365 feet high *Bullwinkle* offshore production platform in the Gulf of Mexico

Groundbreakers

1989	Maersk Oil applies multistage fraccing in horizontal wells in the Dan field in Danish waters
1992	Norsk Hydro develops the inflow control device (ICD)
1992-95	Downhole tractors for tool conveyance in highly deviated and horizontal wells are introduced by companies such as UK's Sondex and Danish Welltec
1993	Aker of Oslo builds the Hibernia, the first iceberg-resistant drilling and production platform that now sits in 80 meters of water 315 kilometers off the coast of Newfoundland
1995	Aker installs the largest Condeep platform ever, the *Troll A*, for Shell's Troll gas field in the Norwegian North Sea
1997	Mitchell Energy starts using slick water as fracturing fluid and achieves major improvements fraccing the Barnett shale
2002	Devon Energy drills seven horizontal wells in the Barnett shale and uses multistage fraccing, marking the beginning of the US shale gas boom
2006	EOG Resources begins using multistage fraccing of shale oil in North Dakota
2007	The Anadarko-operated semisubmersible production platform, *Independence Hub*, starts producing from a record depth of 7,000 feet in the Gulf of Mexico
	FMC Technologies construct the world's longest tie-back, 120 kilometers to an onshore facility, at the Ormen Lange subsea installation in the Norwegian North Sea
2009	Harold Hamm's Continental Resources begin multistage fraccing with 30 stages to extract commercial quantities of crude
2012	Statoil introduces the concept of the Subsea Factory, aiming to place all subsea production and processing elements on the ocean floor
2014	Allseas launches a crane ship with a lifting capacity of 48,000 tons, a world record

Acknowledgements

OUR FIRST thanks must be to Stephen Whittaker, recently retired from Schlumberger, who helped crystallize the idea of this book at the beginning and who, throughout the four years it took to write it, never failed to encourage the authors with his wise words and suggestions.

Second, we owe huge thanks to Schlumberger geophysicist Pascal Edme, who serendipitously occupied the office next to Henry Edmundson's at Schlumberger Gould Research in Cambridge, UK, and introduced Henry to his friend Mark Mau, professional business historian living in Cambridge. Thanks to Pascal, the Edmundson-Mau connection was made and over time grew into a truly enjoyable collaboration without which this book would never have seen the light of day.

Third, it takes financial backing to write such a book, and for this we thank Andrew Gould, at that time Chairman and CEO of Schlumberger, for seeing immediately the value of our enterprise and sponsoring the research and writing. He was also enthusiastic to embrace the notion that the book should represent as fairly as possible the contributions of all players in the oil and gas upstream industry, thereby saving the book from degenerating into a vanity project. When Andrew Gould retired to become Chairman of the BG group, his sponsorship was readily continued by Ashok Belani, currently Chief Technology Officer of Schlumberger, to whom we owe equal thanks.

The book was written in phases—first drilling, then exploration, and finally reservoir and production: the three stages were then combined to provide the book's continuous chronological narrative. For each phase, we relied on a core group of mainly Schlumberger experts to ensure that the first cast of our net would catch enough industry experts to initiate research and ultimately lead us to colleagues of their own that we missed on the first round. For the drilling phase, we are therefore indebted to Stephen Whittaker, Bill Wright and Iain Cooper; for exploration we thank Phil Christie, Stephen Whittaker, Bill Wright and independent geologist Peter Dolan; for reservoir and production we thank Ernie Brown, Stephen Whittaker, Bill Wright, Joseph Ayoub, Kamel Bennaceur, Fikri Kuchuk, Alan Sibbit, Bob Kleinberg and Eduardo Proano.

We greatly appreciate the interviews conducted with the 127 industry experts who shared their observations, experience and insights. They gave generously of their time, knowledge and sometimes even hospitality. Every expert helped us better understand the evolution of upstream technology and science. In addition, they reviewed various drafts and made important comments and corrections. Overall, their contribution was significant and indispensable. Their names are listed in the References.

Mark, being new to the upstream oil and gas business, benefited greatly from a series of introductory courses and field visits. We thank Hooman Sadrpanah, then managing director of Schlumberger's training arm NExT, for making space in a NExT introductory course on petroleum exploration and production, taught in this instance by veteran industry expert Alain Brie. We thank Willem van Adrichem, at the time the Schlumberger manager in the Netherlands, for arranging a tour of a Schlumberger wireline field location, a Smith field location in the Netherlands and to Shell NAM for allowing a visit to a modern, automated rig drilling in its Groningen gas field. We thank Schlumberger colleague Geoff Downton for arranging a tour of the Schlumberger drilling engineering center in Stonehouse, UK, and to colleague Steve Pickering for arranging a tour of the WesternGeco seismic interpretation center in Gatwick, UK. Last, but by no means least, we thank Neil Harbury, founder and managing director of Nautilus training, for providing a space on one of his company's geological field trips to Dorset, UK, with Lance Morrissey as instructor.

The spirit of sharing that connects those museums and universities around the world dedicated to oil and gas equally provided us moral support and key sources of information. Thus, we owe thanks to Ryan Smith, director of the Texas Energy Museum, Beaumont, Texas; Van Romans, president of the Fort Worth Museum of Science and History; Paul Bernhard, curator of the Houston Museum of Natural Science; Finn E. Krogh and Finn Harald Sandberg, respectively director and curator of the Norwegian Petroleum Museum, Stavanger; Christophe de Ceunynck, director of the Schlumberger Museum at Crèvecœur Castle in Normandy; Teresa Tomkins-Walsh, historian and archivist at the University of Houston, M.D. Anderson Library; Mary Kendall, librarian at Churchill College, University of Cambridge; Piotr

Krzywiec, Institute of Geological Sciences, Polish Academy of Sciences, Warsaw; and Robbie Gries, Priority Oil & Gas, Denver, Colorado.

We are also indebted to relevant personnel at the upstream technical professional societies, in particular Amy Esdorn, historian at the Society of Petroleum Engineers (SPE) for sharing transcripts of some recent SPE interviews with industry experts, and Hans Krause, chairman of the History of Petroleum Geology committee, American Association of Petroleum Geologists (AAPG), for inviting us to present this book project at the annual 2014 AAPG international conference and exhibition at Istanbul, and Wolfgang Schollnberger, a member of the same committee.

Throughout the writing, we have been guests of Schlumberger Gould Research. We extend many thanks to Simon Bittleston, current head of Schlumberger research, for tolerating our comings and goings, and allowing us use of guest offices. Of particular note at Schlumberger Gould Research has been the indefatigable and extraordinarily efficient assistance of the library staff, particularly Melanie Atkinson and the manager of Schlumberger's online library Tellus service, Clare Aitken.

No less gratitude goes to our copyeditor, Ginger Oppenheimer, who taught us how to place commas, use American English spelling and, crucially, improved the book's readability.

Finally, we owe stability and success in this endeavor to our wives and families who provided love and support when it mattered most. Therefore, we dedicate this book to them, thanking them for always being there.

References

Interviews conducted for the book

THIS BOOK could not have been written without the personal recollections of over one hundred and twenty well-known industry veterans and experts—see list below. Any good history book relies on its sources. Our prime target was senior or retired engineers and scientists whose career spanned several decades. The furthest back in time that we could rely on personal recollection was about the late 1940s. Prior to that, the book had to rely solely on books and technical papers.

As much as possible we sought to balance interviewees between oil and gas companies, service companies and academia. It was harder to balance contributions globally. Researching technical innovation that took place in the West is easy compared with Russia, with the language barrier and difficulty accessing source material.

In most cases the interview was one-to-one via telephone. Each interview covered five questions related to the interviewee's particular expertise:

1. What upstream innovations have made the biggest impact on the oil and gas industry?
2. How did new technologies, methods and concepts emerge and become a commercial success?
3. Who was responsible for the innovations? What was the balance between oil companies, services companies, research organizations and individual innovators and inventors?
4. What are the future challenges in upstream technology?
5. Who else would be a good person to talk to about the history of upstream technology?

The industry experts were interviewed by Mark Mau, except for those marked with an asterisk that were conducted by Henry Edmundson. The organizational names in parentheses indicate the historical affiliation of the interviewees. The alphabetical list of interviewees is as follows:

- Pierre Adler (Université Pierre et Marie Curie)
- Ram Agarwal (Amoco/BP/Reliance Industries/Petrotel)
- Walt Aldred (Schlumberger)
- Ken Arnold (Shell/Paragon/WorleyParsons)
- Khalid Aziz (Karachi Gas Company/University of Alberta/University of Calgary/Stanford University)
- Milo Backus (GSI/University of Texas at Austin)
- Bert Bally (Shell/Rice University)
- Richard Bateman (Schlumberger/Amoco/Halliburton/Bridas/Gaffney Cline/Texas Tech University)
- Larry Behrmann (Schlumberger)
- Larry Bell (The Atlantic Refining Company/ARCO)
- Guus Berkhout (Shell/TU Delft)
- Jacques Bosio (Elf Aquitaine)
- Ted Bourgoyne (Louisiana State University)
- George Boyadjieff (National Oilwell Varco)
- Al Breitenbach (Scientific Software-Intercomp)
- Alistair Brown (GSI/Consultant)
- Ernie Brown (Schlumberger)
- Murray Burns (Shell/Petro-Marine Engineering/CBS Engineering/Technip)
- Chris Chapman (University of Toronto/University of Cambridge/Schlumberger)
- Roland Chemali (Schlumberger/Baker Hughes/Halliburton/Oxy)
- Jean Chevallier (SEDCO/Schlumberger)
- Phil Christie (Schlumberger)
- Jon Claerbout (Stanford University)
- Brian Clark (Schlumberger)
- Joe Dunn Clegg (Shell)
- George Coates (Schlumberger/Numar)
- John Cook (Schlumberger)
- Claude Cooke (Columbia University/Exxon)
- David Curry (Baker Hughes)
- Marlan Downey (Shell/ARCO/Roxanna Oil)

- Geoff Downton (Schlumberger)
- John Dribus (Mobil/Schlumberger)
- Mike Dyson (Shell/BG Group)
- Darwin Ellis (Schlumberger/Darwin Petrophysics)
- Mike Forrest (Shell)
- Robert Freedman (Shell/Petroleum Physics/Schlumberger)
- Peter Gaffney (Gaffney Cline/Baker Hughes)
- Marvin Gearhart (Gearhart-Owen Industries/Rockbit International)
- Ken Glennie (Shell)
- Andrew Gould (Schlumberger/BG Group)
- Sam Gray (Amoco/Veritas/CGGVeritas/CGG)
- Sid Green (Terratek/Schlumberger/University of Utah)*
- Alain Gringarten (Bureau de Recherches Géologique et Minières/Schlumberger/Scientific Software-Intercomp/Imperial College)
- Larry Harrington (General Dynamics/The Western Company of North America)
- Billy Hendricks (Schlumberger/NL Industries/Sperry Sun/Halliburton)
- Mel Hightower (Chevron/Exxon/Superior Oil/Arco)
- George Hirasaki (Shell/Rice University)
- Roger Hoestenbach (Shell/Paragon)
- Stephen Holditch (Shell/Texas A&M University/S.A. Holditch Associates)
- Jonathan Holmes (ECL/Intera/Schlumberger)
- Jim Hornabrook (BP)
- Roland Horne (University of Auckland/Stanford)
- Ed Horton (Standard Oil of California/Global Marine Development Company/Deep Oil Technology/Horton Wison Deepwater)
- Robert Hunt-Grubbe (Schlumberger/COP-GO/Sondex)
- Charles Ingold (Schlumberger)
- Larry Jacobson (Schlumberger/Halliburton)
- Mike Jellison (National Oilwell Varco)

- David Jenkins (BP)
- Pete Johnson (Project Mohole)
- George King (Amoco/BP/Rimrock Energy/Apache)
- Dave Kingston (Standard Oil of New Jersey/Humble Oil/Exxon)
- Nick Koutsabeloulis (BP/Vector International Processing System/Schlumberger)★
- Gregers Kudsk (Maersk Drilling)
- Jean Laherrère (Total)
- Leif Larsen (Statoil/Kappa Engineering)
- Andy Leonard (BP)
- Roy Lindseth (United Geophysical/Richmond Exploration/Engineering Data Processors)
- David Llewelyn (BP)
- Bryan Lovell (BP/University of Cambridge)
- Stefan Luthi (Schlumberger/TU Delft)
- Dan McKenzie (University of Cambridge)
- Harry O'Neal McLeod (Phillips/Exxon/Dowell/University of Tulsa/Conoco)
- Jan Meek (Shell/SBM Offshore/Heerema)
- Drew Michel (Shell/Halliburton/ROV Technologies)
- Tommy Miller (ECL/Intera/Schlumberger/Ridgeway Kite Software)
- Keith Millheim (Amoco)
- Robert Mitchell (Landmark/Halliburton)
- John Mogford (BP)
- Carl Montgomery (ConocoPhillips/ARCO/Dowell Schlumberger/NSI Technologies)
- Jim Munns (GSI/Texaco/Amoco/UK Department of Trade and Industry/Endeavour Energy/GDF Suez)
- Greg Myers (Integrated Ocean Drilling Program/Consortium for Ocean Leadership)
- Akif Narimanov (State Oil Company of Azerbaijan Republic)
- Erik Nelson (Schlumberger)
- Uri Nooteboom (IntecSea)

- Amos Nur (MIT/Stanford University)
- Alistair Oag (Burmah Oil/BP/Schlumberger)
- Donald Peaceman (Humble Oil/Exxon)
- Bob Peebler (Schlumberger/Landmark/Halliburton/ION)
- Thomas Perkins (Dow Chemical Company/The University of Texas at Austin/The Atlantic Refining Company/ARCO)
- Mike Prats (Shell)
- Stephen Prensky (Texaco/U.S. Geological Survey/Prensky Consulting Services)
- Bill Rehm (Dresser)
- Volker Reichert (Schlumberger)
- David Reid (National Oilwell Varco)
- Leon Robinson (Humble Oil/Exxon)
- Jean-Claude Roegiers (Los Alamos Scientific Laboratory/ University of Toronto/Dowell/Dowell Schlumberger/The University of Oklahoma)★
- Martin Rylance (BP)
- Nansen Saleri (Chevron/Saudi Aramco/QRI Reservoir Impact)
- Dan Scott (Baker Hughes)
- Rudolf Shagiev (Gubkin Russian State University of Oil and Gas/Petroleum Business Institute)
- Peter Sharpe (Shell)
- Brian Skeels (Exxon/FMC Technologies)
- Roger Slatt (Arco/Colorado School of Mines/The University of Oklahoma)
- Hobie Smith (Smith International)
- Michael Smith (Amoco/NSI Technologies)
- Fred Stalkup (Arco/Petrotel)
- Gordon Sterling (Shell)
- Philippe Theys (Schlumberger)
- E.C. Thomas (Shell/Bayou Petrophysics)
- Leon Thomsen (State University of New York/Amoco/BP/University of Houston)
- John Thorogood (BP)

- Scott Tinker (Marathon Oil/The University of Texas at Austin)
- Jay Tittman (Schlumberger)
- Sven Treitel (MIT/Amoco)
- Bernie Vining (Exxon/Baker Hughes)
- Robin Walker (Geco-Prakla/WesternGeco/Schlumberger)
- Koen Weber (Shell/TU Delft)
- Dietrich Welte (TU Aachen/Nuclear Research Centre Jülich/Integrated Exploration Systems/Schlumberger)
- Paul Willhite (Continental Oil/University of Kansas)
- Frank Williford (SEDCO)
- Andy Woods (University of Cambridge)
- Paul Worthington (Park Royd P&P)
- Mario Zamora (M-I Swaco, Schlumberger)
- Yu Zhang (Chinese Academy of Sciences/California Institute of Technology/VeritasDGC/CGGVeritas/CGG/ConocoPhillips)
- Martin Ziegler (Shell/Exxon)
- Walter Ziegler (Exxon/Petrofina)
- Tom Zimmerman (Schlumberger)

Other sourced interviews

Interviews at the Houston History Project, Series 7, Energy Development, University of Houston Libraries (2002-2003)
- Aubrey Bassett (Shell)
- Sam Evans (GSI)
- Robert "Bob" Graebner (GSI)

Interviews at the University of Aberdeen/British Library Project "Lives in the Oil Industry," a project of National Life Stories (2003-2005)
- Ken Glennie (Shell)

Interviews at the University of Cambridge, film interviews of academics and others, https://www.repository.cam.ac.uk/handle/1810/25 (2007)
- Dan McKenzie (University of Cambridge)

Interviews at www.minigeology.com (2010-12)
- Bert Bally (Shell/Rice University)

- Arnold Bouma (Scripps Institution of Oceanography/Utrecht University/Texas A&M University/US Geological Survey/Gulf Oil/Chevron/Louisiana State University)
- Nikolai Lopatin (Moscow State University/Russian Academy of Sciences)
- Emiliano Mutti (University of Milan/Exxon/University of Turin/University of Parma)
- Pete Vail (Exxon/Rice University)

Interviews at the Offshore Energy Center, Oral History Project, University of Houston Libraries (1999-2002)
- Robert "Bob" Bauer (Global Marine Exploration Company)
- Alexander Grigoryan (All-Union Scientific Research Institute for Drilling Technology, Moscow (VNIIBT))
- André Rey-Grange (SEDCO)

Interviews at the American Association of Drilling Engineers (AADE) Drilling Fluids Hall of Fame, Class of 2006 (DVD produced by M-I Swaco)
- Martin Chenevert (Exxon)
- Tommy Mondshine (Baroid)

Interviews at the American Association of Drilling Engineers (AADE) Drilling Fluids Hall of Fame, Class of 2008 (DVD produced by M-I Swaco)
- Ronald Clark (Shell)

Interviews at the Society of Petroleum Engineers, History of Petroleum Technology
- James Brill (University of Tulsa)
- Stephen Holditch (Texas A&M University/S.A. Holditch Associates)
- Fikri Kuchuk (Sohio Petroleum/Schlumbeger)
- John Lee (Exxon/Texas A&M University/University of Houston)
- Ralph Veatch (Stanolind/Amoco/Software Enterprises)

Bibliography

Achterbergh, Niels (2010): *In-Situ Oil Combustion: processes perpendicular to the main gas flow direction*, Delft: ISAPP Knowledge Centre

Adams, Neal and Alfred Eustes III (2011): "Drilling Problems," in: Mitchell, Robert and Stefan Miska (eds.)(2011): *Fundamentals of Drilling Engineering*, Richardson (Texas): Society of Petroleum Engineers, pp. 625-76

Adler, Pierre (2013): *A Lévy Flight through Porous Media*, presentation slides, University of Wyoming, Laramie, Wyoming, September 19-20

Adler, Pierre and Howard Brenner (1992): *Porous Media: geometry and transports*, Stoneham (Massachusetts): Butterworth-Heinemann

Adler, Pierre, Christian Jacquin and Jean Quiblier (1990): "Flow in simulated porous media," *International Journal of Multiphase Flow*, vol. 16, no. 4, pp. 691-712

Afghoul, Ali Chareuf et al. (2004): "Coiled tubing: the next generation," *Oilfield Review*, spring, pp. 38-57

Al-Ali, Zaki et al. (2009): "Looking deep into the reservoir," *Oilfield Review*, vol. 21, no. 2, pp. 38-47

Al-Amer, Abdulhadi et al. (2005): "Tractoring: a new era in horizontal logging for Ghawar Field, Saudi Arabia," conference paper, *SPE Middle East Oil and Gas Show and Conference*, 12-15 March, Kingdom of Bahrain

Al-Asimi, Mohammad et al. (2002): "Advances in well and reservoir surveillance," *Oilfield Review*, winter, pp. 14-35

Al-Shehri, Abdullah et al. (2013): "Illuminating the reservoir: magnetic NanoMappers, conference paper, *SPE* Middle East Oil and Gas Show and Conference, 10-13 March, Manama, Bahrain

Albertin, Uwe et al. (2002): "The time for depth imaging," *Oilfield Review*, spring, pp. 2-15

Albright, James et al. (1988): *The Crosswell Acoustic Surveying Project*, Los Alamos (New Mexico): Los Alamos National Laboratory

Aldred, Walt et al. (2005): "Changing the way we drill," *Oilfield Review*, spring, pp. 42-49

Alekperov, Vagit (2011): *Oil of Russia: past, present & future*, Minneapolis: East View Press

Alfsen, T.E. (1995): "Pushing the limits for extended reach drilling: new world record from platform Statfjord C, Well C2," *SPE Drilling & Completion*, vol. 10, no. 2, pp. 71-6

Alkhelaiwi, F. and D. Davies (2007): "Inflow control devices: application and value quantification of a developing technology," conference paper, *International Oil Conference and Exhibition*, 27-30 June, Veracruz, Mexico

Allaud, Louis and Maurice Martin (1977): *Schlumberger: the history of a technique*, New York: John Wiley and Sons

Allen, David et al. (1987): "Logging while drilling," *Oilfield Review*, vol. 1, no. 1, pp. 4-17

Allen, Frank et al. (1997): "Extended-Reach Drilling: breaking the 10-km barrier," *Oilfield Review*, winter, pp. 32-47

Allen, Frank, Tony Meader and Graham Riley (2000): "The secret of world-class extended-reach drilling," *Journal of Petroleum Technology*, June 2000, pp. 41-3

Allen, Philip and John Allen (2005): *Basin Analysis: principles and applications*, Oxford: Blackwell Publishing

Allen, Thomas and Alan Roberts (1993): *Production Operations*, vol. 1 (Well completions, Workover, and Stimulation), Tulsa (Oklahoma): Pennwell

Almeida, Rob (2011): "Dual gradient technology: a game-changer for offshore drilling," *gCaptain*, December 20

Alvarez, Jose, Raul Moreno and Ronald Sawatzky (2014): "Can SAGD be exported? Potential challenges," conference paper, *SPE Heavy and Extra Heavy Oil Conference: Latin America*, 24-26 September, Medellín, Colombia

American Petroleum Institute (ed.)(1961): *History of Petroleum Engineering*, New York: American Petroleum Institute

Amundsen, Lasse et al. (1999): "Multicomponent seabed seismic data: a tool for improved imaging and lithology fluid prediction," conference paper, *Offshore Technology Conference*, 5 March, Houston

Andersen, Svend Aage et al. (1990): "Exploiting reservoirs with horizontal wells: the Maersk experience," *Oilfield Review*, vol. 2, no. 3, pp. 11-21

Anderson, R.A. et al. (1980): "A production logging tool with simultaneous measurements," *Journal of Petroleum Technology*, vol. 32, no. 2, pp. 191-8

Anderson, Robert (1985): *Fundamentals of the Petroleum Industry*, London: Weidenfeld and Nicolson

Andrew, Robert Newton and Frederick Stone (1982): "MWD field use and results in the Gulf of Mexico," conference paper, *SPE*

Annual Technical Conference and Exhibition, September 26-29, 1982, New Orleans, Louisiana

Angehrn, J. and S. Sie (1987): "A high data rate fiber optic well logging cable," *The Log Analyst*, vol. 28, no. 2

Anstey, Nigel (1982): *Simple seismics for the petroleum geologist, the reservoir engineer, the well-log analyst, the processing technician, and the man in the field*, Boston: International Human Resources Development Corporation

---------- (2008): *Letter to the Conveners*, EAGE Vibroseis Workshop, October 1, Isle of Man

---------- (2013): *History of Exploration*, memo from Nigel Anstey to Mark Mau, Ramsey (Isle of Man), 10 March

Anstey, Nigel and Turhan Taner (1975): *Broad Line Seismic Profiling*, US Patent no. 3885225 patented May 20

Antonov, Yu. N. and D.S. Daev (1965): "Dielectric logging equipment," *Geophysical Equipment*, Nm Nedram Rel 26

Archie, G.E. (1942): "The electrical resistivity log as an aid in determining some reservoir characteristics," *Transactions AIME*, vol. 146, pp. 54-62

---------- (1950): "Introduction to petrophysics of reservoir rocks," *AAPG Bulletin*, vol. 34, May 1950, pp. 943-61

Arnold, Ken and Maurice Stewart (2008): *Surface Production Operations*, vol. 1 (Design of Oil-Handling Systems and Facilities), Oxford: Gulf Professional Publishing

Arps, J.J. (1963): "Continuous logging while drilling," conference paper, *SPE Annual Fall Meeting*, October 6-9, New Orleans

Arps, J.J. and J.L. Arps (1964): "The subsurface telemetry problem: a practical solution," *Journal of Petroleum Technology*, vol. 16, no. 5, pp. 487-93

Auletta, Ken (1984): *The Art of Corporate Success: the story of Schlumberger*, New York: G.P. Putnam's Sons

Badry, Rob et al. (1994): "Downhole optical analysis of formation fluids," *Oilfield Review*, January, pp. 21-8

Bai, Yong and Qiang Bai (2012): *Subsea Engineering Handbook*, Amsterdam: Gulf Professional Publishing

Baibakov, Nikolai Konstantinovich and Aleksandr Rubenovich Garushev (1989): *Thermal Methods of Petroleum Production*, Amsterdam and Oxford: Elsevier (Translated by W.J. Cieslewicz, Colorado School of Mines)

Baker, Alan et al. (1995): "Permanent monitoring: looking at lifetime reservoir dynamics," *Oilfield Review*, Winter, pp. 32-46

Baker, Ron (1998): *A Primer of Offshore Operations*, Austin: The University of Texas at Austin

Baker Hughes (ed.)(2007): "Baker Hughes: 100 years of service," *InDepth* 13, 2 (special issue)

Baker Hughes INTEQ (ed.)(1992): *Advanced Wireline & MWD Procedures Manual*, Houston: Baker Hughes Technical Publications Group

---------- (1997): *Baker Hughes INTEQ's Guide to Measurement While Drilling*, Houston: Baker Hughes INTEQ Technical Communications Group

Bakker, Eelco et al. (2003): "The new dynamics of underbalanced perforating," *Oilfield Review*, winter, pp. 54-67

Baldock, Simon et al. (2012): "Orthogonal wide azimuth surveys: acquisition and imaging," *First Break*, vol. 30, September, pp. 35-41

Ballard, Robert and Will Hively (2000): *The Eternal Darkness: a personal history of deep-sea exploration*, Princeton: Princeton University Press

Bamforth, Steve et al. (1996): "Revitalizing production logging," *Oilfield Review*, vol. 8, no. 4, pp. 44-60

Barclay, Frazer et al. (2008): "Seismic inversion: reading between the lines," *Oilfield Review*, spring 2008, pp. 42-63

Barenblatt, G.I., S.A. Christianovich, Y.P. Zheltov and G.K. Maximovich (1959): "Theoretical principles of hydraulic fracturing of oil strata," conference paper, 5th *World Petroleum Congress*, 30 May-5 June, New York

Barenblatt, G. I., Yu. P. Zheltov, and I.N. Kochina (1960): "Basis concepts in the theory of seepage of homogeneous liquids in fissured rocks," *Journal of Applied Mathematics* (Soviet), vol. 24, no. 5, pp. 1286-1303

Barlett, Donald and James Steele (1979): *Empire: the life, legend, and madness of Howard Hughes*, New York: Norton

Barnes, Kenneth (1931): *A Method for Determining the Effective Porosity of a Reservoir-Rock*, The Pennsylvania State College

Bulletin, Mineral Industries Experiment Station, bulletin 10, State College (Pennsylvania): The Pennsylvania State College

Bateman, Richard (2009): "Petrophysical data acquisition, transmission, recording and processing: a brief history of change from dots to digits," conference paper, *SPWLA 50th Annual Logging Symposium*, 21-24 June, The Woodlands, Texas

---------- (2012): *Openhole Log Analysis and Formation Evaluation*, Richardson (Texas): Society of Petroleum Engineers

Bates, T. and G. Coyle (2005): "Accelerating technology acceptance: nucleating and funding E&P technology," conference paper, *SPE Annual Technical Conference and Exhibition*, October 9-12, Dallas

Bearden, John (2007): "Electrical Submersible Pumps," in: Clegg, Joe Dunn (ed.): *Petroleum Engineering Handbook*, vol. IV (Production operations engineering), Richardson (Texas): Society of Petroleum Engineers, pp. 625-711

Bearden, John, Earl Brookbank and Brown Wilson (2009): *How we did it then: the evolutionary growth of ESPs (A survey of progress and changes in the ESP product during the past 82+ years)*, oral presentation given at the SPE ESP Workshop, The Woodlands, Texas, USA, 29 April-1 May (reprinted in: Noonan, Shauna (2011): *Electric Submersible Pumps*, digital edition, Richardson (Texas): Society of Petroleum Engineers, pp. 1-33)

Beasley, Craig et al. (2012): "Simultaneous sources: the inaugural full-field, marine seismic case history," conference paper, *SEG* Annual Meeting, Las Vegas, November 4-9

Beaton, Kendall (1957): *Enterprise in Oil: a history of Shell in the United States*, New York: Appleton-Century-Crofts

Beaton, Tim, Dan Calnan and Rocky Seale (2007): "Identifying applications for turbodrilling and evaluating historical performances in North America," *Journal of Canadian Petroleum Technology*, vol. 46, no. 6 (June)

Bednar, Bee (2005): *A Brief History of Seismic Migration*, Houston, article manuscript

Beeby-Thompson, Arthur (1904): *The Oil Fields of Russia and The Russian Petroleum Industry: a practical handbook on the exploration, exploitation, and management of Russian oil properties*, London: Crosby Lockwood and Son

Beecher, C.E. and H.C. Fowler (1961): "Production Techniques and Control," in: American Petroleum Institute (ed.): *History of Petroleum Engineering*, New York: American Petroleum Institute, pp. 745-810

Behrmann, Larry (2007): *After 58 years of Perforating, does it still have a future?* presentation slides, internal Schlumberger presentation, Houston, April 24,

Behrmann, Larry et al. (1996): "Quo vadis, extreme overbalance?" *Oilfield Review*, autumn, pp. 18-33

Behrmann, Larry et al. (2000): "Perforating practices that optimize productivity," *Oilfield Review*, spring, pp. 52-74

Bellarby, Jonathan (2009): *Well Completion Design*, Amsterdam: Elsevier

Berkhout, Guus (2008): "Changing the mindset in seismic data acquisition," *The Leading Edge*, vol. 27, July 2008, pp. 924-38

---------- (2013): "Decentralized blended acquisition: are networks the next big step in seismic data collection?" *EAGE* London Meeting, June, talk summary, in: www.earthdoc.org/publication/ publicationdetails/?publication=68347

Berry, James (1959): "Acoustic velocity in porous media," *AIME Petroleum Transactions*, vol. 216, pp. 262-70

Billingham, Matthew et al. (2011): "Conveyance—down an out in the oil field," *Oilfield Review*, vol. 23, no. 2, pp. 18-31

Biot, Maurice Anthony (1952): "Propagation of elastic waves in a cylindrical bore containing a fluid," *Journal of Applied Physics*, vol. 23, no. 9

Blair, P.M. and C.F. Weinaug (1969): "Solution of two-phase flow problems using implicit difference equations," *Society of Petroleum Engineers Journal*, vol. 9, no. 4, pp. 417-24

Bleier, Roger, Arthur Leuterman and Cheryl Stark (1993): "Drilling fluids: making peace with the environment," *Journal of Petroleum Technology* 45, 1 (January 1993), pp. 6-10

Bleier, Roger, Arvind Patel, Raymond McGlothlin and H.N. Brinkley (1993): *Oil Based Synthetic Hydrocarbon Drilling Fluid*, United States Patent no. 5189012, patented Feb. 23

Bokserman, A.A., V.P. Filippov and V.Yu. Filanovskii (1998): "Oil Extraction," in: Krylov, N.A., A.A. Bokserman, and E.R. Stavrosky (ed.): *The Oil Industry of the Former Soviet Union: reserves and prospects, extraction, transporation*, Amsterdam: Gordon and Breach Science Publishers, pp. 69-184

Bommer, Paul (2008): *A Primer of Oilwell Drilling*, Austin (Texas): University of Texas at Austin Petroleum

Bonolis, Luisa (2005): "Bruno Pontecorvo: from slow neutrons to oscillating neutrinos," *American Journal of Physics*, vol. 73, no. 6, pp. 487-99

Bosio, Jacques (2011): "A life led horizontally," *Offshore Engineer*, January

Bosworth, Steve et al. (1998): "Key issues in multilateral technology," *Oilfield Review*, winter, pp. 14-28

Bourdet, Dominique et al. (1983): "A new set of type curves simplifies well test analysis," *World Oil*, May, pp. 95-106

Bowker, Geoffrey (1994): *Science on the Run: information management and industrial geophysics at Schlumberger, 1920-1940*, Cambridge (Massachussetts): MIT Press

Bowker, Kent (2003): "Recent developments of the Barnett Shale play, Fort Worth Basin," *West Texas Geological Society Bulletin*, vol. 42, no. 6, pp. 4-11

Bram, Kurt at al. (1995): "The KTB Borehole: Germany's superdeep telescope into the Earth's crust," *Oilfield Review*, January, pp. 4-22

Brantly, John (1961a): "Percussion-Drilling System," in: American Petroleum Institute (ed.): *History of Petroleum Engineering*, New York: American Petroleum Institute, pp. 133-269

---------- (1961b): "Hydraulic Rotary-Drilling System with addendum on pneumatic rotary drilling," in: American Petroleum Institute (ed.): *History of Petroleum Engineering*, New York: American Petroleum Institute, pp. 271-452

---------- (1971): *History of Oil Well Drilling*, Houston: Gulf Publishing Breitenbach, E.A., G.A. King and K.N.B. Dunlop (1989): "The range of application of reservoir monitoring," conference paper, *SPE Annual Technical Conference and Exhibition*, 8-11 October, San Antonio, Texas

Breitenbach, E.A. et al. (1991): "Monitoring oil/water fronts by direct measurement," *Journal of Petroleum Technology*, vol. 43, no. 5, pp. 596-602

Brekke, Kristian and S. Lien (1994): "New and simple completion methods for horizontal wells improve the production performance in high-permeability, thin oil zones," *SPE Drilling & Completion*, vol. 9, no. 3, pp. 205-9

Bremner, Chad et al. (2006): "Evolving technologies: electrical submersible pumps," *Oilfield Review*, winter, pp. 30-43

Bret-Rouzaut, Nadine and Michel Thom (2005): "Technology strategy in the upstream petroleum supply chain," *Institut Français du Pétrole*, Les cahiers de l'économie, no. 57 (March)

Brewer, John (1975): "The tension leg platform concept", conference paper, *API*, annual meeting papers, division of production, 7-9 April, Dallas

Brewer, Robert (2000): *VSP Data in Comparison to other Borehole Seismic Data*, online presentation: www.searchanddiscovery.com/ documents/geophysical/brewer/images/brewer.pdf

Briggs, Robert (1964): "Development of a downhole television camera," conference paper, *SPWLA* 5th Annual Logging Symposium, 13-15 May, Midland, Texas

Brill, James and Hemanta Mukherjee (1999): *Multiphase Flow in Wells*, Richardson (Texas): Society of Petroleum Engineers

Bromberger, Mary (1954): *Comment ils ont fait fortune*, Paris: Librairie Plon

Brons, Folkert (1940): *Process and Apparatus for Exploring Geological Strata*, US Patent no. 2220509 patented November 5

Brown, Alistair (2011): *Interpretation of Three-Dimensional Data*, Tulsa (Oklahoma): The American Association of Petroleum Geologists and the Society of Exploration Geophysicists

Brown, Anthony Cave (1999): *Oil, God, and Gold: the story of Aramco and the Saudi Kings*, Boston: Houghton Mifflin

Brown, George (2004): "Permanent reservoir monitoring using fiber optic distributed temperature measurements," *SPE* Distinguished Lecturer presentation, 2004-2005

Brown, J., Brown, Lee and Jasper Jackson (1981): "NMR measurements on western gas sands core," SPE/DOE Low Permeability Symposium, 27-29 May, Denver, Colorado

Brown, Kermit and James Lea (1985): "Nodal systems analysis of oil and gas wells," *Journal of Petroleum Technology*, vol. 37, no. 10, pp. 1751-63

Brown, Robert (2001): "The earth's field NML development at Chevron," *Concepts in Magnetic Resonance*, vol. 13, no. 6, pp. 344-66

Buia, Michele et al. (2008): "Shooting seismic surveys in circles," *Oilfield Review*, autumn, pp. 27-8

Burgen, Jack and Hilton Evans (1975): "Direct digital laserlogging," conference paper, 50th annual fall meeting of the *Society of Petroleum Engineers of AIME*, 28 September-1 October, Dallas

Burkhardt, J.A. and T.W. Michie (1979): "Submerged Production System," conference paper, *Offshore Technology Conference*, April 30, 1979, Houston

Burleson, Clyde (1999): *Deep Challenge! The true epic story of our quest for energy beneath the sea*, Houston: Gulf Publishing

Butler, Roger (1994): *Horizontal Wells for the Recovery of Oil, Gas and Bitumen*, Houston: Gulf Publishing Company

Butler, R.M. and D.J. Stephens (1981): "The gravity drainage of steam-heated heavy oil to parallel horizontal wells," *Journal of Canadian Petroleum Technology*, vol. 20, no. 2, pp. 90-6

Butler, R.M., G.S. McNab and H.Y. Lo (1981): "Theoretical studies on the gravity drainage of heavy oil during in-situ steam heating," *The Canadian Journal of Chemical Engineering*, vol. 59, no. 4, pp. 455-60

Calhoun, George and Herbert Allen (1951): "*Equipment for Inserting Small Flexible Tubing into High-Pressure Wells*, US Patent no. 2567009, patented September 4

Campbell-Kelly, Martin and William Aspray (2004): *Computer: a history of the information machine*, London: Basic Books

Canadian Society of Exploration Geophysicists (ed.)(2002): "'Young people in our societies can look forward to many years of progress': an interview with Leon Thomsen," *CSEG Recorder*, June 2002, pp. 26-33

Candler, John et al. (1995): "Seafloor monitoring for synthetic-based mud discharged in the Western Gulf of Mexico," conference paper, *SPE/EPA* Exploration and Production Environmental Conference, March 1995 27-29, Houston

Cannon, George and Thomas Pennington (1971): "Rotary Drag-Type Drilling Bits," in: Brantly, John: *History of Oil Well Drilling*, Houston: Gulf Publishing, pp. 1060-70

Cannon, D.E., C. Cao Minh and R.L. Kleinberg (1998): "Quantitative NMR interpretation," *SPE Annual Technical Conference and Exhibition*, 27-30 September, New Orleans

Carden, Richard and Robert Grace (2007): *Horizontal and Directional Drilling*, seminar catalogue, Tulsa (Oklahoma): Petroskills

Carll, John Franklin (1880): *Second Geological Survey of Pennsylvania: 1875 to 1879: the geology of the oil regions of Warren, Venango, Clarion, and Butler counties*, Harrisburg (Pennsylvania): The Board of Commissioners for The Second Geological Survey

Carlson, Burt (1979): "First OCS subsea completion," *Petroleum Engineer*, August, pp. 98-9

Carnahan, B.D. et al. (1999): "Fiber optic temperature monitoring technology," conference paper, *SPE* Western Regional Meeting, 26-27 May, Anchorage, Alaska

Carstens, Halfdan (2006): "Quantum leap for seismic," *GeoExpro*, May 2006, pp. 22-4

Cartwright, Joe and Mads Huuse (2006): "3D seismic technology: the geological 'Hubble'," *Basin Research*, vol. 17, pp. 1-20

Ceruzzi, Paul (2003): *A History of Modern Computing*, Cambridge (Massuchussetts): MIT Press

Chang, Harry (1978): "Polymer flooding technology yesterday, today, and tomorrow," *Journal of Petroleum Technology*, vol. 30, no. 8, pp. 1113-28

Chang, Harry et al. (2006): "Advances in polymer flooding and alkaline/surfactant polymer processes as developed and applied in the People's Republic of China", *Journal of Petroleum Technology*, vol. 58, no. 2, pp. 84-9

Chenevert, Martin (1970): "Shale control with balanced-activity oil-continuous muds," *Journal of Petroleum Technology*, October 1970, pp. 1309-16

Cheng, J. et al. (2010): "Study on remaining oil distribution after polymer flooding", conference paper, *SPE Annual Technical Conference and Exhibition*, Florence, Italy, 19-22 September

Cheshire, Ian et al. (1980): "An efficient fully implicit simulator," conference paper, *European Offshore Technology Conference and Exhibition*, 21-24 October, London

Cholet, Henri (1997): *Progressing Cavity Pumps*, Paris: Éditions Technip

Christie, Phil et al. (1995): "Borehole seismic data sharpen the reservoir image," *Oilfield Review*, winter, pp. 18-31

Cirigliano, A.J. and R.E Leibach (1967): "Gravel packing in Venezuela," conference paper, 7th *World Petroleum Congress*, 2-9 April, Mexico City

Claerbout, Jon (1973): *Proposal to initiate the Stanford Exploration Project to do fundamental research in reflection seismology*, Stanford (California): Stanford University

---------- (1976): *Fundamentals of Geophysical Data Processing*, New York: McGraw-Hill

---------- (1985): *Imaging the Earth's Interior*, London: Blackwell

Clark, Brian (1983): *Well Logging Apparatus and Method using Transverse Magnetic Mode*, European Patent Application, no. 83401907.7, September 29, 1983

Clark, J.B. (1949): "A hydraulic process for increasing the productivity of wells," *AIME Petroleum Transactions*, 1949, vol. 1, pp. 1-9

Clavier, Christian, George Coates and Jean Dumanoir (1984): "Theoretical and experimental bases for the dual-water model for interpretation of shaly sands," *Society of Petroleum Engineers Journal*, April 1984, pp. 153-68

Clegg, Joe Dunn and Erich Klementich (2007): "Tubing Selection, Design, and Installation," in: Clegg, Joe Dunn (ed.): *Petroleum Engineering Handbook*, vol. IV (Production operations engineering), Richardson (Texas): Society of Petroleum Engineers, pp. 105-48

Coates, George, Christian Clavier and Yves Boutemy (1983): "A study of the dual-water model based on log data," *Journal of Petroleum Technology*, vol. 35, no. 1, pp. 158-66

Coats, Keith (1982): "Reservoir simulation: state of the art," *Journal of Petroleum Technology*, vol. 34, no. 8, pp. 1633-42

Coberly, C.J. (1961): "Production Equipment," in: American Petroleum Institute (ed.): *History of Petroleum Engineering*, New York: American Petroleum Institute, pp. 617-744

Comeaux, Malcolm (1987): *One Hundred and One Years of Geography at Arizona State University*, Tempe: Arizona State University Department of Geography

Congress of the United States Office of Technology Assessment (ed.)(1985): *Technology & Soviet Energy Availability*, Washington (DC): US Government

Conrad, K.M. (1962): "Application of the wireline formation tester," conference paper, *SPE Drilling and Production Practices Conference*, 5-6 April, Beaumont, Texas

Consortium for Ocean Leadership (ed.)(2007): *Ocean Drilling Program: final technical report 1983-2007*, Washington (D.C.): Consortium for Ocean Leadership

Constable, Steven (2010): *Seafloor Electromagnetic Methods Consortium: a research proposal*, La Jolla (California)

Constable, Steven and Leonard Srnka (2007): "An introduction to marine controlled-source electromagnetic methods for hydrocarbon exploration," *Geophysics*, vol. 72, no. 2, pp. WA3-WA12

Cooke, Claude (1977): "Fracturing with a high-strength proppant," *Journal of Petroleum Technology*, vol. 29, no. 10, pp. 1222-6

Cookson, Colter (2014): "Nanotech sensors to reveal reservoir," *The American Oil & Gas Reporter*, July

Coolidge, Robert et al. (2007): "Special report: BP, Baker run first expandable monobore liner extension system," *Oil & Gas Journal*, December 2

Cooper Cameron (ed.)(2003): *Designation ceremony, Cameron first ram-type BOP, an ASME Historic Mechanical Engineering Landmark, July 14, 2003*, Houston: Cooper Cameron Corporation Division

Cotter, W. (1962): "Twenty-three years of gas injection into a highly undersaturated crude reservoir," *Journal of Petroleum Technology*, vo. 14, issue 4, pp. 361-5

Craig, Edward Hubert Cunningham (1912): *Oil-Finding: an introduction to the geological study of petroleum*, London: E. Arnold

Crampin, Stuart and David Taylor (1994): "The potential for monitoring the progress of production fronts across hydrocarbon reservoirs," conference paper, *Rock Mechanics in Petroleum Engineering*, 29-31 August, Delft, Netherlands

Czarniecki, Stanislaw (1993): "Grzybowski and his School: the beginnings of applied micropaleontology in Poland at the turn of the 19th and 20th centuries," in: Kaminski, M.A., S. Geroch, and D.G. Kaminski (eds.): *The Origins of Applied Micropalaeontology: The School of Jozef Grzybowski*, Krakow: The Grzybowski Foundation, pp. 1-15 (also in: www.gf.tmsoc.org/Documents/Origins/ Czarniecki-Origins-1993.pdf)

Daev, D.S., and S.B. Denisou (1970): "About high frequency induction logging," *Geophysical Equipment*, M. Rel 42

Dake, Laurence (2001): *The Practice of Reservoir Engineering*, Amsterdam: Elsevier

Darcy, Henry (Jr.)(2003): "Henry Darcy: Inspecteur Général des Ponts et Chaussées," in: Brown, Glenn, Jürgen and Willi Hager (eds.): *Henry P. G. Darcy and other Pioneers in Hydraulics: contributions in celebration of the 200th birthday of Henry Philibert Gaspard Darcy*, Reston (Virginia): American Society of Civil Engineers, pp. 4-13

Darley, Henry, George Gray and Ryen Caenn (2011): *Composition and Properties of Drilling and Completion Fluids*, Waltham (Massachusetts): Gulf Professional Publishing

Dawe, Richard (ed.)(2000): *Modern Petroleum Technology*, vol. 1 (Upstream), New York: John Wiley

Deffeyes, Kenneth (2009): *Hubbert's Peak: the impending world oil shortage*, Princeton: Princeton University Press

DeGolyer, Everette Lee (1938): "Historical Notes of the Development of the Technique of Prospecting for Petroleum," in: Dunstan, A.E. (1938): *The Science of Petroleum: a comprehensive treatise of the principles and practice of the production, transport and distribution of mineral oil*, London: Oxford University Press, vol. 1, pp. 268-75

---------- (1961): "Concepts on Occurrence of Oil and Gas," in: American Petroleum Institute (ed.)(1961): *History of Petroleum Engineering*, New York: American Petroleum Institute, pp. 15-33

Delamaide, E., P. Corlay and W. Demin (1994): "Daqing oil field: the success of two pilots initiates first extension of polymer injection in a giant oil field," conference paper, *SPE/DOE Improved Oil Recovery Symposium*, Tulsa, 17-20 April

Dellinger, Thomas and Glen Tolle (1986): "Mobil identifies extended reach drilling advantages, possibilities in the North Sea," *Oil & Gas Journal*, May 26

Demaison, Gerald (1984): "The Generative Basin Concept," in: Demaison, Gerald and Roelof Murris (eds.): *Petroleum Geochemistry and Basin Evaluation*, American Association of Petroleum Geologists Memoir 35, pp. 1-14

Denney, Dennis (2014): "Nanotechnology applications for challenges in Egypt," *Journal of Petroleum Technology*, vol. 66, no. 2, pp. 123-6

Deutsch, C.V. and J.A. McLennan (2005): *Guide to SAGD (Steam Assisted Gravity Drainage) Reservoir Characterization Using Geostatistics*, Edmonton: University of Alberta

Devereux, Steve (1999): *Drilling Technology in Nontechnical Language*, Tulsa (Oklahoma): PennWell Books

Dickey, Parke (1959): "The first oil well," *Journal of Petroleum Technology*, vol. 59 (Oil Industry Centennial), January, pp. 14-25

\-\-\-\-\-\-\-\-\-\- (1979): *Petroleum Development Geology*, Tulsa (Oklahoma): Petroleum Publishing Company

Dixon, Dougal (1992): *The Practical Geologist: the introductory guide to the basics of geology and to collecting and identifying rocks*, New York: Simon & Schuster

Dobrin, Milton and Carl Savit (1976): *Introduction to Geophysical Prospecting*, New York and London

Doll, Henri-Georges (1949): "Introduction to induction logging and application to logging of wells drilled with oil base mud," *Journal of Petroleum Technology*, vol. 1, issue 6, pp. 148-62

\-\-\-\-\-\-\-\-\-\- (1953): "Two decades of electrical logging," *Journal of Petroleum Technology*, vol. 5, no. 9 (September), pp. 33-41

Dorozynski, Alexander (2011): *A History of Oil in Russia*, Paris: unpublished manuscript

Dow, Wallace (1974): "Application of oil-correlation and source-rock data to exploration in Williston basin," *American Association of Petroleum Geologists Bulletin*, vol. 58, no. 7, pp. 1253-62

Downton, Geoff et al. (2000): "New directions in rotary steerable drilling," *Oilfield Review*, spring, pp. 18-29

Dragoset, Bill (2005): "A historical reflection on reflections," *The Leading Edge*, vol. 24, supplement, pp. S46-S70

Dria, Dennis (2007): "Mud Logging," in: Holstein, Edward (ed.): *Petroleum Engineering Handbook*, vol. V (a)(Reservoir engineering and petrophysics), Richardson (Texas): Society of Petroleum Engineers, pp. 357-77

Drijkoningen, G.G. (2003): *Seismic Data Acquisition*, course TA3600 document, Delft: Delft Technical University

Duncan, William, Jr. (1996): *Organic Liquid Base Drilling Fluid with Terpene*, US Patent no. 5559085 patented Sep. 24

Dyke, Kate van (1997): *Fundamentals of Petroleum*, Austin: The University of Texas at Austin

EAGE (ed.)(2008): "Virtues of VSP highlighted at Galperin Readings," special topic, *First Break*, vol. 26, August

Eastman, H. John (1971): "Directional Drilling, 1960s," in: Brantly, John: *History of Oil Well Drilling*, Houston: Gulf Publishing, pp. 1182-1209

Eaton, Ben (1969): "Fracture gradient prediction and its application in oilfield operations," *Journal of Petroleum Technology*, vol. 21, no. 10 (October), pp. 1353-60

Eaton, F.M. et al. (1976): "The Cyber Service Unit: an integrated logging system," conference paper, *SPE Annual Fall Technical Conference and Exhibition*, 3-6 October, New Orleans

Edmundson, Henry (1981): "Production logging," *The Technical Review*, vol. 29, no. 2

\-\-\-\-\-\-\-\-\-\- (1986a): "Horizontal fractures: debunking a myth", *The Technical Review*, vol. 34, no. 3 (October), pp. 4-8

\-\-\-\-\-\-\-\-\-\- (1986b): "Basics of failure mechanics," *The Technical Review*, vol. 34, no. 3 (October), pp. 10-19

\-\-\-\-\-\-\-\-\-\- (1987): "Raymond Sauvage: recollections of Schlumberger Wireline's first four years," *The Technical Review*, vol. 35, no. 1, pp. 4-15

\-\-\-\-\-\-\-\-\-\- (1988): "Archie II: electrical conduction in hydrocarbon-bearing rock," *The Technical Review*, vol. 36, no. 4, pp. 12-21

---------- (1989): "Archie III: electrical conduction in shaly sands," *Oilfield Review*, vol. 1, no. 3, pp. 43-53

---------- (2010): *Patents and Other Sundry Matters*, Cambridge, unpublished article

---------- (2010a): *Schlumberger History: old ties arise anew*, Cambridge, unpublished article

Edwards, David et al. (2011): "Reservoir simulation: keeping pace with oilfield complexity", *Oilfield Review*, vol. 23, no. 4 (winter), pp. 4-15

Eichenberger, Ursula and Ursula Markus (2008): *Augusto Gansser: aus dem Leben eines Welt-Erkunders*, Zürich: AS Verlag

Elkins, Lloyd (1961a): "Thermal, Solvent and Improved Gas-Drive Oil-Recovery Methods," in: American Petroleum Institute (ed.): *History of Petroleum Engineering*, New York: American Petroleum Institute, pp. 883-906

Elkins, Lloyd (1961b): "Research," in: American Petroleum Institute (ed.): *History of Petroleum Engineering*, New York: American Petroleum Institute, pp. 1081-1113

Elliott, Dave et al. (2011): "Managed pressure drilling erases the lines," *Oilfield Review*, spring, pp. 14-23

Elliott, John (1925): "The Elliott Core Drills," *Transactions of the AIME*, vol. G-25, no. 1, pp. 58-61

Ellis, Darwin and Julian Singer (2008): *Well Logging for Earth Scientists*, Dordrecht: Springer

Ellis, Richard (1998): "An Overview of Frac Packs: a technical revolution (evolution) process," *Journal of Petroleum Technology*, vol. 50, no. 1, pp. 66-8

Ellis, Tor et al. (2009): "Inflow control devices: raising profiles," *Oilfield Review*, vol. 21, no. 4, pp. 30-37

Embry, Ashton (2009): *Practical Sequence Stratigraphy*, online document, Canadian Society of Petroleum Geologists, www.cspg.org

Emmermann, Rolf and Jörn Lauterjung (1997): "The German Continental Deep Drilling Program KTB: overview and major results," *Journal of Geophysical Research*, 102,B8 (August 10, 1997), pp. 18179-18201

Ertekin, Turgay, Jamal Abou-Kassem and Gregory King (2001): *Basic Applied Reservoir Simulation*, Richardson (Texas): Society of Petroleum Engineers

Eustes III, A.W. (2011): "Drilling Fluids," in: Mitchell, Robert and Stefan Miska (eds.): *Fundamentals of Drilling Engineering*, Richardson (Texas): Society of Petroleum Engineers, pp. 87-138

Eve, A.S. and D.A. Keys (1954): *Applied Geophysics in the Search for Minerals*, Cambridge: Cambridge University Press

Feenstra, Robijn and Anthony Kamp (1984): *Downhole Motor and Method for Directional Drilling of Boreholes*, United States Patent no. 4485879 patented Dec. 4

Filippov, Andrei et al. (1999): "Expandable tubular solutions," conference paper, *SPE Annual Technical Conference and Exhibition* held in Houston, Texas, October 3-6

Finch, David (1985): *Trace through Time: the history of geophysical exploration for petroleum in Canada*, Calgary: Canadian Society of Exploration Geophysicists

Finn, L.D. (1976): "A new deepwater offshore platform: the guyed tower," conference paper, *Offshore Technology Conference*, 3-6 May, Houston

FMC Energy Systems (ed.)(2004): *FMC Surface Welhead Catalog*, Houston: FMC Technologies

Fontenot, John (1986): "Measurement while drilling: a new tool," *Journal of Petroleum Technology*, vol. 38, no. 2, pp. 128-30

Forbes, R.J. and D.R. O'Beirne (1957): *The Technical Development of the Royal Dutch Shell 1890-1940*, Leiden: E.J. Brill

Ford, John (2000): "Drilling Operations," in: Dawe, Richard A. (ed.): *Modern Petroleum Technology*, Chichester: John Wiley, vol. 1 (Upstream), pp. 101-29

Foster, Brandon, Doug Hupp and Hal Martens (2011): *ENI Nikaitchuq: on the edge of the envelope at the end of the earth, an ERD case history*, symposium paper, Schlumberger Drilling Symposium

Fourmann, J.M. (1975): *Seismic Processing: wave equation migration*, CGG technical series, Massy: Compagnie Générale de Géophysique

Frehner, Brian (2011): *Finding Oil: the nature of petroleum geology, 1859-1920*, Lincoln (Nebraska): University of Nebraska Press

Frasch, Hermann (1896): *Increasing the Flow of Oil-Wells*, US Patent no. 556669 patented March 17

Fretwell, James (2007): "Hydraulic Pumping in Oil Wells," in: Clegg, Joe Dunn (ed.): *Petroleum Engineering Handbook*, vol. IV (Production operations engineering), Richardson (Texas): Society of Petroleum Engineers, pp. 713-56

Frink, Philip (2006): "Managed pressure drilling: what's in a name?" *Drilling Contractor*, March/April, pp. 37-9

Fromyr, Eivind (2010): "The role of wide azimuth in subsalt imaging," conference paper, *Offshore Technology Conference*, Houston, May, 3-6

Gaddy, Dean (1998): "Pioneering work, economic factors provide insights into Russian drilling technology," *Oil & Gas Journal* 96, no. 27 (July 6), pp. 67-9

Gadelle, Claude and Gerard Renard (1999): "*Increasing oil production through horizontal and multilateral wells*," paper presented at the workshop on Enhanced Production of Old Oil Fields, Surgut, Russia, March 17-18, 1999

Gala, Deepak and Steve Nas (2009): "Underbalanced Drilling Operations," in: Aadnoy, Bernt et al. (eds.): *Advanced Drilling and Well Technology*, Richardson (Texas): Society of Petroleum Engineers, pp. 678-702

Galperin, Evsei Iosifovich (1985): *Vertical Seismic Profiling and its Exploration Potentials*, Dordrecht: Kluwer

Gammack, William and Donald Knox (1932): *Gas Lift for Oil Wells*, US Patent no. 1789866 patented January 20

Gates, Alexander (2002): *A to Z of Earth Scientists*, New York: Facts on File

Gault, Allen (1996): "Riserless drilling: circumventing the size/cost cycle in deepwater," *Offshore Magazine*, May, pp. 49-54

Gelfgat, Yakov, Mikhail Gelfgat and Yuri Lopatin (2003): *Advanced Drilling Solutions: lessons from the FSU*, 2 vol., Tulsa (Oklahoma): PennWell

Gerding, Mildred (1986): *Fundamentals of Petroleum*, Austin (Texas): The University of Texas at Austin Petroleum

Gerretson, Frederik Carel (1958): *History of the Royal Dutch Shell*, 4 vol., Leiden: E.J. Brill

Giddens, Paul (1955): *Standard Oild Company (Indiana): oil pioneer of the Middle-West*, New York: Appleton-Century-Crofts

Gilbert, W.E. (1954): "Flowing and gas-lift well performance," conference paper, *API*, Spring Meeting of the Pacific Coast District, Division of Prouction, May 6-7, Los Angeles

Gillingham, W.J. (1977): *Schlumberger: the first years*, Paris: Schlumberger

Gilman, James and Chet Ozgen (2013): *Reservoir Simulation: history matching and forecasting*, Richardson (Texas): Society of Petroleum Engineers

Giusti, L.E. (1974): "CSV makes steam soak work in Venezuela field," *Oil & Gas Journal*, November 4, pp. 88-93

Glasby, Geoffrey P. (2006): "Abiogenic origin of hydrocarbons: an historical overview," *Resource Geology* vol. 56, no. 1, pp. 85-98

Glennie, Ken (1998): *Petroleum Geology of the North Sea: basic concepts and recent advances*, London: Blackwell Science

Gluyas, Jon and Richard Swarbrick (2004): *Petroleum Geoscience*, Oxford: Blackwell Science

Gogarty, W. B. and W.C. Tosch (1968): "Miscible-type waterflooding: oil recovery with micellar solutions," *Journal of Petroleum Technology*, vol. 20, no. 12, pp. 1407-14

Goins, W.C., K.R. Webster and S.C. Berry (1965): "A continuous multistage tracing technique," Journal of Petroleum Technology, vol. 17, no. 6, pp. 619-25

Gow, Sandy (2005): *Roughnecks, Rock Bits and Rigs: the evolution of oil well drilling technology in Alberta, 1883-1970*, Calgary: University of Calgary Press

Grace, Robert (1974a): "The Problem of Deviation and Dog Legging in Rotary Boreholes," in: Moore, Preston: *Drilling Practices Manual*, Tulsa (Oklahoma): PennWell Books, pp. 327-53

---------- (1974b): "Rotary Drilling Bits," in: Moore, Preston (1974): *Drilling Practices Manual*, Tulsa (Oklahoma): PennWell Books, pp. 354-80

---------- (2003): *Blowout and Well Control Handbook*, Amsterdam: Gulf Professional Publishing

Grant, Bruce and Stefan Szasz (1954): "Development of an underground heat wave for oil recovery," *AIME Petroleum Transactions*, May 1954, pp. 23-3

Gray, Sam (2012): *A Brief History of Depth and Time Imaging*, SEG presentation slide, Calgary

Gray, Samuel and Michael O'Brian (1996): "Can we image beneath salt?" *The Leading Edge of Exploration*, vol. 15, no. 1, pp. 17-22

Grebe, John and Sylvia Stoesser (1935): "Increasing crude production 20,000,000 Bbl. from established fields," *World Petroleum*, August 1935, pp. 473-82

Green, Cecil (2001): "Reflections on the Field of Exploration Geophysics," in: Lawyer, Lee, Charles Bates, and Robert Rice (2000): *Geophysics in the Affairs of Mankind: A personalized History of Exploration Geophysics*, Tulsa (Oklahoma): Society of Exploration Geophysicists, pp. 371-8

Greenberg, Jerry (2012): "Holistic approach to drilling means going beyond rig equipment: an interview with David Reid," *Drilling Contractor*, May 14

Gringarten, Alain (2008): "From straight lines to deconvolution: the evolution of the state of the art in well test analysis," *SPE Reservoir Evaluation & Engineering*, vol. 11, no. 1, pp. 41-62

Gruffeille, Jean-Paul et al. (2010): "Exploring Oligocene targets using multi-azimuth acquisitions: application of non-linear slope tomography," conference paper, 72nd *EAGE* Conference & Exhibition, Barcelona, June 14-17

Gunning, T.B. (1864): *Oil Ejector for Oil Wells*, US Patent no. 45152 patented November 22

Gustafson, Thane (2012): *Wheels of Fortune: the battle for oil and power in Russia*, Cambridge (Masachussetts): Harvard University Press

Haan, H.J. and L. Schenk (1969): "Drive project in the Tia Juana Field, Western Venezuela," *Journal of Petroleum Technology*, vol. 21, issue 1, pp. 111-9

Hager, Dorsey (1939): *Fundamentals of the Petroleum Industry*, New York and London: McGraw-Hill

Hallundbæk, Jørgen (1995): "Reduction of cost with new well intervention technology, well tractors," conference paper, *Offshore Europe*, September 5-8, Aberdeen

Hallundbæk, Jørgen et al. (1997): "Wireline well tractor: case histories," conference paper, *Offshore Technology Conference*, May 5, Houston

Handren, P.J., T.B. Jupp and J.M. Dees (1993): "Overbalance perforating and stimulation method for wells," conference paper, *SPE Annual Technical Conference and Exhibition*, 3-6 October, Houston

Hannegan, Don (2009): "Managed-Pressure Drilling," in: Aadnoy, Bernt et al. (eds.): *Advanced Drilling and Well Technology*, Richardson (Texas): Society of Petroleum Engineers, pp. 750-64

Hannegan, Don and K. Fisher (2005): "Managed pressure drilling in marine environments," conference paper, *International Petroleum Technology Conference*, 21-23 November, Doha, Qatar

Hannegan, Don and Roger Stave (2006): "The time has come to develop Riserless Mud Recovery technology's deepwater capabilities," *Drilling Contractor*, September/October, pp. 50-4

Harbaugh, John, John Doveton and John Davis (1977): *Probability Methods in Oil Exploration*, New York: John Wiley

Harries, Steve (2012): "Reading Between the Lines: point-receiver data, isometrically sampled in both crossline and inline directions, fully captures the three-dimensional seismic wavefield for the first time," *GeoExpro*, October 2012, pp. 50-52

Harrison, David and Yves Chauvel (2007): "Reservoir Pressure and Temperature," in: Holstein, Edward (ed.): *Petroleum Engineering Handbook*, vol. V a (Reservoir engineering and petrophysics), Richardson (Texas): Society of Petroleum Engineers, pp. 683-717

Hasan, A.R. and C.S. Kabir (2002): *Fluid Flow and Heat Transfer in Wellbores*, Richardson (Texas): Society of Petroleum Engineers

Hassler, Gerald (1944): "*Method and Apparatus for Permeability Measurements*, US Patent no. 2345935, patented April 4

Hatley, Allen (1995): *The Oil Finders: a collection of stories about exploration*, Utopia (Texas): Centex Press

Haugen, Jonny (1998): "Rotary steerable system replaces slide mode for directional drilling applications," *Oil & Gas Journal*, February 3

Heerema, Pieter, Alexandre Horowitz and Henricus Willemsen (1980): *Stabilizing System on a Semi-Submersible Crane Vessel*, US Patent no. 4231313A patented Nov. 4

Helbig, Klaus and Leon Thomsen (2005): "75-plus years of anisotropy in exploration and reservoir seismics: a historical review of concepts and methods," *SEG 75th Anniversary issue*, pp. 9ND-23ND

Henderson, Bob (1975): *Conoco Geophysics: the first fifty years*, Houston: Conoco

Henry, James (1982): "Stratigraphy of the Barnett Shale (Mississippian) and associated reefs in the northern Fort Worth Basin," *Petroleum Geology of the Fort Worth Basin and Bend Arch Area*, Dallas Geological Society, pp. 157-77

Herrick, R.C., S.H. Couturie and D.L. Best (1979): "An improved nuclear magnetism logging system and its application to formation evaluation," conference paper, 54th Annual Fall Technical Conference and Exhibition of the *Society of Petroleum Engineers of AIME*, 23-26 September, Las Vegas

Herron, Susan and Michael Herron (1996): "Quantitative lithology: an application for open and cased hole spectroscopy," conference paper, *SPWLA* 37th Annual Logging Symposium, June 16-19, New Orleans, Louisiana

Hertzog, Russ (1980): "Laboratory and field evaluation of an inelastic neutron scattering and capture gamma ray spectrometry tool," *Society of Petroleum Engineers Journal*, vol. 20, no. 5, pp. 327-40

Hilchie, Douglas (1990): *Wireline: a history of the well logging and perforating business in the oil fields*, Boulder (Colorado): Douglas W. Hilchie, Inc.

Hill, A.D., Ding Zhu and Michael Economides (2008): *Multilateral Wells*, Richardson (Texas): Society of Petroleum Engineers

Hillegeist, Paul (2012): *Subsea Acceleration: fathoming new technologies*, Hong Kong: CLSA

Hirasaki, George, Clarence Miller and Maura Puerto (2011): "Recent advances in surfactant EOR," *SPE Journal*, vol. 16, no. 4

Holcomb, David, Robert Hardy and David Glowka (1997): *Disposable Fiber Optics Telemetry for Measuring While Drillings*, Albuquerque (New Mexico): Sandia National Laboratories

Holditch, Stephen (2007): "Hydraulic Fracturing," in: Clegg, Joe Dunn (ed.): *Petroleum Engineering Handbook*, vol. IV (Production operations engineering), Richardson (Texas): Society of Petroleum Engineers, pp. 323-66

Holly, Christopher, Martin Mader and Jesse Toor (2012): *Alberta Department of Energy: oil sands production profile, 2002-2010*, Edmonton (Alberta): Government of Alberta

Holmes, Arthur (1965): *Principles of Physical Geology*, London and Edinburgh: Thomas Nelson

Horkowitz, John and Darrel Cannon (1997): "Complex reservoir evaluation in open and cased wells," conference paper, *SPWLA* 38th Annual Logging Symposium, June 15-18, Houston

Horne, Roland (2007): "Listening to the reservoir: interpreting data from permanent downhole gauges," *Journal of Petroleum Technology*, vol. 59, no. 12, pp. 78-86

Horner, D.R. (1951): "Pressure build-up in wells," conference paper, 3rd *World Petroleum Congress*, 28 May-6 June, The Hague

Houston Geological Society Bulletin (ed.)(1986): *Dave R. Kingston: biographical sketch*, Houston: Houston Geological Society

Hubbert, Marion King and David Willis (1957): "Mechanics of hydraulic fracturing," *AIME Petroleum Transactions*, vol. 210, pp. 153-68

Huber, T.A. and G.H. Tausch (1953): "Permanent-Type Well Completion", *AIME Petroleum Transactions*, vol. 198, pp. 11-16

Hughes, Richard (1955): "Theories on the Accumulation of Petroleum of Interest to Production Personnel," paper presented at the spring meeting of the Eastern District, Pittsburgh, May 1955, Division of Production, *API Drilling and Production Practice*, pp. 402-11

Hunt, T. Sterry (1862): "Notes on the History of Petroleum or Rock Oil," *Smithsonian Institution Annual Report*, for the year 1861, pp. 319-29, here pp. 325-6

Hunt-Grubbe, Robert (2007): *From little acorns… the story of a fledgling company* (Sondex), Potterne (Wiltshire): unpublished manuscript

Hyne, Norman (2001): *Nontechnical Guide to Petroleum Geology, Exploration, Drilling, and Production*, Tulsa (Oklahoma): PennWell Books

Ikelle, Luc and Lasse Amundsen (2005): *Introduction to Petroleum Seismology*, Tulsa (Oklahoma): Society of Exploration Geophysicists

International Association of Geophysical Contractors (ed.)(2002): *Marine Seismic Operations: an overview*, Houston: International Association of Geophysical Contractors (IAGC)

International Energy Agency (2012): *World Energy Outlook 2012: executive summary*, Paris: International Energy Agency

Jackson, Jasper (2001): "Los Alamos well logging project," *Concepts in Magnetic Resonance*, vol. 13, no. 6, pp. 368-78

Jackson, Warren and John Campbell (1946): "Some practical aspects of radioactivity well logging," *Transactions of the AIME*, vol. 165, no. 1, pp. 241-67

Jacobs, Trent (2014): "Pioneering subsea gas compression offshore Norway," *Journal of Petroleum Technology*, vol. 66, no 2, pp. 58-65

Jahn, Frank, Mark Cook and Mark Graham (2008): *Hydrocarbon Exploration and Production*, Amsterdam and Oxford: Elsevier

James, Huw (2013): "Innovation geophysics: a memoir," *First Break*, vol. 31, issue 6, pp. 161-7

James, R.W. and Bjørn Helland (1992): "The Greater Ekofisk area: addressing drilling fluid challenges with environmental justifications," conference paper, *European Petroleum Conference*, November 16-18, 1992, Cannes, France

Jewett, R.L. and G.F. Schurz (1970): "Polymer flooding: a current appraisal," *Journal of Petroleum Technology*, vol. 22, issue 6, pp. 675-84

Johansen, B., O. Holberg and K. Ovreba (1995): "Sub-sea seismic: impact on exploration and production," conference paper, *Offshore Technology Conference*, 1 May, Houston

Johnson, Hamilton (1961): "A history of well logging," conference paper, *SPWLA* 2nd annual logging symposium, 18-19 May, Dallas

Johnson, Pete (2011): "Some Recollections of the Birth of Deep-Sea Drilling," in: www.nationalacademies.org/moholeacc.html

Jolly, R.N. (1953): "Deep-hole geophone study in Garvin County, Okla.," *Geophysics*, vol. 28, no. 3, pp. 662-70

Jordan, Frank and Arthur White (1932): *Valve Construction for Gas Lift Pumps*, US Patent no. 1789855 patented January 20

Journal of Petroleum Technology (ed.): "People: John Barr," *Journal of Petroleum Technology*, April, p. 108

---------- (2008): "Special section: legends of drilling," *Journal of Petroleum Technology*, December

Karcher, J.C. (1933): *Method and Apparatus for Exploring Bore Holes*, US Patent no. 1927664 patented Sept. 19

Kass-Simon, Gabriele, Patricia Farnes and Deborah Nash (eds.)(1993): *Women of Science: righting the record*, New York: John Wiley and Sons

Kazemi, H. et al. (1976): "Numerical simulation of water-oil flow in naturally fractured reservoirs," *Society of Petroleum Engineers Journal*, vol. 16, no. 6 (December), pp. 317-26

Kelts, Kerry and Michael Arthur (1981): "Turbidites after Ten Years of Deep-Sea Drilling: wringing out the mop?" in: Warme, John, Robert Douglas and Edward Winther (eds.): *The Deep Sea Drilling Project: a decade of progress*, Tulsa (Oklahoma): Society of Economic Paleontologists and Mineralogists, pp. 91-127

Kenyon, W. et al. (1988): "A three-part study of NMR longitudinal relaxation properties of water-saturated sandstones," *SPE Formation Evaluation*, September, pp. 622-36

Kersey, Alan, James Dunphy and Arthur Hay (1998): "Optical reservoir instrumentation system," conference paper, *Offshore Technology Conference*, 5 April, Houston

King, George (2007): "Perforating," in: Clegg, Joe Dunn (ed.): *Petroleum Engineering Handbook*, vol. IV (Production operations engineering), Richardson (Texas): Society of Petroleum Engineers, pp. 149-73

---------- (2010): "Thirty years of gas shale fracturing: what have we learned?" conference paper, *SPE Annual Technical Conference and Exhibition*, 19-22 September, Florence, Italy

Kingston, David (1995): "The Rover Boys and other Stories," in: Hatley, Allen: *The Oil Finders: a collection of stories about exploration*, Utopia (Texas): Centex Press, pp. 1-26

Kleinberg, Robert (1998): "Nuclear Magnetic Resonance," in: Wong, Po-zen (ed.): *Methods in the Physics of Porous Media*, San Diego: Academic Press

---------- (2001): "NMR well logging at Schlumberger," *Concepts in Magnetic Resonance*, vol. 13, no. 6, pp. 396-403

Kleinberg, Robert and Jasper Jackson (2001): "An introduction to the history of NMR well logging," *Concepts in Magnetic Resonance*, vol. 13, no. 6, pp. 340-2

Kleinberg, Robert et al. (1995): "Nuclear magnetic resonance imaging: technology for the 21st century," *Oilfield Review*, autumn, pp. 19-33

Koelemeijer, Paula (2013): "Robert Stoneley and core-mantle boundary Stoneley modes," *Pembroke College Annual Gazette*, pp. 29-33

Kragh, Helge (1999): *Quantum Generations: a history of physics in the twentieth century*, Princeton: Princeton University Press

Krail, P.M. (2010): *Airguns: theory and operation of the marine seismic source*, course notes, University of Texas at Austin

Krebs, Jerome et al. (2009): "Fast full-wavefield seismic inversion using encoded sources," *Geophysics*, vol. 74, no. 6, pp. 177-88

Kuchuk, Fikri, Mustafa Onur and Florian Hollaender (2010): *Pressure Transient Formation and Well Testing: convolution, deconvolution and nonlinear estimation*, Amsterdam: Elsevier

Krehl, Peter (2008): *History of Shock Waves, Explosions and Impacts*, Berlin and New York: Springer

Kunzig, Robert (2000): *Mapping the Deep: the extraordinary story of ocean science*, London: Sort of Books

Kutchin, Joseph (2001): *How Mitchell Energy & Development Corp. got its start and how it grew*, Boca Raton: Universal Publishers

Lanagan, Mike (1882): *Safety Attachment for Oil Wells and Tanks*, US Patent no. 267903 patented Nov. 21

Lankford, Raymond (1971): "Marine Drilling," in: Brantly, John (1971): *History of Oil Well Drilling*, Houston: Gulf Publishing, pp. 1358-1444

Larner, Ken et al. (1978): "Depth migration of complex offshore seismic profiles," conference paper, *Offshore Technology Conference*, 8-11 May, Houston

Larson, Henrietta and Kenneth Porter (1959): *History of Humble Oil & Refining Company: a study in industrial growth*, New York: Harper

Lathim, Rod (1995): *The Spirit of the Big Yellow House: a history of Summerland's founding family*, Santa Barbara (California): Emily Publications

Latimer, Rebecca (2011): "Inversion and Interpretation of Impedance Data," in: Brown, Alistair: *Interpretation of Three-Dimensional Seismic Data*, Tulsa (Oklahoma): The American Association of Petroleum Geologists and the Society of Exploration Geophysicists pp. 309-49

Lawyer, Lee, Charles Bates and Robert Rice (2001): *Geophysics in the Affairs of Mankind: a personalized history of exploration geophysics*, Tulsa (Oklahoma): Society of Exploration Geophysicists

Lawyer, Lee and Ben Giles (2013): "Doodlebugger diary: air gun history," *GSH Journal*, Geophysical Society of Houston, vol. 3, no. 8 (April 2013), pp. 34, 38

Lea, James (2007): "Artificial Lift Selection," in: Clegg, Joe Dunn (ed.): *Petroleum Engineering Handbook*, vol. IV (Production operations engineering), Richardson (Texas): Society of Petroleum Engineers, pp. 411-56

Lebourg, M., R.Q. Fields and C.A. Doh (1957): "A method of formation testing on logging cable," *Society of Petroleum Engineers* paper

Lebourg, M.P. and G.R. Hodgson (1952): "A method of perforating casing below tubing," *Petroleum Transactions, AIME*, vol. 195, pp. 303-10

Lecuyer, Christophe (1992): "The making of a science based technological university: Karl Compton, James Killian, and the reform of MIT, 1930-1957," *Historical Studies in the Physical and Biological Sciences*, 23, 1, pp. 153-80

Lee, John (1982): *Well Testing*, Richardson (Texas): Society of Petroleum Engineers

Lee, John, John Rollins and John Spivey (2003): *Pressure Transient Testing*, Richardson (Texas): Society of Petroleum Engineers

Leffler, William, Richard Pattarozzi and Gordon Sterling (2003): *Deepwater Petroleum Exploration and Production: a nontechnical guide*, Tulsa (Oklahoma): PennWell Books

---------- (2011): *Deepwater Petroleum Exploration & Production*, Tulsa (Oklahoma): PennWell

Lemmens, Herman, Alan Butcher and P.W. Botha (2010): "FIB/SEM and automated mineralogy for core and cuttings analysis," conference paper, *SPE Russian Oil and Gas Conference and Exhibition*, 26-28 October, Moscow

Lenn, Chris, Steve Bamforth, and Hitesh Jariwala (1996): "Flow diagnosis in an extended reach well at the Wytch Farm Oilfield using a new toolstring combination incorporating novel production technology," conference paper, *SPE Annual Technical Conference and Exhibition*, 6-9 October, Denver, Colorado

Léonardon, Eugène (1961): "Logging, Sampling, and Testing," in: American Petroleum Institute (ed.): *History of Petroleum Engineering*, New York: American Petroleum Institute, pp. 493-578

Lesso, William (2009): "Geosteering," in: Aadnoy, Bernt et al. (eds.): *Advanced Drilling and Well Technology*, Richardson (Texas): Society of Petroleum Engineers, pp. 458-73

Leverett, M.C. (1941): "Capillary behavior in porous solids," *Transactions of the AIME*, vol. 142, pp. 152-69

Levesque, Cyrille (2006): "Crosswell electromagnetic resistivity imaging: illuminating the reservoir," *Middle East & Asia Reservoir* Review, no. 7, pp. 24-33

Levorsen, Arville Irving (1967): *Geology of Petroleum*, San Francisco: W.H. Freeman

Lewis, James (1961): "Fluid Injection," in: American Petroleum Institute (ed.): *History of Petroleum Engineering*, New York: American Petroleum Institute, pp. 847-81

Linsley, Judith Walker, Ellen Walker Rienstra and Jo Ann Stiles (2002): *Giant under the Hill: a history of the Spindletop discovery at Beaumont, Texas, in 1901*, Austin (Texas): Texas State Historical Association

Longhurst, Henry (1959): *Adventure in Oil: the story of British Petroleum*, London: Sidgwick and Jackson

Lopatin, Nikolai (1971): "Temperature and geologic time as factors in coalification (in Russian)," *Akad. Nauk SSSR, Izv. Serv.Geol.*, no. 3, pp. 95-106

Lopatin, Nikolai, Dietrich Welte et al. (1999): "Gas generation and accumulation in the West Siberian Basin," *AAPG Bulletin*, vol. 83, October 1999, no. 10, pp. 1642-65

Lund, G.G. et al. (2009): "Advanced flow assurance system for the Ormen Lange subsea gas development," conference paper, *Offshore Technology Conference*, 4-7 May, Houston

Luthi, Stefan (2012): *The Carbon Cycle, Organic Matter and Maturation*, course presentation slides, Delft: TU Delft

Lyons, William and Gary Plisga (2004): *Standard Handbook of Petroleum and Natural Gas Engineering*, Houston: Professional Publishing

M-I Swaco (ed.)(2011): *Offshore TCC Hammermill System*, Houston: M-I Swaco

Ma, Zaitian (1981): *Finite Difference Migration with Higher Order Approximation*, technical report of the China National Oil and Gas Exploration and Development Co.

Mach, Joe, Eduardo Proano and Kermit Brown (1979): "A nodal approach for applying systems analysis to the flowing and artificial lift oil or gas well," *Society of Petroleum Engineers* paper no. 8025

Macini, Paolo and Ezio Mesini (2003): "Darcy's law from water to the petroleum industry: when and who?" in: Brown, Glenn, Jürgen, and Willi Hager (eds.): *Henry P. G. Darcy and other pioneers in hydraulics: contributions in celebration of the 200th birthday of Henry Philibert Gaspard Darcy*, Reston (Virginia): American Society of Civil Engineers, pp. 78-89

Malloy, Kenneth (2007): "Managed pressure drilling: what is it anyway?" *World Oil*, March 2007, pp. 27-34

Margrave, Gary, Robert Ferguson and Chad Hogan (2010): *Full Waveform Inversion with Wave Equation Migration and Well Control*, CREWES research report, vol. 22, Calgary: University of Calgary

---------- (2011): *Full Waveform Inversion using One-Way Migration and Well Calibration*, CREWES convention paper, Calgary: University of Calgary

Marine Technology Society (ed.)(2013): *Advances in Marine Technology*, Washington D.C.: Marine Technology Society

Marion, Bruce (2014): "Cross-well imaging offers higher resolution," *The American Oil & Gas Reporter*, January 2014

Marsh, H.N. and Ward Kelly (1971): "Development of Instruments," (Rotary Drilling Fluid) in: Brantly, John: *History of Oil Well Drilling*, Houston: Gulf Publishing, pp. 1133-45

Matthews, Cam et al. (2007): "Progressing Cavity Pumping Systems," in: Clegg, Joe Dunn (ed.): *Petroleum Engineering Handbook*, vol. IV (Production operations engineering), Richardson (Texas): Society of Petroleum Engineers, pp. 757-837

Maurer, William (1968): *Novel Drilling Techniques*, Oxford and New York: Pergamon Press

Maurer, William et al. (1977): *Downhole Drilling Motors: technical review, final report*, Houston: Maurer Engineering

Maxwell, Shaun et al. (1998): "Microseismic logging of the Ekofisk reservoir," conference paper, *SPE/ISRM Rock Mechanics in Petroleum Engineering*, 8-10 July, Trondheim, Norway

Maxwell, Shaun et al. (2002): "Microseismic imaging of hydraulic fracture complexity in the Barnett Shale," conference paper, *SPE Annual Technical Conference and Exhibition*, 29 September-2 October, San Antonio, Texas

Maynard, Lara (ed.)(1997): *Hibernia: promise of rock and sea*, St. John (Newfoundland): Breakwater Books

Mayne, W. Harry (1989): *50 years of Geophysical Ideas*, Tulsa (Oklahoma): The Society of Exploration Geophysicists

McClellan, James E. and Harold Dorn (2006): *Science and Technology in World History: an introduction*, Washington (D.C.): The Johns Hopkins University Press

McCray, Arthur and Frank Cole (1959): *Oil Well Drilling Technology*, Norman (Oklahoma): University of Oklahoma Press

McDaniel, Robert and Henry Dethloff (1989): *Patillo Higgins and the Search for Texas Oil*, College Station (Texas): Texas A&M University Press

McGregor, Kenneth (1967): *The Drilling of Rock*, London: C.R. Books

McLean, John and Robert Haigh (1954): *The Growth of Integrated Oil Companies*, Boston: Harvard University

McPhee, John (2000): *Annals of the Former World*, New York: Farrar, Straus and Giroux

Meador, Richard (2009): "Logging-while-drilling: a story of dreams, accomplishments, and bright futures," conference paper, *SPWLA 50th Annual Logging Symposium*, 21-24 June, The Woodlands, Texas

Meador, Richard and P.T. Cox (1975): "Dielectric constant logging: a salinity independent estimation of formation water volume," conference paper, fall meeting of the *Society of Petroleum Engineers of AIME*, 28 September-1 October, Dallas

Meissner, Fred (1978): "Petroleum geology of the Bakken Formation Williston Basin, North Dakota and Montana," in: Estelle, D. and R. Miller (eds.): *The Economic Geology of the Williston Basin*, Billings (Montana): Montana Geological Society, pp. 207-30

Meissner, Rolf (1986): *The Continental Crust: a geophysical approach*, London: Academic Press

Mercer, James and Laura Nesbit (1992): *Oil-Base Drilling Fluid Comprising Branched Chain Paraffins such as the Dimer of 1-Decene*, US Patent no. 5096883 patented Mar. 17

Middle East & Asia Reservoir Review (ed.)(2007): "Frac Packing: fracturing for sand control," *Middle East & Asia Reservoir Review*, no. 8, pp. 36-49

Miller, C.C., A.B. Dyes and C.A. Hutchinson (1950): "The estimation of permeability and reservoir pressure from bottom hole pressure build-up characteristics," *Journal of Petroleum Technology*, vol. 2, no. 4, pp. 91-104

Miller, Melvin (2001): "Numar and Numalog overview," *Concepts in Magnetic Resonance*, vol. 13, no. 6, pp. 379-85

Miller, Melvin et al. (1990): "Spin echo magnetic resonance logging: porosity and free fluid index determination, conference paper," *SPE Annual Technical Conference and Exhibition*, 23-26 September, New Orleans, Louisiana

Millikan, Charles (1961): "Cementing," in: American Petroleum Institute (ed.): *History of Petroleum Engineering*, New York: American Petroleum Institute, pp. 453-92

---------- (1971): "Cementing," in: Brantly, John: *History of Oil Well Drilling*, Houston: Gulf Publishing, pp. 1306-41

Mills, R. (1920): "Experimental studies of subsurface relationships in oil and gas fields," *Economic Geology*, vol. 15, no. 5, pp. 398-421

Miska, Stefan (2011): "Directional Drilling," in: Mitchell, Robert and Stefan Miska (eds.)(2011): *Fundamentals of Drilling Engineering*, Richardson (Texas): Society of Petroleum Engineers, pp. 449-583

Mitchell, John, Valerie Marcel and Beth Mitchell (2012): *What Next for the Oil and Gas Industry*, Chatham House, October, published in: www.chathamhouse.org/publications/papers/view/186327

Mitzakis, John (1911): *The Russisan Oil Fields and Petroleum Industry: Being a Practical and Concise Handbook on the Management and Exploitation of Oil Properties, including the History of the Russian Petroleum Industry, Survey Outputs, Tenure of Proliferous Lands, Legislation of English Companies in the Russian Empire, and a List of the Russian Mining Laws*, London: Pall Mall

Montgomery, Carl and Michael Smith (2010): "Hydraulic Fracturing: history of an enduring technology," in: Society of Petroleum Engineers (ed.): *Legends of Hydraulic Fracturing*, Richardson (Texas): Society of Petroleum Engineers, pp. 1-9

Moore, Thomas and George Cannon (1936): *Weighted Oil Base Drilling Fluid*, US Patent no. 2055666 patented Sept. 29

Moore, W.W. (ed.)(1981): *Fundamentals of Rotary Drilling: the rotary drilling system: a professional and practical training guide to its equipment, procedures and technology*, Dallas: Energy Publications

Moreton, Richard (ed.)(1995): *Tales from Early UK Oil Exploration 1960-1979*, London: Petroleum Exploration Society of Great Britain

Morgan, James (1992): "Horizontal drilling applications of petroleum technologies for environmental purposes," *Ground Water Monitoring & Remediation*, summer 1992, pp. 98-102

Morris, R.L., D.R. Grine and T.E. Arkfeld (1964): Using Compressional and Shear Acoustic Amplitudes for the Location of Fractures," *Journal of Petroleum Technology*, vol. 16, no. 6, pp. 623-32

Morton, Andrew et al. (2003): "Evaluation and impact of sparse-grid, wide-Azimuth 4C-3D node data from the North Sea," conference paper, *SEG Annual Meeting*, 26-31 October, Dallas

Moseley, L.M. (1976): "Field evaluation of Direct Digital Well Logging," conference paper, 17th *SPWLA* annual logging symposium, June 9-12, Denver

Muskat, Morris (1949): *Physical Principles of Oil Production*, New York: McGraw-Hill

Mutti, E. and R. Ricchi Lucci (1972): "Le Torbiditi del 'Appennino settentrionale': introduzione all' analisi de facies," *Soc. Geol. Italiana Mem.*, vol. 11, pp. 161-99 (English translation: "Turbidites of the Northern Apennines: introduction to facies analysis," published in *International Geology Review*, vol. 20 (1978), pp. 125-66

Myers, Gary (2007): "Nuclear Logging," in: Holstein, Edward (ed.): *Petroleum Engineering Handbook*, vol. V (a)(Reservoir engineering and petrophysics), Richardson (Texas): Society of Petroleum Engineers pp. 243-87

Myers, Greg (2008): "Ultra-deepwater riserless mud circulation with dual gradient drilling," *Scientific Drilling*, no. 6 (July), pp. 48-51

Nalonnil, Ajay and Bruce Marion (2010): "High resolution reservoir monitoring using crosswell seismic," conference paper, *SPE Asia Pacific Oil and Gas Conference and Exhibition*, 18-20 October, Brisbane, Queensland, Australia

National Oilwell Varco (ed.)(2010): *Drilling Evolution 1985-2010*, Houston: National Oilwell Varco

National Petroleum Council (ed.)(1965): *Impact of New Technology on the U.S. Petroleum Industry 1946-1965*, Washington D.C.: National Petroleum Council

National Research Council (ed.)(1940): *Industrial Research Laboratories of the United States*, 7th edition, Washington (D.C.): National Academy of Sciences

Neal, W. Howard et al. (2007): *Oil and Gas Technology Development*, Global Oil & Gas Study, US National Petroleum Council, July 18, published in: www.npc.org/study_topic_papers/26-ttg-ogtechdevelopment.pdf

Nelson, Erik, Michel Michaux and Bruno Drochon (2006): "Cementing Additives and Mechanisms of Action," in: Nelson, Erik B. and Dominique Guillot (eds.): *Well Cementing*, Sugar Land: Schlumberger, pp. 49-91

Nielsen, R.F. and L.T. Bissey (1966): *Petroleum Production Research at Penn State in Retrospect*, University Park (Pennsylvania): The Pennsylvania State University

Northrop, David and Karl-Heinz Frohne (1990): "The Multiwell Experiment: a field laboratory in tight gas sandstone," *Journal of Petroleum Technology*, vol. 42, no. 6, pp. 772-9

Nur, Amos (1982): "Seismic imaging in enhanced recovery," conference paper, *SPE Enhanced Oil Recovery Symposium*, 4-7 April, Tulsa, Oklahoma

Nyhavn, Fridtjof and Anne Dalager Dyrli (2010): "Permanent tracers embedded in downhole polymers prove their monitoring capabilities in a hot offshore well," conference paper, *SPE Annual Technical Conference and Exhibition*, 19-22 September, Florence, Italy

O'Brian, T.B. and W.C. Goins (1960): *The Mechanics of Blowouts and How to Control Them*, API Drilling and Production Practices P41

Offshore Magazine (ed.)(2000): "Expandable casing program helps operator hit TD with larger tubulars," *Offshore Magazine*, no. 1

Offshore Energy Center (2009): *The Star*, third quarter

Offshore Engineer (ed.)(2003): "Automating the Drill Floor," *Offshore Engineer*, June 10

Oil & Gas Journal (ed.)(1992): "Antiwhirl PDC bit designs reduce vibrations," *Oil & Gas Journal*, November 30

---------- (1995): "News: industry pushes use of PDC bits to speed drilling, cut costs," *Oil & Gas Journal*, August 14

---------- (2006): "Roller Cones vs. Diamonds: a reversal of roles," *Oil & Gas Journal*, February 20

---------- (2012): "Chevron's dual gradient drill ship arrives in Gulf of Mexico," *Oil & Gas Journal*, May 7

Okada, Hakuyu (2005): *The Evolution of Clastic Sedimentology*, Edinburgh: Dunedin Academic Press

O'Neill, Frank (1934): "Formation testers", *Transactions of the AIME*, vol. 107, no. 1, pp. 53-61

Oristaglio, Michael and Dorozynski, Alexander (2007): *A Sixth Sense: the life and science of Henri-Georges Doll: oilfield pioneer and inventor*, Cambridge (Massachussetts): The Hammer Company

Ormsby, George (1974): "Drilling Fluid Solids Removal," in: Moore, Preston: *Drilling Practices Manual*, Tulsa (Oklahoma): PennWell Books, pp. 133-204

Ostrander, William (1984): "Plane-wave reflection coefficients for gas sands at nonnormal angles of incidence," *Geophysics*, vol. 49, no. 10 (October), pp. 1637-48

---------- (2006): "Memoirs of a successful geophysicist," *CSEG Recorder*, June 2006, pp. 38-41

Owen, Edgar Wesley (1975): *Trek of the Oil Finders: a history of exploration for petroleum*, Tulsa (Oklahoma): American Association of Petroleum Geologists

Ozbayoglu, Evren (2011): "Rotary Drilling Bits," in: Mitchell, Robert and Stefan Miska (eds.): *Fundamentals of Drilling Engineering*, Richardson (Texas): Society of Petroleum Engineers, pp. 311-84

Pappas, James (ed.)(2010): "Legends of production and operations," *Journal of Petroleum Technology*, vol. 61, no. 12, pp. 33-47

Parcevaux, Philippe, Bernard Piot and Claude Vercaemer (1985): *Cement Compositions for Cementing Wells, Allowing Pressure Gas-Channeling in the Cemented Annulus to be Controlled*, US patent no. 4537918 patented Aug. 27

Payne, Darwin (1979): *Initiative in Energy: the story of Dresser Industries, 1880-1978*, New York: Simon and Schuster

Peaceman, Donald (1957): "Application of large computers to reservoir engineering problems," *Journal of Petroleum Technology*, vol. 9, no. 10, pp. 14-18

---------- (1990): "A Personal Retrospection of Reservoir Simulation," in: Nash, Stephen (ed.): *A History of Scientific Computing*, New York: ACM Press, pp. 106-29

Peaceman, Donald and Henry Rachford (1955): "The numerical solution of parabolic and elliptic differential equations," *Journal of the Society for Industrial and Applied Mathematics*, vol. 2, no. 1, pp. 28-41

Penberthy, W. and E. Echols (1993): "Gravel placement in wells," *Journal of Petroleum Technology*, vol. 45, no. 7, pp. 612-74

Penberthy, W.L. Jr. and C.M. Shaughnessy (1992): *Sand Control*, Richardson (Texas): Society of Petroleum Engineers

Perkins, Thomas and Loyd Kern (1961): "Widths of hydraulic fractures," *Journal of Petroleum Technology*, vol. 13, no. 9, pp. 937-49

Perrodon, Alain (1980): *Géodynamique Pétrolière: genèse et répartition des gisements d'hydrocarbures*, Paris: Masson and Elf-Aquitaine

Petromin (ed.)(2011): "Blowout Preventers: history, performance and advances," *PetroMin*, July/August 2011

Pettenati-Auzière, C., C. Debouvry and E. Berg (1997): "Node-based sea-bottom seismic: a new way to reservoir management," conference paper, 1997 *SEG Annual Meeting*, 2-7 November, Dallas

Pettitt, Roland (1979): "Completion of hot dry rock geothermal well systems," conference paper, *SPE Annual Technical Conference and Exhibition*, 23-26 September, Las Vegas

Pickett, G.R. (1963): "Acoustic character logs and their application in formation evaluation," *Journal of Petroleum Technology*, vol. 15, no. 6, pp. 659-67

Pirtle, Caleb (2005): *Engineering the World: stories from the first 75 years of Texas Instruments*, Dallas: Southern Methodist University Press

Pledge, Thomas (1998): *Saudi Aramco and its People: a history of training*, Houston: Aramco Services Company

Pohlé, Julius (1892): *Process of Elevating Liquids*, US Patent no. 487639 patented December 6

Pontecorvo, Bruno (1941): "Neutron well logging: a new geological method based on nuclear physics," *Oil and Gas Journal*, vol. 40, pp. 32-33

Pothoven, Boudewijn and Matthijs Dicker (2012): *Our Own Course: 50 years of Heerema Marine Contractors*, Rotterdam: Uitgeverij De Tijdgeest Publishers

Poupon, André, William Hoyle and Arthur Schmidt (1971): "Log analysis in complex formations with complex lithologies," *Journal of Petroleum Technology*, vol. 23, no. 8, pp. 995-1005

Powers, R.W., L.F. Ramirez, C.D. Redmond and E.L. Elberg, Jr. (1966): *Geology of the Arabian Peninsula: sedimentary geology of Saudi Arabia*, Washington (D.C.): United States Geological Survey

Prats, Mike (1961): "Effect of vertical fractures on reservoir behavior: incompressible fluid case," *Society of Petroleum Engineers Journal*, vol. 1, no. 2, pp. 105-18

---------- (1986): *Thermal Recovery*, Richardson (Texas): Society of Petroleum Engineers

Pratt, Joseph, Tyler Priest and Christopher Castaneda (1997): *Offshore Pioneers: Brown & Root and the history of offshore oil and gas*, Houston: Gulf Publishing Company

Prensky, Stephen (1992): "Temperature measurements in boreholes: an overview of engineering and scientific applications," *The Log Analyst*, 1992, May-June, pp. 313-33

---------- (1999): "Advances in Borehole Imaging Technology and Applications," in: Lovell, Mike, Gail Williamson, and Peter Harvey (eds.): *Borehole Imaging: applications and case histories*, London: The Geological Society, pp. 1-43

---------- (2012): "What's new in well logging and formation evaluation, part 2," *World Oil*, July, pp. 107-12

---------- (2013): "What's new in well logging and formation evaluation, part 2," *World Oil*, July, pp. 71-78

Prensky, Stephen and Doug Patterson (2007): "Acoustic Logging," in: Holstein, Edward (ed.): *Petroleum Engineering Handbook*, vol. V (a)(Reservoir engineering and petrophysics), Richardson (Texas): Society of Petroleum Engineers, pp. 167-242

Priest, Tyler (2007): *The Offshore Imperative: Shell Oil's search for petroleum in postwar America*, College Station (Texas): Texas A&M University Press

Proffitt, J.M. (1991): "A history of innovation in marine seismic acquisition," *The Leading Edge*, vol. 10, no. 3, pp. 24-30

Prorfir'ev, V.B. (1974): "Inorganic origin of petroleum," *AAPG Bulletin* 1974, v. 58, pp. 3-33

Proubasta, Dolores (1997): "Hugh W. Hardy," *The Leading Edge of Exploration*, vol. 16, pp. 481-6

Prud'Homme, Alex (2014): *Hydrofracking: what everyone needs to know*, Oxford: Oxford University Press

Pullin, Norm, Larry Matthews and Keith Hirsche (1987): "Techniques applied to obtain very high resolution 3-D seismic imaging at an Athabasca tar sands thermal pilot," *The Leading Edge*, vol. 6, no. 12, pp. 10-15

Purcell, W.R. (1949): "Capillary pressures: their measurement using mercury and the calculation of permeability therefrom," *Journal of Petroleum Technology*, vol. 1, no. 2, pp. 39-48

Pye, David (1964): "Improved secondary recovery by control of water mobility," *Journal of Petroleum Technology*, vol. 16, issue 8, pp. 911-6

Radtke, R.J. et al. (2012): "A new capture and inelastic spectroscopy tool takes geochemical logging to the next level," conference paper, SPWLA 53rd Annual Logging Symposium, June 16-20, Cartagena, Colombia

Ramberg, Rune, Simon Davies and Hege Rognoe (2013): "Steps to the Subsea Factory," conference paper, Offshore Technology Conference, 29-31 October, Rio de Janeiro

Ramey, Hank (1992): "Advances in practical well-test analysis," *Journal of Petroleum Technology*, vol. 44, no. 6, pp. 650-9

Raymond, Martin and William Leffler (2006): *Oil and Gas Production in Nontechnical Language*, Tulsa (Oklahoma): PennWell Books

Redden, Jim (2010): "Dual-gradient drilling promises to change the face of deepwater," *Offshore Magazine*, vol. 70, issue no. 50

Rehm, Bill (2011): *Geological Engineering: how we learned to drill safely in the Gulf of Mexico Miocene shale in the GOM: a story of mud and early drilling problems*, paper given at the Missouri University of Science and Technology, November

Rehm, Bill and Jim Hughes (2008): "Equipment Common to MPD Operations," in: Rehm, Bill et al. (eds.): *Managed Pressure Drilling*, Houston: Gulf Publishing, pp. 227-59

Rehm, Bill et al. (2008): "The Why and Basic Principles of Managed Pressure Drilling," in: Rehm, Bill et al. (eds.): *Managed Pressure Drilling*, Houston: Gulf Publishing, pp. 1-38

Reid, David (2007): "Drilling automation: are we there yet?" *E&P Magazine*, January 16

Reistle, C.E. (1961): "Reservoir Engineering," in: American Petroleum Institute (ed.): *History of Petroleum Engineering*, New York: American Petroleum Institute, pp. 811-46

Retalic, Ian, Andy Laird, and Angus McLeod (2009): "Coiled-Tubing Drilling," in: Aadnoy, Bernt et al. (eds.): *Advanced Drilling and Well Technology*, Richardson (Texas): Society of Petroleum Engineers, pp. 764-85

Rhodes, Richard (1986): *The Making of the Atom Bomb*, New York: Simon & Schuster

Rigmor, M. Elde et al. (2000): "Troll West: reservoir monitoring by 4D seismic," conference paper, *SPE European Petroleum Conference*, 24-25 October, Paris

Roberts, Andrew, Robert Newton and Frederick Stone (1982): "MWD field use and results in the Gulf of Mexico," conference paper, *SPE Annual Technical Conference and Exhibition*, September 26-29, New Orleans, Louisiana

Robinson, Enders (2005): *The MIT Geophysical Analysis Group (GAG) from Inception to 1954*, article manuscript, Newburyport (Massachussetts)

---------- (2013): *Exploration interview answers by Enders Robinson*, unpublished document sent to Mark Mau, March 21

Robinson, Leon and Joe Heilhecker (1973): *Method and Apparatus for Treating a Drilling Fluid*, US patent no. 3766997 patented Oct. 23

Rodengen, Jeffrey (1996): *Legend of Halliburton*, Fort Lauderdale (Florida): Write Stuff Enterprises

Rogers, Walter (1971): "History of Drilling Muds," in: Brantly, John (1971): *History of Oil Well Drilling*, Houston: Gulf Publishing, pp. 1127-32

Roodhart, L.P. et al. (1993): "Frack and pack stimulation: application, design, and field experience from the Gulf of Mexico to Borneo," conference paper, *SPE Annual Technical Conference and Exhibition*, 3-6 October, Houston

Rudnick, Leslie (ed.)(2005): *Synthetics, Mineral Oils, and Bio-Based Lubricants: chemistry and technology*, Boca Raton (Florida): CRC Press

Rumble, R.C. (1955): "A subsurface flowmeter," technical note, *Society of Petroleum Engineers*

Rust, W.M. Jr. (1938): "A historical review of electrical prospecting methods," *Geophysics*, vol. 3, no. 1, pp. 1-6

Rutherford, Steven and Robert Williams (1989): "Amplitude-versus-offset variations in gas sands," *Geophysics*, vol. 54, no. 6 (June), pp. 680-8

Salamy, Salam et al. (2008): "Maximum reservoir contact wells performance update: Shaybah Field, Saudi Arabia," *SPE Production & Operation*, vol. 23, no. 4, November, pp. 439-43

Saleri, Nansen (2000): *Re-engineering reservoir management for the New Millennium*, speech given to the Dhahran Geological Society, Dhahran, February

Saleri, Nansen, Salam Salamy and S.S. Al-Otaibi (2003): "The extending role of the drill bit in shaping the subsurface," *Journal of Petroleum Technology*, December, pp. 53-8

Saleri, Nansen et al. (2004): "Shaybah-220: a maximum-reservoir-contact (MRC) well and its implications for developing tight-facies reservoirs," *SPE Reservoir Evaluation & Engineering*, August, pp. 316-21

Sandia Corporation (ed.)(1988): *Multi-Well Experiment MWX-3: as-built report*, Albuquerque (New Mexico): U.S. Department of Energy

Sandiford, Burton (1964): "Laboratory and field studies of water floods using polymer solutions to increase oil recoveries," *Journal of Petroleum Technology*, vol. 16, issue 8, pp. 917-22

Sandwell, D.T. and W.H.F. (1997): "Marine gravity anomaly from Geosat and ERS-1 satellite altimetry," *Journal of Geophysical Research*, vol. 102, no. 10, pp. 039-10 054

Savage, George and Robert Bradbury (1967): *Marine Drilling Apparatus*, US patent no. 3354951, patented Nov. 28

Savre, Wayland and Jack Burke (1963): "Determination of true porosity and mineral composition in complex lithologies with the use of the sonic, neutron, and density surveys," conference paper, *SPWLA* 4th Annual Logging Symposium, 23-24 May, Oklahoma City, Oklahoma

Schempf, F. Jay (2007): *Pioneering Offshore: the early years*, Tulsa (Oklahoma): PennWell Custom Publishing

Schlumberger (ed.)(1998): *Fifty Years of Schlumberger Research in Ridgefield, 1948-1998*, Paris: Schlumberger

---------- (2001): *A Brief History of Oil in the Middle East*, Dubai: Schlumberger Technical Services

---------- (2007): *80 Years of Innovation*, Paris: Schlumberger

---------- (2013): *Schlumberger: 40 years of permanent downhole monitoring*, internal paper

Schlumberger, Anne Gruner (1982): *The Schlumberger Adventure: two brothers who pioneered in petroleum technology*, New York: Arco Publishing

Schlumberger, Conrad (1920): *Etude sur la Prospection Electrique du Sous-Sol*, Paris: Gauthier-Villars et Cie

Schlumberger, Marcel, Henri-Georges Doll and A. Perebinossoff (1937): "Temperature measurements in oil wells," *Journal of the Institute of Petroleum Technologists*, vol. 23, no. 159

Schneider, William (1998): "3D Seismic: a historical note," *The Leading Edge of Exploration*, vol. 17, pp. 375-80

Schoenberger, Michael (1996): "The Growing Importance of 3-D Seismic Technology," conference paper, *Offshore Technology Conference*, May 6-9, Houston

---------- (2000): "Geophysics," in: Dawe, Richard A. (ed.): *Modern Petroleum Technology*, Chichester: John Wiley, vol. 1 (Upstream), pp. 55-100

Schollnberger, Wolfgang (2007): "Geologie als Basis erfolgreicher Erdöl-Erdgas-Exploration: zu Zeiten Hans Höfers und Heute" (Geology at the Base of Successful Petroleum Exploration: in Hans Höfer's Days and today), *Erdöl Erdgas Kohle*, vol. 123, no. 11, p. 1

Schroeter, Thomas von, Florian Hollaender and Alain Gringarten (2001): "Deconvolution of well test data as a nonlinear total least squares problem," conference paper, *SPE Annual Technical Conference and Exhibition*, 30 September-3 October, New Orleans

Schubert, Jerome and Brandee Elieff (2009): "Deepwater Dual-Gradient Drilling," in: Aadnoy, Bernt et al. (eds.): *Advanced Drilling and Well Technology*, Richardson (Texas): Society of Petroleum Engineers, pp. 621-34

Schultz, A.L., W.T. Bell and H.J. Urbanosky (1975): "Advancements in uncased-hole, wireline formation-tester techniques," *Journal of Petroleum Technology*, vol. 27, no. 11, pp. 1331-6

Science & Technology Review (ed.)(1996): "Exploring oil fields with crosshole electromagnetic induction," *Science & Technology Review*, August 1996, pp. 20-3

Sclater, J. and Phil Christie (1980): "Continental stretching: an explanation of the post-mid-Cretaceous subsidence of the central North Sea," *Journal of Geophysical Research*, vol. 85, pp. 3711-39

Scott, Dan (2006): "The history and impact of synthetic diamond cutters and diamond enhanced inserts on the oil and gas industry," *Industrial Diamond Review*, vol. 66, no. 1

Scott, Floyd (1971): "Hughes Cone Bit," in: Brantly, John: *History of Oil Well Drilling*, Houston: Gulf Publishing, pp. 1070-97

Sédillot, François (1998): "The Hibernia gravity base structure," conference paper, The Eighth *International Offshore and Polar Engineering Conference*, 24-29 May, Montréal

Segré, Emilio (1993): *A Mind Always in Motion: the autobiography of Emilio Segré*, Berkeley: University of California Press

Selley, Richard (1982): *An Introduction to Sedimentology*, London: Academic Press

---------- (1998): *Elements of Petroleum Geology*, San Diego and London: Academic Press

---------- (2000): "Geoscience," in: Dawe, Richard (ed.): *Modern Petroleum Technology*, Chichester: John Wiley, vol. 1, pp. 23-40

Sen, P.N., C. Scala, and M.H. Cohen (1981): "A self-similar model for sedimentary rocks with application to the dielectric constant of fused glass beads," *Geophysics*, vol. 46, no. 5, pp. 781-95

Sengupta, Souvik (2011): "Target detectability of marine controlled source electromagnetic method: insights from 1D modeling," conference presentation, *GEO India*, New Delhi, January 12-14

Sheinman, A. B. et al. (1935): "Gasification of crude oil in reservoir sands," *Neftyance Khozyaistva*, 28 (April 1935), English translation published in *Petroleum Engineer*, 10 (December 1938), p. 27, and (February 1939), p. 91

Shell (ed.)(2009): "Underbalanced drilling offers more," *EP Technology*, no. 2, pp. 14-16

Shen, Jinsong et al. (2008): "Application of 2.5D cross-hole electromagnetic inversion in Gudao Oil Field, East China," *Applied Geophysics*, vol. 5, no. 3, pp. 159-69

Shepard, Francis (1980): *Francis P. Shepard Autobiography*, San Diego: University of California

Sheppard, Mike, Stuart Jardine and Dave Malone (1994): "Putting a damper on drilling's bad vibrations," *Oilfield Review*, January, pp. 15-20

Sheriff, Robert (1991): *Encyclopedic Dictionary of Exploration Geophysics*, Tulsa (Oklahoma): Society of Exploration Geophysicists

Sheriff, Robert and Lloyd Geldart (1995): *Exploration Seismology*, Cambridge: Cambridge University Press

Simmons, Craig (2008): "Henry Darcy (1803-1858): immortalized by his scientific legacy," *Hydrogeology Journal*, vol. 16, pp. 1023-38

Singerman, Philip (1990): *An American Hero: the Red Adair story*, Boston: Little, Brown and Company

Sloss, Lawrence (1963): "Sequences in the Cratonic interior of North Amercia," *Bulletin of the Geological Society of America*, vol. 74, pp. 93-114

Smalley, Craig (2000): "Heavy Oil and Viscous Oil," in: Dawe, Richard (ed.)(2001): *Modern Petroleum Technology*, vol. 1: Upstream, Chichester: Wiley, pp. 409-35

Smith, Ken (2009): "Dual gradient drilling: has its time finally come?" presentation paper at the *AADE* Emerging Technologies Forum, April 22

Smith, Ken et al. (1999): "SubSea MudLift Drilling JIP: achieving dual-gradient technology," *World Oil*, August

Smithson, Tony (2012): "Detonation for delivery: defining perforating," *Oilfield Review*, spring, pp. 55-6

Smits, A.R. et al. (1993): "In-situ optical fluid analysis as an aid to wireline formation sampling," *SPE Formation Evaluation*, vol. 10, no. 2, pp. 91-8

Sonier, F. and P. Chaumet (1974): "A fully implicit three-dimensional model in curvilinear coordinates," *Society of Petroleum Engineers Journal*, vol. 14, no. 4, pp. 361-70

Sonneland, L. et al. (1997): "4D seismic on Gullfaks," conference paper, *Offshore Technology Conference*, 5 May, Houston

Spanne, P. et al. (1994): "Synchrotron computed microtomography of porous media," *Physical Review Letters*, vol. 73, no. 14, pp. 2001-4

Speight, James (2012): *Oil Sand Production Processes*, Oxford: Gulf Professional Publishing

---------- (2013): *Shale Gas Production Processes*, Oxford: Gulf Professional Publishing

Sperling, L.H. (2006): *Introduction to Physical Polymer Science*, New York: John Wiley & Sons

Spinnler, R.F., F.A. Stone and C. Ray Williams (1978): *Mud Pulse Logging while Drilling Telemetry System: design, development, and demonstrations*, Bartlesville (Oklahoma): Teleco Oilfield Services, Inc.

Spotkaeff, Matthew (2007): *Logging While Drilling*, workshop presentation, SPE Queensland

Srnka, Leonard (1986): *Method and Apparatus for Offshore Electromagnetic Sounding Utilizing Wavelength Effects to Determine Optimum Source and Detector Positions*, US Patent no. 4,617,518 patented Oct. 14

St. John, Bill (2013): *Life of an International Petroleum Geologist*, unpublished manuscript, Kerrville (Texas)

Staal, J.J. and J.D. Robinson (1977): "Permeability profiles from acoustic logging," conference paper, *SPE Annual Fall Technical Conference and Exhibition*, 9-12 October, Denver, Colorado

Steen, Øyvind (1993): *På Dypt Vann: Norwegian Contractors 1973-1993*, Oslo: Aker

Stegemeier, George, H.J. Hill and J. Reisberg (1973): "Aqueous surfactant systems for oil recovery," *Journal of Petroleum Technology*, vol. 25, no. 2, pp. 186-194

Stegner, Wallace (1970): *Discovery: the search for Arabian oil*, Beirut: Middle East Export Press

Stiles, David (2006): "Annular Formation Fluid Migration," in: Nelson, Erik and Dominique Guillot (eds.): *Well Cementing*, Sugar Land: Schlumberger, pp. 289-317

Stoneley, Robert (1995): *Introduction to Petroleum Exploration for Non-Geologists*, Oxford: Oxford University Press

Stovall, Smith (1934): "Recovery of oil from depleted sands by means of dry steam," *The Oil Weekly*, August 13, pp. 17-24

Submarine Telecoms Forum (ed.)(2008): "Why we don't automate drilling," *Submarine Telecoms Forum*, issue 40 (Offshore oil & gas issue), pp. 11-14

Suman, John (1961): "Evolution by Companies," in: American Petroleum Institute (ed.): *History of Petroleum Engineering*, New York: American Petroleum Institute, pp. 63-132

Sweatman, Ron (2011): "Cementing," in: Mitchell, Robert and Stefan Miska (eds.): *Fundamentals of Drilling Engineering*, Richardson (Texas): Society of Petroleum Engineers, pp. 139-178

Sweet, George Elliott (1966): *The History of Geophysical Prospecting*, Los Angeles: Science Press

Swigart, Theodore (1961): "Handling of Oil and Gas in the Field" in: American Petroleum Institute (ed.): *History of Petroleum Engineering*, New York: American Petroleum Institute, pp. 907-97

Syed, Ali et al. (2002): "Combined stimulation and sand control", *Oilfield Review*, summer, pp. 30-47

Tait, Samuel (1946): *The Wildcatters: an informal history of oil-hunting in America*, Princeton: Princeton University Press

Takcs, Gabor (2009): *Electrical Submersible Pumps Manual: design, operations, and maintenance*, Houston: Gulf Professional Publishing

Tauxe, Lisa (2010): *Essentials of Paleomagnetism*, Berkeley: University of California Press

Teekay Petrojarl (ed.)(2011): *Petrojarl I*, Trondheim: Teekay Petrojarl

Telford, W. M. et al. (1976): *Applied Geophysics*, Cambridge: Cambridge University

Teufel, Lawrence, Douglas Rhett and Helen Farrell (1991): "Effect of reservoir depletion and pore pressure drawdown on in situ stress and deformation in the Ekofisk Field, North Sea," conference paper, 32nd US Symposium on Rock Mechanics, *American Rock Mechanics Association*, 10-12 July, Norman, Oklahoma

The Geological Society of America (ed.)(1999): *Memorial to Lawrence L. Sloss*, Boulder (Colorado): The Geological Society of America

The Texas Ranger Dispatch (ed.)(2002): "Drilling on the slant," *The Texas Ranger Dispatch*, no. 7 (summer)

Thedy, E.A. et al. (2013): "Jubarte Permanent Reservoir Monitoring: installation and first results," conference paper, 13th

International Congress of the *Brazilian Geophysical Society*, August 26-29, Rio de Janeiro

Theys, Philippe (2010): "Marcel Schlumberger," *Petrophysics*, vol. 51, no. 4 (August)

Thomas, E.C. (1992): "50th anniversary of the Archie Equation: Archie left more than just an equation," *The Log Analyst*, May-June 1992, pp. 199-205

---------- (2007): "Petrophysics," in: Holstein, Edward (ed.): *Petroleum Engineering Handbook*, vol. V (a)(Reservoir engineering and petrophysics), Richardson (Texas): Society of Petroleum Engineers, pp. 77-87

Thomsen, Leon (1997): "Seismic anisotropy: from constipation to exploration effectiveness," *CSEG Recorder*, December, pp. 4-5

Timur, Aytekin (1968): "Effective porosity and permeability of sandstones investigated through nuclear magnetic resonance principles," conference paper, SPWLA 9th Annual Logging Symposium, 23-26 June, New Orleans, Louisiana

Tinkle, Lon (1970): *Mr. De: a biography of Everette Lee DeGolyer*, Toronto: Little, Brown and Company

Tinsley, John (1980): "Study of factors causing annular gas flow following primary cementing," *Journal of Petroleum Technology*, vol. 32, no. 8, pp. 1427-37

Tisot, Jean-Paul (ed.)(2008): *Understanding the Future: geosciences serving society*, Strasbourg: Editions Hirlé

Tissot, Bernard and Dietrich Welte (1978): *Petroleum Formation and Occurrence: a new approach to oil and gas exploration*, Berlin and New York: Springer-Verlag

Tixier, M.P., D. Meunier and J.L. Bonnet (1971): "The production combination tool: a new system for production monitoring," *Journal of Petroleum Technology*, vol. 23, no. 5, pp. 603-13

Tolf, Robert (1976): *The Russian Rockefellers: the saga of the Nobel family and the Russian oil industry*, Stanford (California): Hoover Institution Press

Treitel, Sven (2005): *The MIT Geophysical Analysis Group (GAG), 1954 and beyond*, article manuscript

Underdown, D., K. Das and H. Nguyen (1985): "Gravel packing highly deviated wells with a crosslinked polymer system," *Journal of Petroleum Technology*, vol. 37, no. 12, pp. 2197-202

Uren, Lester (1945): *Apparatus for Placing Gravel in Wells*, US Patent no. 2372461 patented March 27

---------- (1946-53): *Petroleum Production Engineering*, 3 vol., New York: McGraw-Hill

---------- (1971): "History of Drilling Fluid," in: Brantly, John: *History of Oil Well Drilling*, Houston: Gulf Publishing, pp. 1122-7

Uren, Lester and E.H. Fahmy (1927): "Factors influencing the recovery of petroleum from unconsolidated sands by waterflooding, *Transactions of the AIME*, vol. 77, no 1, pp. 318-35

Vail, Peter et al. (1977): "Seismic Stratigraphy and Global Changes of Sea Level," *AAPG Memoir*, vol. 26, pp. 49-212

Vail, Peter, Bob Mitchum and John Sangree (2002): "Sequence stratigraphy: evolution and effects," conference paper, 22nd Annual Gulf Coast Section *SEPM* Foundation Bob F. Perkins Research Conference, March 10, 2002, Houston

Vance, Harold (1961a): "Completion Methods," in: American Petroleum Institute (ed.): *History of Petroleum Engineering*, New York: American Petroleum Institute, pp. 579-616

---------- (1961b): "Evaluation", in: American Petroleum Institute (ed.): *History of Petroleum Engineering*, New York: American Petroleum Institute, pp. 999-1080

Varhaug, Matt (2014): "Big reels at the wellsite," *Oilfield Review*, summer, pp. 63-4

Varian, Russell (1951): *Method and Means for Correlating Nuclear Properties of Atoms and Magnetic Fields*, US Patent no. 2561490A patented October 21

Vassiliou, Marius (2009): *Historical Dictionary of the Petroleum Industry*, Lanham (Maryland): Scarecrow Press

Veatch, Ralph, Zissis Moschovidis and Robert Fast (1989): "An Overview of Hydraulic Fracturing," in: Holditch, Stephen et al. (eds.): *Recent Advances in Hydraulic Fracturing*, Richardson (Texas): Society of Petroleum Engineers, pp. 1-38

Veldman, Hans and George Lagers (1997): *50 Years Offshore*, Delft: Foundation for Offshore Studies

Vidrine, D.J. and E.J. Benit (1968): "Field verification of the effect of differential pressure on drilling rate," *Journal of Petroleum Technology*, vol. 20, no. 7, pp. 675-82

Vincent, R.P., R.M. Leibrock and C.W. Ziemer (1948): "Well flowmeter for logging producing ability of gas sands," *Transactions of the AIME*, vo. 174, no. 1, pp. 305-14

Wade, R.T. et al. (1965): "Production logging: the key to optimum well performance," *Journal of Petroleum Technology*, vol. 17, no. 2, pp. 137-44

Walker, James (1985): *Weld-On Casing Connector*, US patent no. 4509777 patented Apr. 9

Walther, René (2010): "Pechelbronn from 1918 to 1962, or Constitution of a National Oil Company Based on a Local Deposit," in: Beltran, Alain (ed.): *A Comparative History of National Oil Companies*, Bruxelles: P.I.E. Peter Lang, pp. 199-214

Walton, G.G. (1972): "Three-dimensional seismic method," *Geophysics*, vol. 37, pp. 417-30

Wang, Zhijing, Amos Nur and Michael Batzle (1988): "Effect of different pore fluids on velocities of rocks," conference paper, SEG Annual Meeting, 30 October-3 November, Anaheim, California

Wankui, G. et al. (2000): "Commercial pilot test of polymer flooding in Daqing Oil field", conference paper, *SPE/DOE Improved Oil Recovery Symposium*, Tulsa, 3-5 April

Warren, J.E. and P.J. Root (1963): "The behavior of naturally fractured reservoirs," *Society of Petroleum Engineers Journal*, vol. 3, no. 3, pp. 245-55

Warren, Kenneth (2007): "Emulsion Treating," in: Arnold, Kenneth (ed.): *Petroleum Engineering Handbook*, vol. III (Facilities and Construction Engineering), Richardson (Texas): Society of Petroleum Engineers, pp. 61-122

Warren, Tommy (2006): "Steerable motors hold out against rotary steerables," paper prepared for the 2006 *SPE Annual Technical Conference and Exhibition* held in San Antonio (Texas), September 24-27, 2006

Warren, Tommy et al. (1990): "Bit whirl: a new theory of PDC bit failure," *SPE Drilling Engineering*, vol. 5, no. 4 (December), pp. 275-281

Warren, Tommy at al. (1993): Field testing of low-friction gauge PDC bits, *SPE Drilling Engineering*, vol. 8, no. 1 (March), pp. 21-27

Weaver, Bobby (2010): *Oilfield Trash: life and labor in the oil patch*, College Station (Texas): Texas A&M University Press

Weber, Koenraad (1997): "A Historical Overview of the Efforts to Predict and Quantify Hydrocarbon Trapping Features in the Exploration Phase and in Field Development Planning," in: Møller-Pedersen, P. and A.G. Koestler (ed.): *Hydrocarbon Seals: Importance for Exploration and Production*, Singapore: Elsevier, pp. 1-13

Weber, Koenraad and R. Eijpe (1971): "Geological note: mini-permeameters for consolidated rock and unconsolidated sand," *AAPG Bulletin*, vol. 55, no. 2 (February), pp. 307-9

Weber, Koenraad et al. (1978): "Simulation of water Injection in a barrier-bar-type, oil-rim reservoir in Nigeria", *Journal of Petroleum Technology*, vol. 30, no. 11, pp. 1555-66

Weinbrandt, R.M. and Irving Fatt (1969): "Scanning electron microscope study of the pore structure of sandstone," conference paper, 11th *U.S. Symposium on Rock Mechanics*, 16-19 June, Berkeley, California

Wharton, Russell (1983): *Well Logging Fiber Optic Communication System*, US patent no 4,389,645 patented June 21

White, Gerald (1962): *Formative Years in the Far West: a history of Standard Oil Company of California and predecessors through 1919*, New York: Apple-Century-Crofts

Willhite, Paul (1986): *Waterflooding*, Richardson (Texas): Society of Petroleum Engineers

Willhite, Paul and Don Green (1998): *Enhanced Oil Recovery*, Richardson (Texas): Society of Petroleum Engineers

Willhite, Paul and Randall Seright (2011): *Polymer Flooding*, Richardson (Texas): Society of Petroleum Engineers

Williams, Mike (2004): "Better turns for rotary steerable drilling," *Oilfield Review*, spring, pp. 4-9

Williamson, Bob et al. (2004): "Offshore hammermill process meets OSPAR discharge limit," *Oil & Gas Journal*, October 5

Williamson, Harold and Arnold Daum (1959): *The American Petroleum Industry: the age of illumination 1859-1899*, Evanston (Illinois): Northwestern University Press

Wilt, Michael et al. (1995): "Crosshole electromagnetic tomography: a new technology for oil field characterization," *The Leading Edge*, March, pp. 173-7

Winchester, Simon (2001): *The Map that Changed the World*, London: Penguin

Winkler, Kenneth W., Hsui-Lin Liu and David Linton Johnson (1989): "Permeability and borehole Stoneley waves: comparison between experiment and theory," *Geophysics*, vol. 54, no. 1, pp. 66-75

Womack, J. E. (1988): "Simultaneous Vibroseis encoding techniques," conference paper, *SEG Annual Meeting*, Anaheim, California, October 30-November 3

Works, Madden (1971): "Blowout Preventers," in: Brantly, John (1971): *History of Oil Well Drilling*, Houston: Gulf Publishing, pp. 1290-1305

Worthington, Paul (1985): "The evolution of shaly-sand concepts in reservoir evaluation," *The Log Analyst*, vol. 26, no. 1, pp. 23-40

Worthington, Paul, K. Boyle and X.D. Jing (2000): "Petrophysics," in: Dawe, Richard (ed.)(2000): *Modern Petroleum Technology*, vol. 1 (Upstream), New York: John Wiley, pp. 131-206

Wright, Russell and Alexander Sas-Jaworsky (1998): *Coiled Tubing Handbook*, Houston: Gulf Publishing Company

Wyckoff, R.D., H.G. Botset, M. Muskat and D.W. Reed (1933): "The measurement of the permeability of porous media for homogeneous fluids," *Review of Scientific Instruments*, vol. 4, no. 7, pp. 394-405

Yergin, Daniel (1991): *The Prize: the epic quest for oil, money and power*, New York: Simon & Schuster

---------- (2011): *The Quest: energy, security, and the remaking of the modern world*, London: Allen Lane

Youmans, Arthur et al. (1964): "The Neutron Lifetime Log," conference paper, *SPWLA 5th Annual Logging Symposium*, 13-15 May, Midland, Texas

Zamora, Mario (2001): *The 100th Birthday of Drilling Mud*, Power Point Presentation, Houston: American Association of Drilling Engineers

Zanden, Jan Luiten van et al. (2007): *A History of Royal Dutch Shell*, 3 vol., Oxford: Oxford University Press

Zemanek, Joseph et al. (1969): "The borehole televiewer: a new logging concept for fracture location and other types of borehole inspection," *Journal of Petroleum Technology*, vol. 21, issue 6, pp. 762-74

Zieglar, Donald (2000): "A historic review of petroleum exploration and production ideas, concepts and technologies," *Oil-Industry History*, vol. 1, no. 1, pp. 10-24

Zuckerman, Gregory (2013): *The Frackers: the outrageous inside story of the new energy revolution*, London: Portfolio Penguin

Internet sources

American Association of Petroleum Geologists (www.aapg.org)
American Oil & Gas Historical Society (www.aoghs.org)
American Petroleum Institute (www.api.org)
Aramco Expats: Saudi Aramco Expats (www.aramcoexpats.com)
Azerbaijan's Oil History: a chronology leading up to the Soviet era (www.azer.com/aiweb/categories/magazine/ai102_folder/102_articles/102_oil_chronology.html)
Baker Hughes (www.bakerhughes.com)
BP Statistical Review of World Energy (www.bp.com/statisticalreview)
CGG (www.cgg.com)
CSEG Recorder (http://cseg.ca/resources/recorder)
DeGolyer Library (http://smu.edu/cul/degolyer/)
Directional Driller (www.directional-driller.com)
Drilling Contractor (www.drillingcontractor.org)
E&P magazine (www.epmagcom)
Engineering and Technology Wiki (www.ethw.org)
Funding Universe (www.fundinguniverse.com)
Google Patents (www.google.com/patents)
Halliburton (www.halliburton.com)
History of Petroleum Technology (www.spe.org/industry/history)
International Tungsten Industry Association (www.itia.info)
Maersk Oil (www.maerskoil.com)
National Ocean Industries Association (www.noia.org)
National Oilwell Varco (www.nov.com)
Offshore Energy Center's Ocean Star (www.oceanstaroec.com)
Offshore Magazine (www.offshore-mag.com)
Oil & Gas Journal (www.ogj.com)
OnePetro (www.onepetro.org)
Petroleum Geo-Services (www.pgs.com)
Petroleum History Institute (www.petroleumhistory.org)
R.C. Baker Memorial Museum (www.rcbakermuseum.com)
Reference for Business (www.referenceforbusiness.com)
Schlumberger (www.slb.com)
SEG Virtual Museum (http://virtualmuseum.seg.org)
Society of Petroleum Engineers (www.spe.org)
Strange Science: the rocky road to modern paleontology and biology (www.strangescience.net)

The American Institute of Mining, Metallurgical, and Petroleum Engineers (www.aimehq.org)

The Encyclopedia of Earth (www.eoerath.org)

The Geological Society of London (www.geolsoc.org.uk)

The Handbook of Texas Online (www.tshaonline.org)

The History of the Oil Industry (www.sjvgeology.org/history)

Trademarkia (www.trademarkia.com)

Transocean (www.deepwater.com)

Weatherford (www.weatherford.com)

Wikipedia (www.wikipedia.org)

World Oil (www.worldoil.com)

Glossary of acronyms used for organizations and companies

AAPG	American Association of Petroleum Geologists
AEC	Advanced Energy Consortium
AOSTRA	Alberta Oil Sands Technology and Research Authority
API	American Petroleum Institute
ARCO	Atlantic Richfield Company
CASOC	California-Arabian Standard Oil Company
CGG	Compagnie Générale de Géophysique
CMG	Computer Modelling Group
CUSS	A consortium of Continental Oil Company, Union Oil Company of California, Superior Oil Company and Shell Oil Company
DOE	US Department of Energy
ECL	Exploration Consultants Limited
EMI	ElectroMagnetics Instruments Inc.
GAG	Geophysical Analysis Group
GRC	Geophysical Research Corporation
GSI	Geophysical Service Incorporated
IEA	International Energy Agency
IFP	Institut Français du Pétrole
MIT	Massachusetts Institute of Technology
OCT	Oil Center Tool company
ODI	Oil Dynamics Inc.
ODECO	Ocean Drilling and Exploration Company
OPEC	Organization of the Petroleum Exporting Countries
PGS	Petroleum Geo-Services
SEG	Society of Exploration Geophysicists
SOCAL	Standard Oil of California
SPE	Society of Petroleum Engineers
SSC	Seismograph Service Corporation
USGS	US Geological Survey

Mark Mau and Henry Edmundson

Glossaries of oilfield technical terms

We recommend:

The Schlumberger Oilfield Glossary at www.glossary.oilfield.slb.com, also available as a free application for some mobile phones

The Society of Petroleum Engineers' (SPE) PetroWiki at http://petrowiki.org/PetroWiki.

Endnotes

A note on marks

The book does not feature any trade mark or copyright symbols as it has been written as a narrative of oilfield technology development and not as a marketing text. The majority of the technologies mentioned now carry legacy marks. However, those that are current, or date from the recent past, are the property of the companies identified in the relevant sections of the text.

A note on units

We have tried to avoid any unit conversions for a smoother text flow. Whenever the narrative describes actions taking place in North America or Asia, the units are in pounds, feet and miles; for Europe, except the UK, Russia and South America and Africa, the units are in kilograms, meters and kilometers. For the UK, the units are in pounds, meters (for depth) and miles.

Chapter 1: Beginnings

[1] Marco Polo, travel report, 13th century, in: "The Book of Sir Marco Polo the Venetian" (1871), quoted in: Hager, Dorsey (1939): *Fundamentals of the Petroleum Industry*, New York and London: McGraw-Hill, pp. 26-7

[2] Dorozynski, Alexander (2011): *A History of Oil in Russia*, Paris: unpublished manuscript, p. 20

[3] Bommer, Paul (2008): *A Primer of Oilwell Drilling*, Austin (Texas): University of Texas at Austin Petroleum, pp. 7-8; Gelfgat, Yakov, Mikhail Gelfgat and Yuri Lopatin (2003): *Advanced Drilling Solutions: lessons from the FSU*, Tulsa (Oklahoma): PennWell, vol. 1, p. 1; Vassiliou, Marius (2009): *Historical dictionary of the petroleum industry*, Lanham (Maryland): Scarecrow Press, p. 531; and www.azer.com/aiweb/categories/magazine/ai102_folder/102_articles/102_oil_chronology.html

[4] Beeby-Thompson, Arthur (1904): *The Oil Fields of Russia and The Russian Petroleum Industry: a practical handbook on the exploration, exploitation, and management of Russian oil properties*, London: Crosby Lockwood and Son, p. 3; Brantly, John (1971): *History of Oil Well Drilling*, Houston: Gulf Publishing, pp. 29-152; Dorozynski, Alexander (2011), p. 40; and Mitzakis, John (1911): *The Russisan Oil Fields and Petroleum Industry: Being a Practical and Concise Handbook on the Management and Exploitation of Oil Properties, including the History of the Russian Petroleum Industry, Survey Outputs, Tenure of Proliferous Lands, Legislation of English Companies in the Russian Empire, and a List of the Russian Mining Laws*, London: Pall Mall, pp. 43-49

[5] Dickey, Parke (1959): "The first oil well," *Journal of Petroleum Technology*, vol. 59 (Oil Industry Centennial), January, pp. 14-25, here p. 24; Raymond, Martin and William

Leffler (2006): *Oil and Gas Production in nontechnical language*, Tulsa (Oklahoma): PennWell Books, p. 1; Williamson, Harold and Arnold Daum (1959): *The American Petroleum Industry: the age of illumination 1859-1899*, Evanston (Illinois): Northwestern University Press, p. 80; and www.petroleumhistory.org/OilHistory/pages/drake/firstproduction.html

[6] Brantly, John (1961a): "Percussion-Drilling System" in: American Petroleum Institute (ed.)(1961): *History of Petroleum Engineering*, Dallas: American Petroleum Institute, pp. 133-269, here pp. 137-8; and Brantly, John (1971), p. 1257

[7] Tolf, Robert (1976): *The Russian Rockefellers: the saga of the Nobel family and the Russian oil industry*, Stanford (California): Hoover Institution Press, p. 50

[8] Beeby-Thompson, Arthur (1904), pp. 139-174; Mitzakis, John (1911), pp. 43-49; and Tolf, Robert (1976), pp. 65-6

[9] Beecher, C.E. and H.C. Fowler (1961): "Production techniques and control," in: American Petroleum Institute (ed.): *History of Petroleum Engineering*, New York: American Petroleum Institute, pp. 745-810, here p. 751; Brantly, John (1971), pp. 165, 174, 1280; Coberly, C.J. (1961): "Production equipment," in: American Petroleum Institute (ed.): *History of Petroleum Engineering*, New York: American Petroleum Institute, pp. 617-744, here p. 626-7; Vance, Harold (1961a): "Completion Methods" in: American Petroleum Institute (ed.): *History of Petroleum Engineering*, New York: American Petroleum Institute, pp. 579-616, here pp. 581-2; Williamson, Harold and Arnold Daum (1959), p. 99; http://petrowiki.org/Wellhead_systems; and http://petrowiki.org/Tubing

[10] Coberly, C.J. (1961), p. 627; Payne, Darwin (1979): *Initiative in Energy: The Story of Dresser Industries, 1880-1978*, New York: Simon and Schuster, pp. 15, 52-3; Raymond, Martin and William Leffler (2006), p. 4; and Williamson, Harold and Arnold Daum (1959), p. 99

[11] Payne, Darwin (1979), pp. 14, 40, 43, 52-3, 59-60, 79-80, 87-88, 100; and http://en.wikipedia.org/wiki/Solomon_Robert_Dresser

[12] Coberly, C.J. (1961), p. 624, 636, 639; Dyke, Kate van (1997): *Fundamentals of Petroleum*, Austin: The University of Texas at Austin, p. 146; and Williamson, Harold and Arnold Daum (1959), p. 99

[13] Beeby-Thompson, Arthur (1904), pp. 195, 213-44; and www.branobelhistory.com/themes/society/the-pivotal-role-of-azerbaijan-oil-and-baku/

[14] Dorozynski, Alexander (2011), p. 2

Chapter 2: The Birth of a Science

[1] www.strangescience.net/hutton.htm

[2] Owen, Edgar Wesley (1975): *Trek of the Oil Finders: a history of exploration for petroleum*, Tulsa (Oklahoma): American Association of Petroleum Geologists, p. 17

[3] Holmes, Arthur (1965): *Principles of Physical Geology*, London and Edinburgh: Thomas Nelson, pp. 43, 54; McPhee, John (2000): *Annals of the Former World*, New York: Farrar, Straus and Giroux, p. 72; Owen, Edgar Wesley (1975), p. 19; and Selley, Richard (1998): *Elements of Petroleum Geology*, San Diego and London: Academic Press, p. 1

[4] Owen, Edgar Wesley (1975), pp. 19, 22

[5] Interview with Bert Bally; Winchester, Simon (2001): *The Map that Changed the World*, London: Penguin; and http://en.wikipedia.org/wiki/Nicolas_Steno

[6] http://en.wikipedia.org/wiki/Geologic_time_scale

[7] William Logan, cited in: Hunt, T. Sterry (1862): "Notes on the History of Petroleum or Rock Oil," *Smithsonian Institution Annual Report for 1861*, pp. 319-29, here pp. 325-6; and Owen, Edgar Wesley (1975), pp. 51, 67, 71

[8] Interview with Walter Ziegler; DeGolyer, Everette Lee (1938): "Historical notes of the development of the technique of prospecting for petroleum" in: Dunstan, A.E. (1938): *The Science of Petroleum: a comprehensive treatise of the principles and practice of the production, transport and distribution of mineral oil*, London: Oxford University Press, vol. 1, pp. 268-75, here p. 270; Forbes, R.J. and D.R. O'Beirne (1957): *The Technical Development of the Royal Dutch/Shell 1890-1940*, Leiden: E.J. Brill, p. 61; Hager, Dorsey (1939), p. 136; Hunt, T. Sterry (1862); Owen, Edgar Wesley (1975), pp. 34, 51, 54, 61-3, 119-21; and Selley, Richard (2000): "Geoscience," in: Dawe, Richard (ed.): *Modern Petroleum Technology*, Chichester: John Wiley, vol. 1, pp. 23-40, here p. 35

[9] Hager, Dorsey (1939), p. 138; Longhurst, Henry (1959): *Adventure in Oil: the story of British Petroleum*, London: Sidgwick and Jackson, p. 32; and Schlumberger (ed.)(2001): *A Brief History of Oil in the Middle East*, Dubai: Schlumberger Technical Services, p. 36

[10] Anderson, Robert (1985): *Fundamentals of the Petroleum Industry*, London: Weidenfeld and Nicolson, p. 10; Linsley, Judith Walker, Ellen Walker Rienstra and Jo Ann Stiles (2002): *Giant under the Hill: a history of the Spindletop discovery at Beaumont, Texas, in 1901*, Austin (Texas): Texas State Historical Association, p. 40; McDaniel, Robert and Henry Dethloff (1989): *Patillo Higgins and the Search for Texas Oil*, College Station (Texas): Texas A&M University Press, p. 40; Owen, Edgar Wesley (1975), p. 119; and Yergin, Daniel (1991): *The Prize: the epic quest for oil, money and power*, New York: Simon & Schuster, pp. 36, 67

[11] Anstey, Nigel (2013): *History of Exploration*, memo from Nigel Anstey to Mark Mau, Ramsey (Isle of Man), 10 March 2013; Gerding, Mildred (1986): *Fundamentals of Petroleum*, Austin (Texas): University of Texas at Austin Petroleum, p. 19; Hyne, Norman (2001): *Nontechnical Guide to Petroleum Geology, Exploration, Drilling and Production*, Tulsa (Oklahoma): PennWell Books, p. 193; and Linsley, Judith Walker, Ellen Walker Rienstra and Jo Ann Stiles (2002), pp. 56-9

[12] Hyne, Norman (2001), p. 193; and Linsley, Judith Walker, Ellen Walker Rienstra and Jo Ann Stiles (2002), pp. 60-9

Chapter 3: Drilling Gets Established

[1] Anderson, Robert (1985), p. 10; Brantly, John (1971), pp. 805, 216-20; Weaver, Bobby (2010): *Oilfield Trash: life and labor in the oil patch*, College Station (Texas): Texas A&M University Press, p. 3; and Yergin, Daniel (1991), p. 67

[2] White, Gerald (1962): *Formative Years in the Far West: a history of Standard Oil Company of California and predecessors through 1919*, New York: Apple-Century-Crofts, p. 351

[3] *Bakersfield Californian*, May 13, 1909, quoted in: White, Gerald (1962), pp. 353-4

[4] Moore, W.W. (ed.)(1981): *Fundamentals of Rotary Drilling: the rotary drilling system: a professional and practical training guide to its equipment, procedures and technology*, Dallas: Energy Publications, p. 59

[5] Brantly, John (1971), p. 885

[6] Weaver, Bobby (2010), pp. 3-4

[7] Interview with Keith Millheim; and Brantly, John (1961b): "Hydraulic Rotary-Drilling System with addendum on pneumatic rotary drilling," in: American Petroleum Institute (ed.): *History of Petroleum Engineering*, New York: American Petroleum Institute, pp. 271-452, here pp. 325, 404

[8] Hu Harris's letter to John Edward Brantly, July 1, 1954, quoted in: Brantly, John (1971), p. 818

[9] Brantly, John (1961b), pp. 397-8
[10] Baker Hughes (ed.)(2007): "Baker Hughes: 100 years of service," *InDepth* 13, 2 (special issue), pp. 11-12
[11] Baker Hughes (ed.)(2007), pp. 11-12; and Barlett, Donald and James Steele (1979): *Empire: the life, legend and madness of Howard Hughes*, New York: Norton, p. 27
[12] Baker Hughes (ed.)(2007), p. 17; Barlett, Donald and James Steele (1979), p. 52; and Scott, Floyd (1971): "Hughes Cone Bit," in: Brantly, John (1971): *History of Oil Well Drilling*, Houston: Gulf Publishing, pp. 1070-93, here p. 1077
[13] Interview with Hobie Smith; Baker Hughes (ed.)(2007), p. 25; and Scott, Floyd (1971), p. 1084
[14] www.spe.org/about/history.php

Chapter 4: Geophysics Enters the Fray

[1] Eve, A.S. and D.A. Keys (1954): *Applied Geophysics in the Search for Minerals*, Cambridge: Cambridge University Press, pp. 25-8; Lawyer, Lee, Charles Bates and Robert Rice (2001): *Geophysics in the Affairs of Mankind: a personalized history of exploration geophysics*, Tulsa (Oklahoma): Society of Exploration Geophysicists, p. 4; Sweet, George Elliott (1966): *The History of Geophysical Prospecting*, Los Angeles: Science Press, pp. 243-4; and Telford, W. M., L.P. Geldart and R.E. Sheriff (1990): *Applied Geophysics*, Cambridge: Cambridge University Press, p. 25

[2] Dixon, Dougal (1992): *The Practical Geologist: the introductory guide to the basics of geology and to collecting and identifying rocks*, New York: Simon & Schuster, p. 17; www.britannica.com/EBchecked/topic/600183/torsion-balance; and http://en.wikipedia.org/wiki/Torsion_spring

[3] www.magnet.fsu.edu/education/tutorials/pioneers/eotvos.html

[4] Beaton, Kendall (1957): *Enterprise in Oil: a history of Shell in the United States*, New York: Appleton-Century-Crofts, p. 201; DeGolyer, Everette Lee (1938), p. 272; Forbes, R.J. and D.R. O'Beirne (1957), p. 113; Lawyer, Lee, Charles Bates and Robert Rice (2001), p. 4; Owen, Edgar Wesley (1975), pp. 397, 433, 516; Sweet, George Elliott (1966), pp. 97, 183; Tinkle, Lon (1970): *Mr. De: a biography of Everette Lee DeGolyer*, Toronto: Little, Brown and Company, pp. 37-98, 159;
www.tshaonline.org/handbook/online/articles/fdevm;
http://en.wikipedia.org/wiki/Everette_Lee_DeGolyer;
http://en.wikipedia.org/wiki/Mexican_Eagle_Petroleum_Company; www-history.mcs.st-andrews.ac.uk/Biographies/Eotvos.html; www.answers.com/topic/j-zsef-e-tv-s; and www.britannica.com/EBchecked/topic/189421/Roland-baron-von-Eotvos

[5] Sweet, George Elliott (1966), p. 94; and Tinkle, Lon (1970), p. 377

[6] Allaud, Louis and Maurice Martin (1977): *Schlumberger: the history of a technique*, New York: John Wiley & Sons, pp. 13-91; Bromberger, Mary (1954): *Comment ils ont fait fortune*, Paris: Librairie Plon, pp. 7-27; DeGolyer, Everette Lee (1938), p. 272; Schlumberger (ed.)(2007): *80 Years of Innovation*, Paris: Schlumberger, p. 11; Schlumberger, Anne Gruner (1982): *The Schlumberger Adventure: two brothers who pioneered in petroleum technology*, New York: Arco Publishing, pp. 6-7; Schlumberger, Conrad (1920): *Etude sur la Prospection Electrique du Sous-Sol*, Paris: Gauthier-Villars et Cie; Theys, Philippe (2010): "Marcel Schlumberger," *Petrophysics*, vol. 51, no. 4 (August); and www.slb.com/about/history/1920s.aspx

[7] Holmes, Arthur (1965), p. 911; www.uh.edu/engines/epi324.htm; and http://en.wikipedia.org/wiki/Zhang_Heng

[8] Lawyer, Lee, Charles Bates and Robert Rice (2001), p. 4; and www.glossary.oilfield.slb.com/en/Terms.aspx?LookIn=term%20name&filter=seismograph

[9] Interview with Albert Bally; Hager, Dorsey (1939), pp. 150-1; Lawyer, Lee, Charles Bates and Robert Rice (2001), p. 8; Sheriff, Robert and Lloyd Geldart (1995): *Exploration Seismology*, Cambridge: Cambridge University Press, p. 4; Schlumberger, Anne Gruner (1982), p. 6; www.tshaonline.org/handbook/online/articles/doo15; and www.s-cool.co.uk/gcse/physics/properties-of-waves/revise-it/refraction-of-waves

[10] Anderson, Robert (1985), p. 99; Beaton, Kendall (1957), pp. 203-4; Forbes, R.J. and D.R. O'Beirne (1957), p. 126; and Yergin, Daniel (1991), pp. 201-2

[11] Beaton, Kendall (1957), p. 204; Telford, W. M., L.P. Geldart and R.E. Sheriff (1990), p. 137; Sheriff, Robert and Lloyd Geldart (1995), p. 6; Sweet, George Elliott (1966), pp. 236, 287; and www.tshaonline.org/handbook/online/articles/doo15

[12] Lawyer, Lee, Charles Bates and Robert Rice (2001), p. 11

[13] Anderson, Robert (1985), p. 99; Dyke, Kate van (1997); and Sweet, George Elliott (1966), pp. 60-2

[14] Sheriff, Robert and Lloyd Geldart (1995), p. 5; and Tinkle, Lon (1970), pp. 162-3

[15] Forbes, R.J. and D.R. O'Beirne (1957), p. 123; and Meissner, Rolf (1986): *The Continental Crust: a geophysical approach*, London: Academic Press, p. 58

[16] Beaton, Kendall (1957), p. 204; Owen, Edgar Wesley (1975), p. 566; Sheriff, Robert and Lloyd Geldart (1995), p. 8; Tinkle, Lon (1970), p. 204; and Vassiliou, Marius (2009), pp. 192-3

[17] Lawyer, Lee, Charles Bates and Robert Rice (2001), p. 268; Oristaglio, Michael and Alexander Dorozynski (2007): *A Sixth Sense: the life and science of Henri-Georges Doll: oilfield pioneer and inventor*, Cambridge (Massachussetts): The Hammer Company, pp. 134-5; Pirtle, Caleb (2005): *Engineering the World: stories from the first 75 years of Texas Instruments*, Dallas: Southern Methodist University Press, p. 2; Sweet, George Elliott (1966), pp. 236-7, 325; and www.cmc.edu/salvatori/about/

[18] www.aapg.org; http://aoghs.org/energy-education-resources/aapg-geology-pros-since-1917/; and www.seg.org/seg/seg-facts/history

Chapter 5: Discovering the Reservoir

[1] Anderson, Robert (1985), p. 105; Darcy, Henry (Jr.)(2003): "Henry Darcy: Inspecteur Général des Ponts et Chaussées," in: Brown, Glenn, Jürgen and Willi Hager (eds.): *Henry P. G. Darcy and other Pioneers in Hydraulics: contributions in celebration of the 200th birthday of Henry Philibert Gaspard Darcy*, Reston (Virginia): American Society of Civil Engineers, pp. 4-13, here p. 11; Macini, Paolo and Ezio Mesini (2003): "Darcy's Law from Water to the Petroleum Industry: When and Who?" in: Brown, Glenn, Jürgen and Willi Hager (eds.): *Henry P. G. Darcy and other pioneers in hydraulics: contributions in celebration of the 200th birthday of Henry Philibert Gaspard Darcy*, Reston (Virginia): American Society of Civil Engineers, pp. 78-89, here pp. 78-80; Reistle, C.E. (1961): "Reservoir Engineering," in: American Petroleum Institute (ed.): *History of Petroleum Engineering*, New York: American Petroleum Institute, pp. 811-46, here p. 815; and http://petrowiki.org/Darcy

[2] Macini, Paolo and Ezio Mesini (2003), pp. 78-81; Reistle, C.E. (1961), p. 815-6; and Simmons, Craig (2008): "Henry Darcy (1803-1858): immortalized by his scientific legacy," *Hydrogeology Journal*, vol. 16, pp. 1023-38

[3] Macini, Paolo and Ezio Mesini (2003), pp. 81-2, 85; Reistle, , C.E. (1961), pp. 817-8, 820; Vance, Harold (1961b): "Evaluation," in: American Petroleum Institute (ed.): *History of Petroleum Engineering*, New York: American Petroleum Institute, pp. 999-1080,

here p. 1027; http://en.wikipedia.org/wiki/Perley_G._Nutting; and http://en.wikipedia.org/wiki/Jean_Léonard_Marie_Poiseuille

[4] Macini, Paolo and Ezio Mesini (2003), p. 82; Wyckoff, R.D., H.G. Botset, M. Muskat and D.W. Reed (1933): "The measurement of the permeability of porous media for homogeneous fluids," *Review of Scientific Instruments*, vol. 4, no. 7, pp. 394-405; Selley, Richard (1998), p. 250; www.aimehq.org/programs/award/bio/morris-muskat; www.aimehq.org/programs/award/bio/ralph-d-wyckoff; http://www.glossary.oilfield.slb.com/en/Terms/d/darcy.aspx; and http://en.wikipedia.org/wiki/Darcy_(unit)

[5] Dria, Dennis (2007): "Mud Logging," in: Lake, Larry (ed.): *Petroleum Engineering Handbook*, vol. V (a)(Reservoir engineering and petrophysics), Richardson (Texas): Society of Petroleum Engineers, pp. 357-77, here p. 357; Léonardon, Eugène (1961): "Logging, Sampling, and Testing," in: American Petroleum Institute (ed.): *History of Petroleum Engineering*, New York: American Petroleum Institute, pp. 493-578, here pp. 505, 515; and http://petrowiki.org/Mud_Logging

[6] Czarniecki, Stanislaw (1993): "Grzybowski and his School: the beginnings of applied micropaleontology in Poland at the turn of the 19th and 20th centuries," in: Kaminski, M.A., S. Geroch, and D.G. Kaminski (eds.): *The Origins of Applied Micropalaeontology: The School of Jozef Grzybowski*, Krakow: The Grzybowski Foundation, pp. 1-15; Owen, Edgar Wesley (1975), p. 522; and www.gf.tmsoc.org/Documents/Origins/Czarniecki-Origins-1993.pdf

[7] Kass-Simon, Gabriele, Patricia Farnes and Deborah Nash (eds.)(1993): *Women of Science: righting the record*, New York: John Wiley and Sons, p. 63; Larson, Henrietta and Kenneth Porter (1959): *History of Humble Oil & Refining Company: a study in industrial growth*, New York: Harper, pp. 115-6; www.jsg.utexas.edu/npl/history/geologists/alva-c-ellisor-1892-1964/; www.jsg.utexas.edu/npl/history/geologists/esther-applin-1895-1972/; www.jsg.utexas.edu/npl/history/geologists/hedwig-t-kniker-1891-1985/; and http://archives.datapages.com/data/bull_memorials/057/057003/pdfs/596.htm

[8] Beecher, C.E. and H.C. Fowler (1961), p. 793; Brantly, John (1971), pp. 9, 369, 1108-12; Elliott, John (1925): "The Elliott Core Drills," *Transactions of the AIME*, vol. G-25, no. 1, pp. 58-61; Forbes, R.J. and D.R. O'Beirne (1957), p. 174; Gow, Sandy (2005): *Roughnecks, Rock Bits and Rigs: the evolution of oil well drilling technology in Alberta, 1883-1970*, Calgary: University of Calgary Press, pp. 40-1; Hyne, Norman (2001), p. 299; McCray, Arthur and Frank Cole (1959): *Oil Well Drilling Technology*, Norman (Oklahoma): University of Oklahoma Press, p. 359; Williamson, Harold and Arnold Daum (1959), p. 289; Zieglar, Donald (2000): "A historic review of petroleum exploration and production ideas, concepts and technologies," *Oil-Industry History*, vol. 1, no. 1, pp. 10-24, here pp. 18-9; www.aimehq.org/programs/award/bio/john-e-elliott; http://directional-driller.com/page46/page13/page13.html; and http://archives.datapages.com/data/bull_memorials/065/065003/pdfs/544.htm

[9] Carll, John Franklin (1880) : *Second Geological Survey of Pennsylvania: 1875 to 1879: the geology of the oil regions of Warren, Venango, Clarion, and Butler counties*, Harrisburg (Pennsylvania): The Board of Commissioners for The Second Geological Survey, p. 278; Elkins, Lloyd (1961b): "Research," in: American Petroleum Institute (ed.): *History of Petroleum Engineering*, New York: American Petroleum Institute, pp. 1081-1113, here p. 1092; Nielsen, R.F. and L.T. Bissey (1966): *Petroleum Production Research at Penn State in Retrospect*, University Park (Pennsylvania): The Pennsylvania State University, p. 4; Vance, Harold (1961b), p. 1027; and www.waterhistory.org/histories/newell/

[10] Barnes, Kenneth (1931): *A Method for Determining the Effective Porosity of a Reservoir-Rock*, The Pennsylvania State College Bulletin, Mineral Industries Experiment Station, bulletin 10, State College (Pennsylvania): The Pennsylvania State College, p. 1; Elkins, Lloyd

(1961b), p. 1092; Reistle, C.E. (1961), pp. 821-2, 826; and Vance, Harold (1961b), p. 1027

[11] Vance, Harold (1961b), p. 1027; and http://archives.datapages.com/data/bull_memorials/067/067004/pdfs/716.htm

[12] Elkins, Lloyd (1961b), p. 1092; Barnes, Kenneth (1931), p. 1; and Nielsen, R.F. and L.T. Bissey (1966), pp. 1-2

[13] Reistle, C.E. (1961), pp. 827, 1092; Barnes, Kenneth (1931), pp. Nielsen, R.F. and L.T. Bissey (1966), p.4; and www.clays.org/journal/archive/volume%201/1-1-306.pdf

[14] Allaud, Louis and Maurice Martin (1977), pp. 101-3; Auletta, Ken (1984): *The Art of Corporate Success: the story of Schlumberger*, New York: G.P. Putnam's Sons, p. 25; Gillingham, W.J. (1977): *Schlumberger: the first years*, Paris: Schlumberger, pp. 7-8; Oristaglio, Michael and Alexander Dorozynski (2007), p. 56; and Walther, René (2010): "Pechelbronn from 1918 to 1962, or Constitution of a National Oil Company Based on a Local Deposit," in: Beltran, Alain (ed.): *A Comparative History of National Oil Companies*, Bruxelles: P.I.E. Peter Lang, pp. 199-214, here p. 202

[15] Allaud, Louis and Maurice Martin (1977), pp. 101-4; and Oristaglio, Michael and Alexander Dorozynski (2007), pp. 24, 52

[16] Interview with Philippe Theys; Allaud, Louis and Maurice Martin (1977), pp. 101-4; Léonardon, Eugène (1961), p. 523; Oristaglio, Michael and Alexander Dorozynski (2007), pp. 24, 52; Rust, W.M. Jr. (1938): "A historical review of electrical prospecting methods," *Geophysics*, vol. 3, no. 1, pp. 1-6; and www.igep.tu-bs.de/geschichte/ambronnb.htm

[17] Léonardon, Eugène (1961), p. 524; and Schlumberger (ed.)(2007), p. 11

[18] Edmundson, Henry (1987): "Raymond Sauvage: recollections Schlumberger Wireline's first years," *The Technical Review*, vol. 35, no. 1, pp. 4-15, here p. 10; Forbes, R.J. and D.R. O'Beirne (1957), p. 205; Johnson, Hamilton (1961): "A history of well logging," conference paper, *SPWLA* 2nd annual logging symposium, 18-19 May, Dallas, p. iii; Léonardon, Eugène (1961), pp. 524, 538; Oristaglio, Michael and Alexander Dorozynski (2007), pp. 66-9; Schlumberger, Anne Gruner (1982); Walther, René (2010), p. 202; and www.brothersreunited.com/brothers2/brothers2/newsletter200501.html

[19] Forbes, R.J. and D.R. O'Beirne (1957), p. 205; Léonardon, Eugène (1961), p. 525; http://petrowiki.org/Spontaneous_(SP)_log; and http://en.wikipedia.org/wiki/Spontaneous_potential

[20] Gillingham, W.J. (1977), pp. 8-9; Johnson, Hamilton (1961), p. 4; Léonardon, Eugène (1961), p. 534; and Oristaglio, Michael and Alexander Dorozynski (2007), p. 90

[21] Allaud, Louis and Maurice Martin (1977), pp. 179-82; Doll, Henri-Georges (1953): "Two decades of electrical logging," *Journal of Petroleum Technology*, vol. 5, no. 9 (September), pp. 33-41, here p. 33; Edmundson, Henry (1987), pp. 13-4; Johnson, Hamilton (1961), pp. iv, 7; Oristaglio, Michael and Alexander Dorozynski (2007), pp. 115-8; and Schlumberger (ed.)(1998): *Fifty Years of Schlumberger Research in Ridgefield, 1948-1998*, Paris: Schlumberger

[22] Allaud, Louis and Maurice Martin (1977), pp. 182-8; Léonardon, Eugène (1961), p. 521; and McCray, Arthur and Frank Cole (1959), p. 368

Chapter 6: First Steps Offshore

[1] Baker, Ron (1998): *A Primer of Offshore Operations*, Austin: The University of Texas at Austin, p. 1; Brantly, John (1971), p. 1364; Lankford, Raymond (1971): "Marine Drilling," in: Brantly, John (1971): *History of Oil Well Drilling*, Houston: Gulf Publishing, pp. 1358-1444, here p. 1365; Lathim, Rod (1995): *The Spirit of the Big Yellow House: a history of Summerland's founding family*, Santa Barbara (California): Emily Publications, pp.

32-47; Leffler, William, Richard Pattarozzi and Gordon Sterling (2003): *Deepwater Petroleum Exploration and Production: a nontechnical guide*, Tulsa (Oklahoma): PennWell Books, p. 1; Schempf, F. Jay (2007): *Pioneering Offshore: the early years*, Tulsa (Oklahoma): PennWell Custom Publishing, pp. 9-10; and www.noia.org/website/article.asp?id=123

[2] Lathim, Rod (1995), p. 59

[3] Lathim, Rod (1995), p. 68; and Schempf, F. Jay (2007), p. 10

[4] Lankford, Raymond (1971), p. 1369; Priest, Tyler (2007): *The Offshore Imperative: Shell Oil's search for petroleum in postwar America*, College Station (Texas): Texas A&M University Press, p. 30; Raymond, Martin and William Leffler (2006), p. 17; and Schempf, F. Jay (2007), p. 13

[5] Brantly (1971), p. 1372; Lankford, Raymond (1971), p. 1372; and Schempf, F. Jay (2007), pp. 10, 28-9

[6] Schempf, F. Jay (2007), pp. 28-29

[7] Interview with Ed Horton; and Raymond, Martin and William Leffler (2006), p. 107

[8] Schempf, F. Jay (2007), pp. 2-6, 22

[9] Priest, Tyler (2007), pp. 57-8; and Schempf, F. Jay (2007), pp. 34, 36, 39

[10] Samuel Lewis in *Scientific American* (1869), here quoted in: Brantly (1971), p. 1362

[11] Priest, Tyler (2007), p. 58; and Schempf, F. Jay (2007), p. 57

[12] Interview of Aubrey Bassett by Tyler Priest (Houston History Project); and Priest, Tyler (2007), pp. 45-6

[13] Dobrin, Milton and Carl Savit (1976): *Introduction to Geophysical Prospecting*, New York and London, pp. 63-4; Lawyer, Lee, Charles Bates and Robert Rice (2001), p. 68; Priest, Tyler (2007), pp. 42, 46; Proffitt, J.M. (1991): "A History of Innovation in Marine Seismic Acquisition," *The Leading Edge*, vol. 10, no. 3, pp. 24-30; Sheriff, Robert (1991): *Encyclopedic Dictionary of Exploration Geophysics*, Tulsa (Oklahoma): Society of Exploration Geophysicists, p. 152; Telford, W. M., L.P. Geldart and R.E. Sheriff (1990), pp. 334-5; and Zanden, Jan Luiten van et al. (2007): *A History of Royal Dutch Shell*, Oxford: Oxford University Press, vol. 1, p. 329

[14] Lawyer, Lee, Charles Bates and Robert Rice (2001), p. 68; Priest, Tyler (2007), pp. 42; and Proffitt, J.M. (1991)

Chapter 7: The Service Industry Takes Hold

[1] Baker Hughes (ed.)(2007), pp. 8-9; Brantly, John (1961b), p. 277; and Brantly, John (1971), pp. 264, 1035, 1273, 1275

[2] www.rcbakermuseum.com/RCBaker.html

[3] Millikan, Charles (1961): "Cementing," in: American Petroleum Institute (ed.)(1961): *History of Petroleum Engineering*, New York: American Petroleum Institute, pp. 453-492, here p. 455

[4] John R. Hill, quoted in: Millikan, Charles (1961), p. 455; Vassiliou, Marius (2009), p. 478; and www.elsmerecanyon.com/picocanyon/history/history.htm

[5] Sweatman, Ron (2011): "Cementing," in: Mitchell, Robert and Stefan Miska (eds.)(2011): *Fundamentals of Drilling Engineering*, Richardson (Texas): Society of Petroleum Engineers, pp. 139-178, here p. 140; en.wikipedia.org/wiki/Joseph_Aspdin; and en.wikipedia.org/wiki/Portland_cement

[6] Elkins, Lloyd (1961b), pp. 1098-1099; Leffler, William, Richard Pattarozzi and Gordon Sterling (2003), p. 5; and Millikan, Charles (1961), p. 456

[7] Rodengen, Jeffrey (1996): *Legend of Halliburton*, Fort Lauderdale (Florida): Write Stuff Enterprises, p. 15

[8] Erle Halliburton, quoted in: *The Cementer*, July/August 1949, here taken from: Rodengen, Jeffrey (1996), p. 16

[9] Rodengen, Jeffrey (1996), pp. 19-20

[10] Millikan, Charles (1961), p. 488; National Research Council (ed.)(1940): *Industrial Research Laboratories of the United States*, 7th edition, Washington (D.C.): National Academy of Sciences, p. 130; and Rodengen, Jeffrey (1996), p. 46

[11] Rodengen, Jeffrey (1996), p. 45

[12] Works, Madden (1971): "Blowout Preventers," in: Brantly, John (1971): *History of Oil Well Drilling*, Houston: Gulf Publishing, pp. 1290-1305, here p. 1291

[13] Lanagan, Mike (1882): *Safety Attachment for Oil Wells and Tanks*, US patent no 267,903, patented Nov. 21; and Works, Madden (1971), p. 1292

[14] Cooper Cameron (ed.)(2003): *Designation ceremony, Cameron first ram-type BOP, an ASME Historic Mechanical Engineering Landmark, July 14, 2003*, Houston: Cooper Cameron Corporation Division; and Works, Madden T. (1971), pp. 1295-7

[15] N.H. Leroy quoted in: Brantly, John (1961b), pp. 379-380

[16] Uren, Lester (1971): "History of Drilling Fluid," in: Brantly, John (1971): *History of Oil Well Drilling*, Houston: Gulf Publishing, pp. 1122-7, here p. 1123

[17] Eustes III, A.W. (2011): "Drilling Fluids," in: Mitchell, Robert and Stefan Miska (eds.)(2011): *Fundamentals of Drilling Engineering*, Richardson (Texas): Society of Petroleum Engineers, pp. 87-138, here pp. 87-88

[18] Darley, Henry, George Gray and Ryen Caenn (2011): *Composition and Properties of Drilling and Completion Fluids*, Waltham (Massachusetts): Gulf Professional Publishing, pp. 44, 63-64; www.glossary.oilfield.slb.com/Display.cfm?Term=bentonite; www.glossary.oilfield.slb.com/Display.cfm?Term=montmorillonite; and en.wikipedia.org/wiki/Bentonite

[19] Darley, Henry, George Gray and Ryen Caenn (2011), pp. 63-5; Moore, Thomas and George Cannon (1936): *Weighted Oil Base Drilling Fluid*, US patent no. 2,055,666, patented Sept. 29; and Uren, Lester (1971), p. 1126

[20] Darley, Henry, George Gray and Ryen Caenn (2011), p. 58; Marsh, H.N. and Ward Kelly (1971): "Development of Instruments," (Rotary Drilling Fluid) in: Brantly, John (1971): *History of Oil Well Drilling*, Houston: Gulf Publishing, pp. 1133-45, here pp. 1141-2; and www.glossary.oilfield.slb.com/display.cfm?term=shale%20shaker

Chapter 8: Beginnings of Production Engineering

[1] Brantly, John (1971), p. 1241-3, 1245; Forbes, R.J. and D.R. O'Beirne (1957), p. 198; Lee, John (1982): *Well Testing*, Richardson (Texas): Society of Petroleum Engineers, p. 97; Lee, John, John Rollins and John Spivey (2003): *Pressure Transient Testing*, Richardson (Texas): Society of Petroleum Engineers, p. 151; Léonardon, Eugène (1961), pp. 561-2; McCray, Arthur and Frank Cole (1959), p. 371; Rodengen, Jeffrey (1996), pp. 34-5; Uren, Lester (1953): *Petroleum Production Engineering*, New York: McGraw-Hill, vol. 3, p. 561; and www.slb.com/services/characterization/testing/drillstem/first_dst.aspx

[2] Brantly, John (1971), pp. 1243, 1245; Léonardon, Eugène (1961), pp. 561-2; O'Neill, Frank (1934): "Formation testers," *Transactions of the AIME*, vol. 107, no. 1, pp. 53-61; and Rodengen, Jeffrey (1996), pp. 34-5

[3] Bowker, Geoffrey (1994): *Science on the Run: information management and industrial geophysics at Schlumberger, 1920-1940*, Cambridge (Massachussetts): MIT Press, pp. 115, 120-5; Edmundson, Henry (2010): *Patents and Other Sundry Matters*, Cambridge, unpublished

article; Oristaglio, Michael and Alexander Dorozynski (2007), pp. 115-8, 172-84; and Rodengen, Jeffrey (1996), pp. 36-7, 41-3

[4] Beecher, C.E. and H.C. Fowler (1961): "Production Techniques and Control," in: American Petroleum Institute (ed.): *History of Petroleum Engineering*, New York: American Petroleum Institute, pp. 745-810, here pp. 794-5; and Harrison, David and Yves Chauvel (2007): "Reservoir Pressure and Temperature," in: Lake, Larry (ed.): *Petroleum Engineering Handbook*, vol. V a (Reservoir engineering and petrophysics), Richardson (Texas): Society of Petroleum Engineers, pp. 683-717, here p. 683

[5] Léonardon, Eugène (1961), pp. 557-8; and Prensky, Stephen (1992): "Temperature measurements in boreholes: an overview of engineering and scientific applications," *The Log Analyst*, 1992, May-June, pp. 313-33, here p. 314

[6] Beecher, C.E. and H.C. Fowler (1961), p. 798; Léonardon, Eugène (1961), p. 558; Prensky, Stephen (1992), p. 314; and Reistle, C.E. (1961), p. 825

[7] Beecher, C.E. and H.C. Fowler (1961), pp. 794, 796, 798; Forbes, R.J. and D.R. O'Beirne (1957), p. 253; Hager, Dorsey (1939), p. 246; Suman, John (1961): "Evolution by Companies," in: American Petroleum Institute (ed.): *History of Petroleum Engineering*, New York: American Petroleum Institute, pp. 63-132, here pp. 76, 82, 85; www.aimehq.org/programs/award/bio/charles-v-millikan; and http://archives.datapages.com/data/bull_memorials/073/073005/pdfs/672.htm

[8] Anderson, Robert (1985), p. 25; Beecher, C.E. and H.C. Fowler (1961), pp. 797, 799; Larson, Henrietta and Kenneth Porter (1959), pp. 436-7; Léonardon, Eugène (1961), pp. 558, 560; and Suman, John (1961), pp. 76, 100, 115

[9] Beecher, C.E. and H.C. Fowler (1961), p. 798; and Prensky, Stephen (1992), p. 314

[10] Leonardon (1961), p. 558; and Smithson, Tony (2012): "Detonation for delivery: defining perforating," *Oilfield Review*, spring, pp. 55-6

[11] Hager, Dorsey (1939), p. 246-7; Raymond, Martin and William Leffler (2006), p. 5; and Vance, Harold (1961a), pp. 587, 589-90

[12] Interview with George King; Forbes, R.J. and D.R. O'Beirne (1957), p. 207; King, George (2007): "Perforating," in: Clegg, Joe Dunn (ed.): *Petroleum Engineering Handbook*, vol. IV (Production operations engineering), Richardson (Texas): Society of Petroleum Engineers, pp. 149-73, here p. 151; Payne, Darwin (1979), p. 224; Vance, Harold (1961a), pp. 590-1; and www.epmag.com/EP-Magazine/archive/Shoot-steel-rock_579

[13] www.epmag.com/EP-Magazine/archive/Shoot-steel-rock_579

[14] Baker Hughes (ed.)(2007), pp. 23, 27; and Vance, Harold (1961a), pp. 590-1

[15] Allaud, Louis and Maurice Martin (1977), pp. 190, 207-8; Edmundson, Henry (2010); Forbes, R.J. and D.R. O'Beirne (1957), p. 207; Millikan, Charles (1961), p. 481; Payne, Darwin (1979), p. 224; and www.epmag.com/EP-Magazine/archive/Shoot-steel-rock_579

[16] Beeby-Thompson, Arthur (1904), pp. 247, 251, 272; Coberly, C.J. (1961), pp. 678-9, 684; Hager, Dorsey (1939), p. 262; Pohlé, Julius (1892): *Process of Elevating Liquids*, US Patent no. 487639 patented December 6; and http://en.wikipedia.org/wiki/Alexander_Mantashev

[17] Interview with Joe Dunn Clegg; Allen, Thomas and Alan Roberts (1993): *Production Operations*, vol. 1 (Well completions, Workover, and Stimulation), Tulsa (Oklahoma): Pennwell, p. 527; Coberly, C.J. (1961), p. 684; Forbes, R.J. and D.R. O'Beirne (1957), pp. 233, 235; Suman, John (1961), p. 127; and Swigart, Theodore (1961): "Handling Oil and Gas in the Field," in: American Petroleum Institute (ed.): *History of Petroleum Engineering*, New York: American Petroleum Institute, pp. 907-97, here p. 921

[18] Coberly, C.J. (1961), pp. 680-1, 683-6

[19] Coberly, C.J. (1961), pp. 686, 688; Gammack, William and Donald Knox (1932): *Gas Lift for Oil Wells*, US Patent no. 1789866 patented January 20; Hager, Dorsey (1939), pp. 262-3; Jahn, Frank, Mark Cook and Mark Graham (2008): *Hydrocarbon Exploration and Production*, Amsterdam and Oxford: Elsevier, p. 259; and Jordan, Frank and Arthur White (1932): *Valve Construction for Gas Lift Pumps*, US Patent no. 1789855 patented January 20

[20] Bearden, John, Earl Brookbank, and Brown Wilson (2009): *How we did it then: the evolutionary growth of ESPs (A survey of progress and changes in the ESP product during the past 82+ years)*, oral presentation given at the SPE ESP Workshop, The Woodlands, Texas, US, 29 April-1 May (reprinted in: Noonan, Shauna (2011): *Electric Submersible Pumps*, digital edition, Richardson (Texas): Society of Petroleum Engineers, pp. 1-33); Coberly, C.J. (1961), p. 691; and http://esppump.com

[21] Bearden, John, Earl Brookbank and Brown Wilson (2009); and http://esppump.com

[22] Bearden, John (2007): "Electrical Submersible Pumps," in: Clegg, Joe Dunn (ed.): *Petroleum Engineering Handbook*, vol. IV (Production operations engineering), Richardson (Texas): Society of Petroleum Engineers, pp. 625-711, here pp. 625-6; Bearden, John, Earl Brookbank and Brown Wilson (2009); and http://petrowiki.org/Electrical_submersible_pumps

[23] Bearden, John, Earl Brookbank and Brown Wilson (2009); Coberly, C.J. (1961), pp. 703-4; and Takcs, Gabor (2009): *Electrical Submersible Pumps Manual: design, operations, and maintenance*, Houston: Gulf Professional Publishing, p. 5

[24] Bearden, John (2007), pp. 625-6; Bearden, John, Earl Brookbank and Brown Wilson (2009); and http://petrowiki.org/Electrical_submersible_pumps

[25] Interview with Joe Dunn Clegg; Bearden, John, Earl Brookbank and Brown Wilson (2009); Coberly, C.J. (1961), pp. 625, 693; Fretwell, James (2007): "Hydraulic Pumping in Oil Wells," in: Clegg, Joe Dunn (ed.): *Petroleum Engineering Handbook*, vol. IV (Production operations engineering), Richardson (Texas): Society of Petroleum Engineers, pp. 713-56, here p. 713; Gerding, Mildred (1986), p. 196; Hager, Dorsey (1939), p. 262; Hyne, Norman (2001), p. 358; Suman, John (1961), p. 86; and http://esppump.com

[26] Beecher, C.E. and H.C. Fowler (1961), p. 757; Frasch, Hermann (1896): *Increasing the Flow of Oil-Wells*, US Patent no. 556669 patented March 17; Vance, Harold (1961a), p. 598; Vassiliou, Marius (2009), p. 198; http://en.wikipedia.org/wiki/Herman_Frasch; and http://en.wikisource.org/wiki/The_Cyclop%C3%A6dia_of_American_Biography/Frasch,_Herman

[27] Grebe, John and Sylvia Stoesser (1935): "Increasing crude production 20,000,000 Bbl. from established fields," *World Petroleum*, August 1935, pp. 473-82; Vance, Harold (1961a), pp. 598-9; and Suman, John (1961), p. 90

[28] Giddens, Paul (1955): *Standard Oild Company (Indiana): oil pioneer of the Middle-West*, New York: Appleton-Century-Crofts, p. vii, Hager, Dorsey (1939), p. 237; and Suman, John (1961), pp. 86, 100, 112

[29] Coberly, C.J. (1961), pp. 715-8, 724; Gerretson, Frederik Carel (1958): *History of the Royal Dutch Shell*, Leiden: E.J. Brill, vol. 1, p. 247; and Swigart, Theodore (1961), p. 920

[30] Hager, Dorsey (1939), pp. 243-4; and Swigart, Theodore (1961), p. 921

[31] Arnold, Ken and Maurice Stewart (2008): *Surface Production Operations*, vol. 1 (Design of Oil-Handling Systems and Facilities), Oxford: Gulf Professional Publishing, p. 150; Warren, Kenneth (2007): "Emulsion Treating," in: Lake, Larry (ed.): *Petroleum Engineering Handbook*, vol. III (Facilities and Construction Engineering), Richardson (Texas): Society of Petroleum Engineers, pp. 61-122, here p. 76; http://petrowiki.org/Emulsion_Treating; and http://en.wikipedia.org/wiki/API_oil-water_separator

[32] Arnold, Ken and Maurice Stewart (2008), p. 403; Baker Hughes (ed.)(2007), p. 12; Coberly, C.J. (1961), p. 727; Warren, Kenneth (2007), p. 66 and http://en.wikipedia.org/wiki/Melvin_De_Groote

Chapter 9: Exploring the World

[1] http://foundation.aapg.org/Pratt.cfm

[2] Owen, Edgar Wesley (1975), pp. 221-2, 375

[3] Anderson, Robert (1985), p. 91; Bommer, Paul (2008), p. 62; Dixon, Dougal (1992), p. 52; Holmes, Arthur (1965), p. 454; Selley, Richard (2000), p. 35; Stoneley, Robert (1995): *Introduction to Petroleum Exploration for Non-Geologists*, Oxford: Oxford University Press, p. 50; www.glossary.oilfield.slb.com/en/Terms.aspx?LookIn=term%20name&filter=biostratigraphy; and www.glossary.oilfield.slb.com/en/Terms/s/stratigraphy.aspx

[4] Frehner, Brian (2011): *Finding Oil: the nature of petroleum geology, 1859-1920*, Lincoln (Nebraska): University of Nebraska Press, pp. 60, 76-7; and Owen, Edgar Wesley (1975), p. 107

[5] Owen, Edgar Wesley (1975), pp. 459-60, 566-7

[6] Powers, R.W., L.F. Ramirez, C.D. Redmond and E.L. Elberg, Jr. (1966): *Geology of the Arabian Peninsula: sedimentary geology of Saudi Arabia*, Washington (D.C.): United States Geological Survey, p. D2; Stegner, Wallace (1970): *Discovery: the search for Arabian oil*, Beirut: Middle East Export Press, p. 70; http://en.wikipedia.org/wiki/Max_Steineke; and www.aramcoexpats.com/obituaries/1952/04/max-steineke.aspx

[7] Comeaux, Malcolm (1987): *One Hundred and One Years of Geography at Arizona State University*, Tempe: Arizona State University Department of Geography, p. 16; and Owen, Edgar Wesley (1975), pp. 1328-9

[8] Brown, Anthony Cave (1999): *Oil, God, and Gold: the story of Aramco and the Saudi Kings*, Boston: Houghton Mifflin, pp. 69-70; and Owen, Edgar Wesley (1975), pp. 1328-30

[9] Beaton, Kendall (1957), p. 204; Hager, Dorsey (1939), p. 148; Owen, Edgar Wesley (1975), pp. 1328-30; Pledge, Thomas (1998): *Saudi Aramco and its People: a history of training*, Houston: Aramco Services Company, p. 6; and Stegner, Wallace (1970), p. 32

[10] Brown, Anthony Cave (1999), p. 74; Owen, p. 1330; Pledge, Thomas (1998), p. 10; and http://en.wikipedia.org/wiki/Max_Steineke

[11] Green, Cecil (2001): "Reflections on the Field of Exploration Geophysics" in: Lawyer, Lee, Charles Bates and Robert Rice: *Geophysics in the Affairs of Mankind: A personalized History of Exploration Geophysics*, Tulsa (Oklahoma): Society of Exploration Geophysicists, pp. 371-8, here p. 377; and Pirtle, Caleb (2005), pp. 10-1

[12] www.aramcoexpats.com/obituaries/1952/04/max-steineke.aspx; and http://en.wikipedia.org/wiki/Max_Steineke

[13] Canadian Society of Exploration Geophysicists (ed.)(2002): "'Young people in our societies can look forward to many years of progress': an interview with Leon Thomsen," *CSEG Recorder*, June 2002, pp. 26-33, here p. 33

[14] Interview with Leon Thomsen; Burleson, Clyde (1999): *Deep Challenge! The true epic story of our quest for energy beneath the sea*, Houston: Gulf Publishing, p. 34; Finch, David (1985): *Trace through Time: the history of geophysical exploration for petroleum in Canada*, Calgary: Canadian Society of Exploration Geophysicists, p. 18; Lawyer, Lee, Charles Bates and Robert Rice (2001), p. 24; www.epmag.com/Production-Drilling/I-married-doodlebugger_33793; www.ogj.com/articles/print/volume-100/issue-31/regular-features/journally-speaking/the-doodlebugger.html; www.readersandrootworkers.org/wiki/Category:Dowsing,_Doodlebugging,_and_Water_Witching; and http://en.wikipedia.org/wiki/Dowsing

[15] Interview with Dave Kingston; Hyne, Norman (2001), p. 455; Kingston, David (1995): "The Rover Boys and other Stories," in: Hatley, Allen: *The Oil Finders: a collection of stories about exploration*, Centex Press: Utopia (Texas), pp. 1-26, here p. 1; and Selley, Richard (1998), p. 363

[16] Interview with Dave Kingston; and Kingston, David (1995), p. 1

[17] Dixon, Dougal (1992), p. 46; DeGolyer (1938), p. 271; and Kingston, David (1995), p. 1

[18] Interviews with Bert Bally and Dave Kingston; and http://en.wikipedia.org/wiki/Brunton_compass

[19] Eichenberger, Ursula and Ursula Markus (2008): *Augusto Gansser: aus dem Leben eines Welt-Erkunders*, Zürich: AS Verlag; and http://en.wikipedia.org/wiki/Augusto_Gansser-Biaggi

[20] Interview with David Jenkins

[21] Owen, Edgar Wesley (1975), pp. 369-70, 521

[22] Interview with Dave Kingston; and www.bbc.co.uk/history/worldwars/wwtwo/aerial_recon_gallery_07.shtml

[23] Interview of Ken Glennie by Hugo Manson (Lives in the Oil Industry project); interview with Martin Ziegler; Anderson, Robert (1985), p. 96; and Forbes, R.J. and D.R. O'Beirne (1957), p. 138

[24] Interview of Ken Glennie by Hugo Manson (Lives in the Oil Industry project); interview with Dave Kingston; and Selley, Richard (1998), p. 6

[25] Interview with Dave Kingston; Anderson, Robert (1985), p. 96; and Lawyer, Lee, Charles Bates and Robert Rice (2001), pp. 166-7

[26] McCray, A.W. and Frank Cole (1959): *Oil Well Drilling Technology*, Norman (Oklahoma): University of
Oklahoma Press, p. 36; Owen, Edgar Wesley (1975), p. 519; and Sweet, George Elliott (1966), pp. 244-5

[27] Lawyer, Lee, Charles Bates and Robert Rice (2001), pp. 64, 166; and www.izmiran.ru/info/history

[28] Okada, Hakuyu (2005): *The Evolution of Clastic Sedimentology*, Edinburgh: Dunedin Academic Press, pp. 121-35; Selley, Richard (1982): *An Introduction to Sedimentology*, London: Academic Press, pp. 1-3, 265; Shepard, Francis (1980): *Francis P. Shepard Autobiography*, San Diego: University of California, p. 45; and www.api.org/globalitems/globalheaderpages/about-api/api-history

[29] Interview with Walter Ziegler; Holmes, Arthur (1965), pp. 346-85; Tauxe, Lisa (2010): *Essentials of Paleomagnetism*, Berkeley: University of California Press http://geomaps.wr.usgs.gov/parks/gtime/radiom.html; http://en.wikipedia.org/wiki/Arthur_Holmes; and http://en.wikipedia.org/wiki/Paleomagnetism

[30] www.slb.com/services/technical_challenges/carbonates.aspx

[31] Dickey, Parke (1979): *Petroleum Development Geology*, Tulsa (Oklahoma): Petroleum Publishing Company, p. 158; Forbes, R.J. and D.R. O'Beirne (1957), pp. 605-6; Priest, Tyler (2007), p. 39-41; http://en.wikipedia.org/wiki/Clastic_rocks; http://en.wikipedia.org/wiki/Dunham_classification; and http://en.wikipedia.org/wiki/Limestone

[32] Interview with Martin Ziegler

[33] Hughes, Richard (1955): "Theories on the Accumulation of Petroleum of Interest to Production Personnel," paper presented at the spring meeting of the Eastern District, Pittsburgh, May 1955, Division of Production, *API Drilling and Production Practice*, pp. 402-11, here p. 402-4; and Owen, Edgar Wesley (1975), pp. 219-20

[34] Yergin, Daniel (2011): *The Quest: energy, security, and the remaking of the modern world*, London: Penguin, p. 233; www.mkinghubbert.com; http://en.wikipedia.org/wiki/M._King_Hubbert; and www.tshaonline.org/handbook/online/articles/fhu85

[35] Yergin, Daniel (2011), p. 234; and http://news.rice.edu/2007/05/10/five-recognized-for-meritorious-service-to-rice/

[36] Deffeyes, Kenneth S. (2009): *Hubbert's Peak: the impending world oil shortage*, Princeton: Princeton University Press, p. 43; Hughes, Richard (1955), p. 405; Priest, Tyler (2007), p. 119; and http://en.wikipedia.org/wiki/M._King_Hubbert

[37] Dixon, Dougal (1992), p. 40; and Holmes, Arthur (1965), pp. 17, 905-6

[38] Interview with Walter Ziegler; interview of Ken Glennie by Hugo Manson (Lives in the Oil Industry project); Ballard, Robert and Will Hively (2000): *The Eternal Darkness: a personal history of deep-sea exploration*, Princeton: Princeton University Press, p. 118; Holmes, Arthur (1965), pp. 1199-1204; Lawyer, Lee, Charles Bates and Robert Rice (2001), pp. 5-6; McPhee, John (2000), p. 131; www.geolsoc.org.uk/en/Plate-Tectonics/Chap1-Pioneers-of-Plate-Tectonics/Alfred-Wegener; and www.geolsoc.org.uk/Plate-Tectonics/Glossary/A-C

[39] Okada, Hakuyu (2005), pp. 136-9; and http://en.wikipedia.org/wiki/Challenger_expedition

[40] Ballard, Robert and Will Hively (2000), pp. 120-4; Lawyer, Lee, Charles Bates and Robert Rice (2001), p. 149; and www.geolsoc.org.uk/en/Plate-Tectonics/Chap1-Pioneers-of-Plate-Tectonics/Harry-Hess

[41] Ballard, Robert and Will Hively (2000), p. 123; Holmes, Arthur (1965), p. 1205; Lawyer, Lee, Charles Bates and Robert Rice (2001), p. 150; McPhee, John (2000), p. 129; and www.geolsoc.org.uk/en/Plate-Tectonics/Chap1-Pioneers-of-Plate-Tectonics/Vine-and-Matthews

[42] McPhee, John (2000), p. 129

[43] Interview with Walter Ziegler; https://www.e-education.psu.edu/earth520/content/l2_p7.html; and www.geolsoc.org.uk/en/Plate-Tectonics/Chap1-Pioneers-of-Plate-Tectonics/John-Tuzo-Wilson

[44] St. John, Bill (2013): *Life of an International Petroleum Geologist*, unpublished manuscript, Kerrville (Texas); and www.aapg.org/explorer/special/tectonicstheory.html

Chapter 10: The Proliferation of Well Logging

[1] Edmundson, Henry (1987), pp. 12-3

[2] Allaud, Louis and Maurice Martin (1977), p. 165; Edmundson, Henry (1987); Edmundson, Henry (1988): "Archie II: electrical conduction in hydrocarbon-bearing rock," *The Technical Review*, vol. 36, no. 4, pp. 12-21, here pp. 12-3; Edmundson, Henry (2010a): *Schlumberger History: old ties arise anew*, Cambridge, unpublished article; Léonardon, Eugène (1961), p. 532

[3] Allaud, Louis and Maurice Martin (1977), pp. 165-6; Edmundson, Henry (1988), pp. 13-4; and www.ornl.gov/ornl/about-ornl/history-2/profiles/miles-c-leverett

[4] Interviews with George Coates, Roland Chemali, Robert Freedman and E.C. Thomas; Bateman, Richard (2012): *Openhole Log Analysis and Formation Evaluation*, Richardson (Texas): Society of Petroleum Engineers, p. 116, 119; Oristaglio, Michael and Alexander Dorozynski (2007), pp. 235-6; Selley, Richard (1998), p. 60; Thomas, E.C. (2007): "Petrophysics," in: Léonardon, Eugène (1961), p. 546 77-87, here p. 83; and www.slb.com/about/history/1940s.aspx

[5] Archie, G.E. (1942): "The electrical resistivity log as an aid in determining some reservoir characteristics," *Transactions AIME*, vol. 146, pp. 54-62; Archie, G.E. (1950): "Introduction to petrophysics of reservoir rocks," *AAPG Bulletin*, vol. 34, May 1950, pp. 943-61; and Thomas, E.C. (1992): "50th anniversary of the Archie Equation: Archie left more than just an equation," *The Log Analyst*, May-June 1992, pp. 199-205, here p. 199

[6] Allaud, Louis and Maurice Martin (1977), pp. 161, 249; Bateman, Richard (2012), p. 213; Doll, Henri-Georges (1949): "Introduction to induction logging and application to logging of wells drilled with oil base mud," *Journal of Petroleum Technology*, vol. 1, issue 6, pp. 148-62, here p. 149; Oristaglio, Michael and Alexander Dorozynski (2007), p. 228; and Schlumberger (ed.)(1998)

[7] Allaud, Louis and Maurice Martin (1977), pp. 200-1; Léonardon, Eugène (1961), p. 541; and Oristaglio, Michael and Alexander Dorozynski (2007), pp. 14, 185-99, 228-30

[8] Interview with Robert Hunt-Grubbe; Allaud, Louis and Maurice Martin (1977), pp. 244-53; and Oristaglio, Michael and Alexander Dorozynski (2007), pp. 230-1

[9] Allaud, Louis and Maurice Martin (1977), pp. 249-53; Oristaglio, Michael and Alexander Dorozynski (2007), p. 158; and Schlumberger (ed.)(1998)

[10] Allaud, Louis and Maurice Martin (1977), pp. 268, 270-3; Baker Hughes (ed.)(2007), pp. 31, 33; Lawyer, Lee, Charles Bates and Robert Rice (2001), p. 268; Léonardon, Eugène (1961), pp. 542-9; and Oristaglio, Michael and Alexander Dorozynski (2007), p. 242

[11] Baker Hughes (ed.)(2007), p. 31; Jackson, Warren and John Campbell (1946): "Some practical aspects of radioactivity well logging," *Transactions of the AIME*, vol. 165, no. 1, pp. 241-67; Léonardon, Eugène (1961), p. 546; Pontecorvo, Bruno (1941): "Neutron well logging: a new geological method based on nuclear physics," *Oil and Gas Journal*, vol. 40, pp. 32-33; Vassiliou, Marius (2009), p. 525; and http://petrowiki.org/Casing_collar_locator

[12] Kragh, Helge (1999): *Quantum Generations: a history of physics in the twentieth century*, Princeton: Princeton University Press, p. 239; Oristaglio, Michael and Alexander Dorozynski (2007), p. 161; Rhodes, Richard (1986): *The Making of the Atom Bomb*, New York: Simon & Schuster, pp. 217-9; Segré, Emilio (1993): *A Mind Always in Motion: the autobiography of Emilio Segré*, Berkeley: University of California Press, p. 160; http://en.wikipedia.org/wiki/Bruno_Pontecorvo; and http://home.web.cern.ch/about/updates/2013/08/centenary-bruno-pontecorvo

[13] Interviews with Robert Hunt-Grubbe and E.C. Thomas; Baker Hughes (ed.)(2007), p. 33; Bonolis, Luisa (2005): "Bruno Pontecorvo: from slow neutrons to oscillating neutrinos," *American Journal of Physics*, vol. 73, no. 6, pp. 487-99, here p. 489; Brons, Folkert (1940): *Process and Apparatus for Exploring Geological Strata*, US Patent no. 2220509 patented November 5; and Pontecorvo, Bruno (1941)

[14] Interviews with Robert Freedman, Robert Hunt-Grubbe and Jay Tittman; Allaud, Louis and Maurice Martin (1977), p. 273; Oristaglio, Michael and Alexander Dorozynski (2007), p. 221; and Schlumberger (ed.)(1998)

[15] Interviews with Robert Hunt-Grubbe and Jay Tittman; Allaud, Louis and Maurice Martin (1977), p. 305; Gluyas, Jon and Richard Swarbrick (2004): *Petroleum Geoscience*, Oxford: Blackwell Science, p. 32; and Schlumberger (ed.)(1998)

[16] Interview with Robert Hunt-Grubbe; Bonolis, Luisa (2005), p. 494; Ellis, Darwin and Julian Singer (2008): *Well Logging for Earth Scientists*, Dordrecht: Springer, p. 351; Kragh, Helge (1999), pp. 203-4; http://en.wikipedia.org/wiki/Bruno_Pontecorvo; http://home.web.cern.ch/about/updates/2013/08/centenary-bruno-pontecorvo; and http://en.wikipedia.org/wiki/Montreal_Laboratory

[17] Ellis, Darwin and Julian Singer (2008), p. 480; Léonardon, Eugène (1961), p. 550; and Prensky, Stephen and Doug Patterson (2007): "Acoustic Logging," in: Lake, Larry (ed.):

Petroleum Engineering Handbook, vol. V (a)(Reservoir engineering and petrophysics), Richardson (Texas): Society of Petroleum Engineers, pp. 167-242, here p. 176

[18] Allaud, Louis and Maurice Martin (1977), p. 310; Ellis, Darwin and Julian Singer (2008), pp. 480-1; Jolly, R.N. (1953): "Deep-hole geophone study in Garvin County, Okla.," *Geophysics*, vol. 28, no. 3, pp. 662-70; Léonardon, Eugène (1961), p. 552; National Petroleum Council (ed.)(1965): *Impact of New Technology on the U.S. Petroleum Industry 1946-1965*, Washington D.C.: National Petroleum Council, p. 69; Oristaglio, Michael and Alexander Dorozynski (2007), p. 264; and http://csegrecorder.com/articles/view/the-sound-of-sonic-a-historical-perspective-and-intro-to-acoustic-logging

[19] Allaud, Louis and Maurice Martin (1977), p. 309; Ellis, Darwin and Julian Singer (2008), p. 480; Oristaglio, Michael and Alexander Dorozynski (2007), p. 263; and http://csegrecorder.com/articles/view/the-sound-of-sonic-a-historical-perspective-and-intro-to-acoustic-logging

[20] Allaud, Louis and Maurice Martin (1977), p. 310; Ellis, Darwin and Julian Singer (2008), pp. 480-1; National Petroleum Council (ed.)(1965), p. 69; and Oristaglio, Michael and Alexander Dorozynski (2007), pp. 264-5

[21] Interview with E.C. Thomas; Bateman, Richard (2012), pp. 276-7; and http://csegrecorder.com/articles/view/the-sound-of-sonic-a-historical-perspective-and-intro-to-acoustic-logging

[22] Allaud, Louis and Maurice Martin (1977), pp. 310-1; Berry, James (1959): "Acoustic velocity in porous media," *AIME Petroleum Transactions*, vol. 216, pp. 262-70, here p. 262; Ellis, Darwin and Julian Singer (2008), pp. 481-2; Léonardon, Eugène (1961), p. 554; National Petroleum Council (ed.)(1965), p. 69; and Oristaglio, Michael and Alexander Dorozynski (2007), pp. 264-5

[23] Allaud, Louis and Maurice Martin (1977), p. 312; and Oristaglio, Michael and Alexander Dorozynski (2007), p. 266

[24] Interview with Robert Freedman; Biot, Maurice Anthony (1952): "Propagation of elastic waves in a cylindrical bore containing a fluid," *Journal of Applied Physics*, vol. 23, no. 9; Ellis, Darwin and Julian Singer (2008), pp. 527, 559; Koelemeijer, Paula (2013): "Robert Stoneley and core-mantel boundary Stoneley modes," *Pembroke College Annual Gazette*, pp. 29-33; Léonardon, Eugène (1961), p. 552; Morris, R.L., D.R. Grine and T.E. Arkfeld (1964): Using compressional and shear acoustic amplitudes for the location of fractures," *Journal of Petroleum Technology*, vol. 16, no. 6, pp. 623-32; Pickett, G.R. (1963): "Acoustic character logs and their application in formation evaluation," *Journal of Petroleum Technology*, vol. 15, no. 6, pp. 659-67; Staal, J.J. and J.D. Robinson (1977): "Permeability profiles from acoustic logging," conference paper, *SPE Annual Fall Technical Conference and Exhibition*, 9-12 October, Denver, Colorado; Winkler, Kenneth W., Hsui-Lin Liu and David Linton Johnson (1989): "Permeability and borehole Stoneley waves: comparison between experiment and theory," *Geophysics*, vol. 54, no. 1, pp. 66-75; and www.glossary.oilfield.slb.com/en/Terms/s/stoneley_wave.aspx

Chapter 11: Production Engineering Matures

[1] Gow, Sandy (2005), p. 134; Hilchie, Douglas (1990): *Wireline: a history of the well logging and perforating business in the oil fields*, Boulder (Colorado): Douglas W. Hilchie, Inc.; Vance, Harold (1961a), p. 591; www.epmag.com/Production-Drilling/From-bazookas-well-perforating_45570; http://aoghs.org/tag/henry-mohaupt/; www.logwell.com/tech/shot/perforator_history.html; and http://en.wikipedia.org/wiki/Shaped_charge

[2] Interview with Bill Rehm; Hilchie, Douglas (1990); McLean, John and Robert Haigh (1954): *The Growth of Integrated Oil Companies*, Boston: Harvard University, p. 261;

www.logwell.com/tech/shot/perforator_history.html;
www.immigrantentrepreneurship.org/entry.php?rec=28; and
http://en.wikipedia.org/wiki/Timken_Roller_Bearing_Company

[3] Interview with George King; King, George (2007), p. 151;
www.logwell.com/tech/shot/perforator_history.html; www.epmag.com/Production-Drilling/From-bazookas-well-perforating_45570; www.thefreedictionary.com/bazooka; www.encyclopediaofarkansas.net/encyclopedia/entry-detail.aspx?entryID=2185; and http://en.wikipedia.org/wiki/Bazooka_(instrument)

[4] Clark, J.B. (1949): "A hydraulic process for increasing the productivity of wells," *AIME Petroleum Transactions*, 1949, vol. 1, pp. 1-9; Montgomery, Carl and Michael Smith (2010): "Hydraulic Fracturing: history of an enduring technology," in: Society of Petroleum Engineers (ed.): *Legends of Hydraulic Fracturing*, Richardson (Texas): Society of Petroleum Engineers, pp. 1-9, here pp. 1-2; National Petroleum Council (ed.)(1967), p. 113; and http://frackingresource.org/what-is-fracking/hydraulic-fracturing-overview/the-history-of-fracking/

[5] Clark, J.B. (1949), p. 1; Holditch, Stephen (2007): "Hydraulic Fracturing," in: Clegg, Joe Dunn (ed.): *Petroleum Engineering Handbook*, vol. IV (Production operations engineering), Richardson (Texas): Society of Petroleum Engineers, pp. 323-66, here p. 323; Montgomery, Carl and Michael Smith (2010), pp. 1-2; National Petroleum Council (ed.)(1967), p. 113; Veatch, Ralph, Zissis Moschovidis and Robert Fast (1989): "An Overview of Hydraulic Fracturing," in: Holditch, Stephen et al. (eds.): *Recent Advances in Hydraulic Fracturing*, Richardson (Texas): Society of Petroleum Engineers, pp. 1-38, here p. 1; and http://frackingresource.org/what-is-fracking/hydraulic-fracturing-overview/the-history-of-fracking/

[6] Clark, J.B. (1949); Montgomery, Carl and Michael Smith (2010), p. 2; National Petroleum Council (ed.)(1967), p. 113; and Rodengen, Jeffrey (1996), p. 59

[7] Brantly, John (1971), pp. 1280-8; and Linsley, Judith Walker, Ellen Walker Rienstra and Jo Ann Stiles (2002), pp. 122-6

[8] FMC Energy Systems (ed.)(2004): *FMC Surface Wellhead Catalog*, Houston: FMC Technologies; http://en.wikipedia.org/wiki/FMC_Technologies; and www.fmctechnologies.com/en/AboutUs/History/FMCtimeline.aspx

Chapter 12: Offshore Goes Deeper

[1] Burleson, Clyde (1999), p. 86; Lankford, Raymond (1971), p. 1426; and interview with Pete Johnson

[2] Burleson, Clyde (1999), pp. 99; and John Steinbeck, cited in: Johnson, Pete (2011): "Some Recollections of the Birth of Deep-Sea Drilling," in: www.nationalacademies.org/moholeacc.html

[3] Ballard, Robert and Will Hively (2000), p. 280; Johnson, Pete (2011); and Lankford, Raymond (1971), p. 1403

[4] Interview with Ted Bourgoyne; Frank Williford's written answer to Mark Mau's interview question; Schempf, F. Jay (2007), p. 167

[5] Interview with Greg Myers; interview of Robert "Bob" Bauer by Tyler Priest (Offshore Energy Center); Burleson, Clyde (1999), pp. 126-7, 132; Lankford, Raymond (1971), pp. 1404-5; and Schempf, F. Jay (2007), pp. 116, 118

[6] Anderson, Robert (1985), p. 150; Gerding, Mildred (1986), p. 159; and Schempf, F. Jay (2007), pp. 123-6

[7] Interview with John Thorogood; Anderson, Robert (1985), p. 150; and Priest, Tyler (2007), pp. 81-2, 85

[8] Schempf, F. Jay (2007), p. 129

[9] Interview of Andre Rey-Grange by Joseph Pratt (Offshore Energy Center)
[10] Interview with Gregers Kudsk
[11] Frank Williford's written answer to Mark Mau's interview question
[12] Ford, John (2000): "Drilling Operations," in: Dawe, Richard A. (ed.)(2000): *Modern Petroleum Technology*, Chichester: John Wiley, vol. 1 (Upstream), pp. 101-29, here p. 108; and Petromin (ed.)(2011): "Blowout preventers: history, performance and advances," *PetroMin*, July/August 2011
[13] Interview with David Llewelyn; Frank Williford's written answer to Mark Mau's interview question; and Schempf, F. Jay (2007), p. 156
[14] Interviews with Jean Chevallier and Ted Bourgoyne; Frank Williford's written answer to Mark Mau's interview questions; Burleson, Clyde (1999), p. 44; Ford, John (2000), p. 104; Savage, George and Robert Bradbury (1967): *Marine Drilling Apparatus*, US patent no. 3354951 patented Nov. 28; and Schempf, F. Jay (2007), p. 156
[15] Interview with David Llewelyn; and Raymond, Martin and William Leffler (2006), p. 101

Chapter 13: The Digital Revolution

[1] Lawyer, Lee, Charles Bates and Robert Rice (2001), p. 20; Sheriff, Robert and Lloyd Geldart (1995), pp. 9, 18, 21; and www.youtube.com/watch?v=OjKQNykc_04
[2] Lawyer, Lee, Charles Bates and Robert Rice (2001), p. 20; and http://en.wikipedia.org/wiki/Magnetic_tape
[3] Mayne, W. Harry (1989): *50 years of Geophysical Ideas*, Tulsa (Oklahoma): The Society of Exploration Geophysicists, pp. 93-4
[4] Interview with Milo Backus; Hyne, Norman (2001), p. 219; Mayne, W. Harry (1989), p. 93; Priest, Tyler (2007), pp. 43, 94; Schempf, F. Jay (2007), p. 77; and Vassiliou, Marius (2009), p. 211
[5] McClellan, James E. and Harold Dorn (2006): *Science and Technology in World History: an introduction*, Washington (D.C.): The Johns Hopkins University Press, p. 406; Tinkle, Lon (1970), p. 372; and http://digital.library.okstate.edu/encyclopedia/entries/D/DE007.html
[6] Lecuyer, Christophe (1992): "The making of a science based technological university: Karl Compton, James Killian, and the reform of MIT, 1930-1957," *Historical Studies in the Physical and Biological Sciences*, vol. 23, no. 1, pp. 153-80
[7] Interview with Sven Treitel; Robinson, Enders (2013): *Exploration interview answers by Enders Robinson*, unpublished document sent to Mark Mau, March 21; Treitel, Sven (2005): *The MIT Geophysical Analysis Group (GAG), 1954 and beyond*, article manuscript; and www.ieeeghn.org/wiki/index.php/Oral-History:Enders_Robinson
[8] Interviews with Milo Backus and Sven Treitel; interview of Robert "Bob" Graebner by Tyler Priest and Joseph Pratt (Houston History Project); Pirtle, Caleb (2005), p. 16; and Treitel, Sven (2005)
[9] Interviews with Milo Backus and Sven Treitel; interview of Robert "Bob" Graebner by Tyler Priest and Joseph Pratt; Finch, David (1985), p. 81; Glennie, Ken (1998): *Petroleum Geology of the North Sea: basic concepts and recent advances*, London: Blackwell Science, pp, 1-29. Lawyer, Lee, Charles Bates and Robert Rice (2001), pp. 113-4; Mayne, W. Harry (1989), pp. 42, 75; and Pirtle, Caleb (2005), p. 16
[10] Moreton, Richard (ed.)(1995): *Tales from Early UK Oil Exploration 1960-1979*, London: Petroleum Exploration Society of Great Britain, pp.34, 38
[11] Interviews with Milo Backus and Roy Lindseth; Bednar, Bee (2005): *A Brief History of Seismic Migration*, Houston, article manuscript, p. 16; Lawyer, Lee, Charles Bates and

Robert Rice (2001), pp. 117, 279-80; Moreton, Richard (ed.)(1995), p. 36; Mayne, W. Harry (1989), p. 41; Robinson, Enders (2013); and www.cgg.com/default.aspx?cid=6035&lang=1

[12] Interviews with Jim Hornabrook and Roy Lindseth; Lawyer, Lee, Charles Bates and Robert Rice (2001), p. 117; Moreton, Richard (ed.)(1995), p. 57; Sheriff, Robert and Lloyd Geldart (1995), pp. 20-1; and Stoneley, Robert (1995), p.77

[13] Interview with Milo Backus; interview of Sam Evans by Tyler Priest and Joseph Pratt; Hyne, Norman (2001), p. 471; Jahn, Frank, Mark Cook and Mark Graham (2008), pp. 30, 34; Robinson, Enders (2013); and Sheriff, Robert and Lloyd Geldart (1995), pp.18, 20-1

[14] Interview with Roy Lindseth; Interview of Sam Evans by Tyler Priest and Joseph Pratt; and Moreton, Richard (ed.)(1995), p. 39

[15] Interview with Sam Gray; Fourmann, J.M. (1975): *Seismic Processing: wave equation migration*, CGG technical series, Massy: Compagnie Générale de Géophysique, p. 2; and Treitel, Sven (2005)

[16] Dobrin, Milton and Carl Savit (1976), p. 238; Hyne, Norman (2001), p. 227; Jahn, Frank, Mark Cook and Mark Graham (2008), pp. 37-8; Sheriff, Robert and Lloyd Geldart (1995), pp. 5-6, 326-7; Schoenberger, Michael (2000): "Geophysics," in: Dawe, Richard A. (ed.): *Modern Petroleum Technology*, Chichester: John Wiley, vol. 1 (Upstream), pp. 55-100, here pp. 82-3; Selley, Richard (1998), pp. 103-5; Stoneley, Robert (1995), p. 78 and Telford, W. M., L.P. Geldart and R.E. Sheriff (1990), pp. 357-9

[17] Interviews with Milo Backus and Sam Gray; Albertin, Uwe et al. (2002): "The time for depth imaging," *Oilfield Review*, spring, pp. 2-15, here p. 2; Bednar, Bee (2005), p. 14; Dobrin, Milton and Carl Savit (1976), p. 39; Dragoset, Bill (2005): "A historical reflection on reflections," *The Leading Edge*, vol. 24, supplement, pp. S46-S70, here p. S47; Gray, Sam (2012): *A Brief History of Depth and Time Imaging*, SEG presentation slide, Calgary; Hyne, Norman (2001), p. 227; Jahn, Frank, Mark Cook and Mark Graham (2008), pp. 37-8; Sheriff, Robert and Lloyd Geldart (1995), pp. 5-6, 44, 244, 326-7

[18] Interviews with Jon Claerbout and Sam Gray; and Bednar, Bee (2005), pp. 17-9

[19] Interviews with Milo Backus, Guus Berkhout, Sam Gray and Sven Treitel; Bednar, Bee (2005), pp. 17-9; Claerbout, Jon (1976): *Fundamentals of Geophysical Data Processing*, New York: McGraw-Hill; Fourmann, J.M. (1975); Gray, Sam (2012); Larner, Ken et al. (1978): "Depth migration of complex offshore seismic profiles," conference paper, *Offshore Technology Conference*, 8-11 May, Houston; http://news.stanford.edu/news/2000/may31/claerbout-531.html; and http://geophysics.geoscienceworld.org/content/48/5/627.abstract

[20] Interview with Jon Claerbout; Bednar, Bee (2005), pp. 17-9; and Claerbout, Jon (1973): *Proposal to initiate the Stanford Exploration Project to do fundamental research in reflection seismology*, Stanford (California): Stanford University

[21] Interviews with Milo Backus, Phil Christie, Sam Gray and Yu Zhang; Bednar, Bee (2005), p. 13; James, Huw (2013): "Innovation geophysics: a memoir," *First Break*, vol. 31, issue 6, pp. 161-7, here p. 163; www.cgg.com; and www.slb.com/services/westerngeco/services/dp/technologies/depth/prestackdepth/rtm.aspx

[22] Interview with David Jenkins; and www.oceanstaroec.com/fame/2006/brightspot.html

[23] Interviews with Alistair Brown and Leon Thomsen; interview of Sam Evans by Tyler Priest and Joseph Pratt; Lawyer, Lee, Charles Bates and Robert Rice (2001), p. 178; Priest, Tyler (2007), p. 130; and www.oceanstaroec.com/fame/2006/brightspot.html

[24] Interview with Leon Thomsen; and www.oceanstaroec.com/fame/2006/brightspot.html

[25] Interviews with Mike Forrest and Leon Thomsen; Priest, Tyler (2007), pp. 129-30, 132; www.oceanstaroec.com/fame/2006/brightspot.html; www.searchanddiscovery.com/documents/forrest/; and www.aapg.org/explorer/wildcat/2000/wildcat05.cfm

[26] Interview with David Jenkins; Priest, Tyler (2007), p. 106; and www.oceanstaroec.com/fame/2006/brightspot.html

[27] Interview with Jim Hornabrook

[28] Lawyer, Lee, Charles Bates and Robert Rice (2001), pp. 87-8; Lawyer, Lee and Ben Giles (2013): "Doodlebugger diary: air gun history," *GSH Journal*, Geophysical Society of Houston, vol. 3, no. 8 (April 2013), pp. 34, 38; http://virtualmuseum.seg.org/bio_stephen_chelminski.html; and www.zoominfo.com/#!search/profile/person?personId=119027121&targetid=profile

[29] Krail, P.M. (2010): *Airguns: theory and operation of the marine seismic source*, course notes, University of Texas at Austin, p. 6; Lawyer, Lee, Charles Bates and Robert Rice (2001), pp. 87-8; www.bolt-technology.com/pages/products.htm; www.geol.lsu.edu/jlorenzo/ReflectSeismol97/eczimmermann/www/eczimmermann.html; and http://virtualmuseum.seg.org/bio_stephen_chelminski.html

[30] Dragoset, Bill (2005), p. S54; Krail, P.M. (2010), p. 6; Lawyer, Lee and Ben Giles (2013), p. 38; www.bolt-technology.com/pages/products.htm; and http://virtualmuseum.seg.org/bio_stephen_chelminski.html

[31] Drijkoningen, G.G. (2003): *Seismic Data Acquisition*, course TA3600 document, Delft: Delft Technical University, p. 35; and http://seismosblog.blogspot.co.uk/2009/12/bell-choir-seismology.html

[32] Lawyer, Lee, Charles Bates and Robert Rice (2001), p. 88; www.kshs.org/kansapedia/burton-mccollum/18255; http://virtualmuseum.seg.org/bio_burton_mccollum.html; and http://en.wikipedia.org/wiki/Seismic_source#Thumper_truck

[33] Dragoset, Bill (2005), pp. S52-3; Krehl, Peter (2008): *History of Shock Waves, Explosions and Impacts*, Berlin and New York: Springer, p. 120; and Robinson, Enders (2005): *The MIT Geophysical Analysis Group (GAG) from Inception to 1954*, article manuscript, Newburyport (Massachussetts), pp. 25-6

[34] Dragoset, Bill (2005), pp. S52-3; Drijkoningen, G.G. (2003), p. 27; Henderson, Bob (1975): *Conoco Geophysics: the first fifty years*, Houston: Conoco; Krehl, Peter (2008), p. 120; and Robinson, Enders (2005), pp. 25-26

[35] Anstey, Nigel (2008): *Letter to the Conveners,* EAGE Vibroseis Workshop, October 1, Isle of Man; Drijkoningen, G.G. (2003), p. 23; Drijkoningen, G.G. (2003), pp. 34-5; Lawyer, Lee, Charles Bates and Robert Rice (2001), p. 269; and http://seismosblog.blogspot.co.uk/2009/12/bell-choir-seismology.html

Chapter 14: Petroleum Geology Changes the Game

[1] Interview with Martin Ziegler

[2] Embry, Ashton (2009): *Practical Sequence Stratigraphy*, online document, Canadian Society of Petroleum Geologists, www.cspg.org; Gates, Alexander (2002): *A to Z of Earth Scientists*, New York: Facts on File , p. 240; Okada, Hakuyu (2005), pp. 66-83; Sloss, Lawrence (1963): "Sequences in the Cratonic interior of North Amercia," *Bulletin of the Geological Society of America*, vol. 74, pp. 93-114; www.earth.northwestern.edu/alumni/profiles/vail-peter.html; www.geotimes.org/july03/profiles.html; and www.glossary.oilfield.slb.com/en/Terms/s/sequence_stratigraphy.aspx

[3] Interview with Walter Ziegler; Embry, Ashton (2009); and Gates, Alexander (2003), pp. 271-2; www.mcz.harvard.edu/Departments/InvertPaleo/Trenton/Intro/GeologyPage/Sedimentary%20Geology/sequencestrat.htm#intro

[4] Interview with Dave Kingston; and Selley, Richard (1998), p. 109

[5] Vail, Peter et al. (1977): "Seismic Stratigraphy and Global Changes of Sea Level," *AAPG Memoir*, vol. 26, pp. 49-212; and Walter Ziegler's email to Mark Mau, August 21, 2013

[6] Embry, Ashton (2009); and Vail, Peter, Bob Mitchum and John Sangree (2002): "Sequence stratigraphy: evolution and effects," conference paper, 22nd Annual Gulf Coast Section *SEPM* Foundation Bob F. Perkins Research Conference, March 10, 2002, Houston

[7] Interview with Bryan Lovell; interview of Ken Glennie by Hugo Manson (Lives in the Oil Industry project); Deffeyes, Kenneth (2009), p. 56; Levorsen, Arville Irving (1967): *Geology of Petroleum*, San Francisco: W.H. Freeman, p. 63; and www.marin.edu/~jim/ring/rturb.html

[8] Gates, Alexander (2003), p. 26; www.minigeology.com/speaker/8361-Arnold-Bouma; http://soundwaves.usgs.gov/2012/02/staff.html; and http://en.wikipedia.org/wiki/Bouma_sequence

[9] Interviews with Dave Kingston and Walter Ziegler; Mutti, E. and R. Ricchi Lucci (1972): "Le Torbiditi del 'Appennino settentrionale': introduzione all' analisi de facies," *Soc. Geol. Italiana Mem.*, vol. 11, pp. 161-99 (English translation: "Turbidites of the Northern Apennines: introduction to facies analysis," published in *International Geology Review*, vol. 20 (1978), pp. 125-66; www.minigeology.com/speaker/11008-Emiliano-Mutti/video/550879-Dreaming-about-turbidites-for-real-full-interview; and http://en.wikipedia.org/wiki/Emiliano_Mutti

[10] Interview with Bryan Lovell; and Okada, Hakuyu (2005), pp. 50-66

[11] Interviews with Bryan Lovell and Walter Ziegler; Interview of Ken Glennie by Hugo Manson (Lives in the Oil Industry project); Gates, Alexander (2003), p. 27; Hyne, Norman (2001), p. 131; Kelts, Kerry and Michael Arthur (1981): "Turbidites after Ten Years of Deep-Sea Drilling: wringing out the mop?" in: Warme, John, Robert Douglas and Edward Winther (eds.): *The Deep Sea Drilling Project: a decade of progress*, Tulsa (Oklahoma): Society of Economic Paleontologists and Mineralogists, pp. 91-127; and Selley, Richard (1982), p. 2

[12] Allen, Philip and John Allen (2005): *Basin Analysis: principles and applications*, Oxford: Blackwell Publishing, p. 4; Gates, Alexander (2003), pp. 136-7; The Geological Society of America (ed.)(1999): *Memorial to Laurence L. Sloss*, Boulder (Colorado): The Geological Society of America; www.mcz.harvard.edu/Departments/InvertPaleo/Trenton/Intro/GeologyPage/Sedimentary%20Geology/sequencestrat.htm; http://en.wikipedia.org/wiki/Cratonic_sequence; and http://en.wikipedia.org/wiki/Marshall_Kay

[13] Interviews with Dave Kingston, Bryan Lovell, Dan McKenzie and Andy Woods; interview of Dan McKenzie by Alan MacFarlane (University of Cambridge); Allen, Philip and John Allen (2005), pp. 4, 13-8; and Gates, Alexander (2003), p. 162

[14] Interviews with Phil Christie, Bryan Lovell and Dan McKenzie; interview of Dan McKenzie by Alan MacFarlane (University of Cambridge); Gates, Alexander (2003), p. 162; and Sclater, J. and Phil Christie (1980): "Continental stretching: an explanation of the post-mid-Cretaceous subsidence of the central North Sea," *Journal of Geophysical Research*, vol. 85, pp. 3711-39

[15] Interviews with Dave Kingston, Bryan Lovell and Walter Ziegler; Hatley, Allen: *The Oil Finders: a collection of stories about exploration*, Utopia (Texas): Centex Press, p. xvi; Houston Geological Society Bulletin (ed.)(1986): *Dave R. Kingston: biographical sketch*,

Houston: Houston Geological Society; and www.geofacets.com/info/elsevier-oil-and-gas-advisory-board

[16] Sandwell, D.T. and W.H.F. (1997): "Marine gravity anomaly from Geosat and ERS-1 satellite altimetry," *Journal of Geophysical Research*, vol. 102, no. 10, pp. 039-10 054

[17] Schollnberger, Wolfgang (2007): "Geologie als Basis erfolgreicher Erdöl-Erdgas-Exploration: zu Zeiten Hans Höfers und Heute" (Geology at the Base of Successful Petroleum Exploration: in Hans Höfer's Days and today), *Erdöl Erdgas Kohle*, vol. 123, no. 11, p. 1

[18] Interview with Dietrich Welte; Allen, Philip and John Allen (2005), pp. 407-8; and Dow, Wallace (1974): "Application of oil-correlation and source-rock data to exploration in Williston basin," *American Association of Petroleum Geologists Bulletin*, vol. 58, no. 7, pp. 1253-62

[19] Interviews with Bert Bally and Dietrich Welte; Campbell, Colin (2012), p. 188; Demaison, Gerald (1984): "The Generative Basin Concept," in: Demaison, Gerald and Roelef Murris (eds.): *Petroleum Geochemistry and Basin Evaluation*, American Association of Petroleum Geologists Memoir 35, pp. 1-14; Perrodon, Alain (1980): *Géodynamique Pétrolière: genèse et répartition des gisements d'hydrocarbures*, Paris: Masson and Elf-Aquitaine; and Tissot, Bernard and Dietrich Welte (1978): *Petroleum Formation and Occurrence: a new approach to oil and gas exploration*, Berlin and New York: Springer-Verlag; and www.glossary.oilfield.slb.com/en/Terms/p/petroleum_system.aspx

[20] Interviews with David Jenkins, Dietrich Welte and Martin Ziegler; Ceruzzi, Paul (2003): *A History of Modern Computing*, Cambridge (Massuchussetts): MIT Press, pp. 177-206; Campbell-Kelly, Martin and William Aspray (2004): *Computer: a history of the information machine*, London: Basic Books, p. 198; and Harbaugh, John, John Doveton and John Davis (1977): *Probability Methods in Oil Exploration*, New York: John Wiley

[21] Interviews with David Jenkins and Dietrich Welte; and www.glossary.oilfield.slb.com/en/Terms/p/petroleum_system.aspx

[22] Interview with Bernie Vining; Glasby, Geoffrey (2006): "Abiogenic origin of hydrocarbons: an historical overview," *Resource Geology* vol. 56, no. 1, pp. 85-98; Prorfir'ev, V.B. (1974): "Inorganic origin of petroleum," *AAPG Bulletin* 1974, v. 58, pp. 3-33; http://encyclopedia2.thefreedictionary.com/Petroleum+Oil; and http://en.wikipedia.org/wiki/Abiogenic_petroleum_origin

[23] Interviews with Guus Berkhout, David Jenkins and Walter Ziegler; interview of Nikolai Lopatin by Daniel Minisini (www.minigeology.com); Walter Ziegler's emails to Mark Mau, January 24 and June 12, 2013; Lopatin, Nikolai (1971): "Temperature and Geologic Time as Factors in Coalification," (in Russian) *Akad. Nauk SSSR, Izv. Serv. Geol.*, no. 3, pp. 95-106; Luthi, Stefan (2012): *The Carbon Cycle, Organic Matter and Maturation*, course presentation slides, Delft: TU Delft, p. 34; and www.landforms.eu/orkney/Geology/Oil/OIL%20orkney%20Basin%20modeling.htm

[24] Bokserman, A.A., V.P. Filippov and V.Yu. Filanovskii (eds.)(1998): "Oil Extraction," in: Krylov, N.A., A.A. Bokserman and E.R. Stavrosky (eds.): *The Oil Industry of the Former Soviet Union: reserves and prospects, extraction, transportation*, Amsterdam: Gordon and Breach Science Publishers, pp. 69-184, here p. 77; and Lopatin, Nikolai, Dietrich Welte et al. (1999): "Gas generation and accumulation in the West Siberian Basin," *AAPG Bulletin*, vol. 83, October 1999, no. 10, pp. 1642-65

[25] Interviews with Jon Claerbout and Bernie Vining; http://en.wikipedia.org/wiki/Tengiz_Field; www.oilru.com/or/38/774/; and http://ocw.tudelft.nl/courses/petroleum-engineering-and-geosciences/petroleum-geology/lectures/lecture-1-introduction/

[26] Interview with Yu Zhang; Claerbout, Jon (1985): *Imaging the Earth's Interior*, London: Blackwell, p. 121; Ma, Zaitian (1981): *Finite Difference Migration with Higher Order*

Approximation, technical report of the China National Oil and Gas Exploration and Development Co.; and
www.tongji.edu.cn/english/themes/10/template/Faculty/MAZaitian.shtml

Chapter 15: Making the Reservoir Work

[1] Lewis, James (1961): "Fluid Injection," in: American Petroleum Institute (ed.): *History of Petroleum Engineering*, New York: American Petroleum Institute, pp. 847-81, here p. 854

[2] Carll, John Franklin (1880), p. 263; and Willhite, Paul (1986): *Waterflooding*, Richardson (Texas): Society of Petroleum Engineers, pp. 1, 5

[3] Dickey, Parke (1979), p. 337; and Lewis, James (1961), p. 865

[4] Lewis, James (1961), pp. 850-1, 863-71; and Willhite, Paul (1986), pp. 1-2

[5] Interview with Fred Stalkup; Bokserman, A.A., V.P. Filippov and V.Yu. Filanovskii (1998), pp. 84-5; and Willhite, Paul (1986), pp. 1-2

[6] Bokserman, A.A., V.P. Filippov and V.Yu. Filanovskii (1998), pp. 84-5; Willhite, Paul (1986), pp. 1-2; Dickey, Parke (1979), p. 338; and www.ogj.com/articles/print/volume-99/issue-9/exploration-development/slower-reserve-growth-rates-observed-in-volga-ural-province-russia.html

[7] Dickey, Parke (1979), pp. 346-7; and Prats, Mike (1986): *Thermal Recovery*, Richardson (Texas): Society of Petroleum Engineers, pp. 1-2

[8] Elkins, Lloyd (1961a): "Thermal, Solvent and Improved Gas-Drive Oil-Recovery Methods," in: American Petroleum Institute (ed.): *History of Petroleum Engineering*, New York: American Petroleum Institute, pp. 883-906, here p. 890; Prats, Mike (1986), pp. 1-2; and www.glossary.oilfield.slb.com/en/Terms/i/in-situ_combustion.aspx

[9] Elkins, Lloyd (1961a), p. 890; Grant, Bruce and Stefan Szasz (1954): "Development of an underground heat wave for oil recovery," *AIME Petroleum Transactions*, May 1954, pp. 23-3, here p. 23; Prats, Mike (1986), p. 2; and Sheinman, A. B. et al. (1935): "Gasification of crude oil in reservoir sands," *Neftyance Khozyaistva*, 28 (April 1935), English translation published in *Petroleum Engineer*, 10 (December 1938), p. 27, and (February 1939), p. 91

[10] Achterbergh, Niels (2010): *In-Situ Oil Combustion: processes perpendicular to the main gas flow direction*, Delft: ISAPP Knowledge Centre; Elkins, Lloyd (1961a), p. 890; and Prats, Mike (1986), p. 2

[11] Interview with Mike Prats; Prats, Mike (1986), p. 3; and Stovall, Smith (1934): "Recovery of oil from depleted sands by means of dry steam," *The Oil Weekly*, August 13, pp. 17-24

[12] Interviews with Mike Prats and Paul Willhite; Giusti, L.E. (1974): "CSV makes steam soak work in Venezuela field," *Oil & Gas Journal*, November 4, pp. 88-93; Haan, H.J. and L. Schenk (1969): "Drive Project in the Tia Juana Field, Western Venezuela," *Journal of Petroleum Technology*, vol. 21, issue 1, pp. 111-9; Prats, Mike (1986), p. 3; and http://encyclopedia.thefreedictionary.com/Huff+and+puff

[13] http://encyclopedia.thefreedictionary.com/Huff+and+puff

[14] Baibakov, Nikolai Konstantinovich and Aleksandr Rubenovich Garushev (1989): *Thermal Methods of Petroleum Production*, Amsterdam and Oxford: Elsevier (Translated by W.J. Cieslewicz, Colorado School of Mines); Bokserman, A.A., V.P. Filippov and V.Yu. Filanovskii (1998), pp. 124, 130-1; and Dickey, Parke (1979), p. 349

[15] Cotter, W. (1962): "Twenty-three years of gas injection into a highly undersaturated crude reservoir," *Journal of Petroleum Technology*, vo. 14, issue 4, pp. 361-5; Lewis, James (1961), pp. 854-6, 858-9; Muskat, Morris (1949): *Physical Principles of Oil Production*, New

York: McGraw-Hill, pp. 470-502;
http://petrowiki.org/Immiscible_gas_injection_in_oil_reservoirs; and
http://tatweerpetroleum.com/en/oilfield/global/oil-field-title.html

[16] Interview with Fred Stalkup

[17] Interviews with George Hirasaki and Paul Willhite; Dickey, Parke (1979), p. 347; Dyke, Kate van (1997), p. 168; Willhite, Paul and Don Green (1998): *Enhanced Oil Recovery*, Richardson (Texas): Society of Petroleum Engineers, p. 186; www.kindermorgan.com/business/co2/flood.cfm; and www.epmag.com/Production-Field-Development/Impact-miscible-CO2-flooding-HCGOR_3850

[18] Dyke, Kate van (1997), p. 168; Sperling, L.H. (2006): *Introduction to Physical Polymer Science*, New York: John Wiley & Sons, pp. 1, 21

[19] Interviews with George Hirasaki and Paul Willhite; Chang, Harry (1978): "Polymer flooding technology yesterday, today, and tomorrow," *Journal of Petroleum Technology*, vol. 30, no. 8, pp. 1113-28; Dyke, Kate van (1997), p. 168; Jewett, R.L. and G.F. Schurz (1970): "Polymer flooding: a current appraisal," *Journal of Petroleum Technology*, vol. 22, issue 6, pp. 675-84; Pye, David (1964): "Improved secondary recovery by control of water mobility," *Journal of Petroleum Technology*, vol. 16, issue 8, pp. 911-6; Sandiford, Burton (1964): "Laboratory and field studies of water floods using polymer solutions to increase oil recoveries," *Journal of Petroleum Technology*, vol. 16, issue 8, pp. 917-22; and Willhite, Paul and Randall Seright (2011): *Polymer Flooding*, Richardson (Texas): Society of Petroleum Engineers, p. 3

[20] Bokserman, A.A., V.P. Filippov and V.Yu. Filanovskii (1998), p. 102

[21] Interviews with Roland Horne, Koen Weber and Paul Willhite; Chang, H.L. et al. (2006): "Advances in polymer flooding and alkaline/surfactant polymer processes as developed and applied in the People's Republic of China," *Journal of Petroleum Technology*, vol. 58, no. 2, pp. 84-9; Cheng, J. et al. (2010): "Study on remaining oil distribution after polymer flooding," conference paper, *SPE Annual Technical Conference and Exhibition*, Florence, Italy, 19-22 September; Delamaide, E., P. Corlay and W. Demin (1994): "Daqing Oil Field: the success of two pilots initiates first extension of polymer injection in a giant oil field," conference paper, *SPE/DOE Improved Oil Recovery Symposium*, Tulsa, 17-20 April; Wankui, G. et al. (2000): "Commercial pilot test of polymer flooding in Daqing Oil field," conference paper, *SPE/DOE Improved Oil Recovery Symposium*, Tulsa, 3-5 April; www.eia.gov/countries/cab.cfm?fips=CH; and http://en.wikipedia.org/wiki/Daqing_Field

[22] Interview with George Hirasaki; Gogarty, W. B. and W.C. Tosch (1968): "Miscible-type waterflooding: oil recovery with micellar solutions," *Journal of Petroleum Technology*, vol. 20, no. 12, pp. 1407-14; Hirasaki, George, Clarence Miller and Maura Puerto (2011): "Recent advances in surfactant EOR," *SPE Journal*, vol. 16, no. 4; Stegemeier, George, H.J. Hill and J. Reisberg (1973): "Aqueous surfactant systems for oil recovery," *Journal of Petroleum Technology*, vol. 25, no. 2, pp. 186-194; and Uren, Lester and E.H. Fahmy (1927): "Factors influencing the recovery of petroleum from unconsolidated sands by waterflooding," *Transactions of the AIME*, vol. 77, no 1, pp. 318-35

Chapter 16: Logging Breakthroughs

[1] Doll, Henri-Georges (1953), p. 41; Edmundson, Henry (1989): "Archie III: electrical conduction in shaly sands," *Oilfield Review*, vol. 1, no. 3, pp. 43-53; Worthington, Paul (1985): "The evolution of shaly-sand concepts in reservoir evaluation," *The Log Analyst*, vol. 26, no. 1, pp. 23-40

[2] Edmundson, Henry (1989); and Ellis, Darwin and Julian Singer (2008), pp. 658-61

[3] Interviews with Georg Coates, Robert Freedman and E.C. Thomas; Clavier, Christian, George Coates and Jean Dumanoir (1984): "Theoretical and experimental bases for the dual-water model for interpretation of shaly sands," *Society of Petroleum Engineers Journal*, April 1984, pp. 153-68; Ellis, Darwin and Julian Singer (2008), p. 662; Edmundson, Henry (1989); and http://en.wikipedia.org/wiki/Cation_exchange_capacity

[4] Interview with George Coates and E.C. Thomas

[5] Interviews with George Coates and E.C. Thomas; and Ellis, Darwin and Julian Singer (2008), pp. 662-3

[6] Interviews with Robert Freedman and E.C. Thomas; Coates, George, Christian Clavier and Yves Boutemy (1983): "A study of the dual-water model based on log data," *Journal of Petroleum Technology*, vol. 35, no. 1, pp. 158-66; and Ellis, Darwin and Julian Singer (2008), pp. 662-3

[7] www.slb.com/services/technical_challenges/carbonates.aspx

[8] Poupon, André, William Hoyle and Arthur Schmidt (1971): "Log analysis in complex formations with complex lithologies," *Journal of Petroleum Technology*, vol. 23, no. 8, pp. 995-1005; and Savre, Wayland and Jack Burke (1963): "Determination of true porosity and mineral composition in complex lithologies with the use of the sonic, neutron, and density surveys," conference paper, *SPWLA 4th Annual Logging Symposium*, 23-24 May, Oklahoma City, Oklahoma

[9] Lawyer, Lee, Charles Bates and Robert Rice (2001), p. 269; and Prensky, Stephen (1999): "Advances in Borehole Imaging Technology and Applications," in: Lovell, Mike, Gail Williamson and Peter Harvey (eds.): *Borehole Imaging: applications and case histories*, London: The Geological Society, pp. 1-43

[10] Briggs, Robert (1964): "Development of a downhole television camera," conference paper, *SPWLA 5th Annual Logging Symposium*, 13-15 May, Midland, Texas; and Prensky, Stephen (1999)

[11] Interview with Robert Hunt-Grubbe; Prensky, Stephen (1999); Zemanek, Joseph et al. (1969): "The borehole televiewer: a new logging concept for fracture location and other types of borehole inspection," *Journal of Petroleum Technology*, vol. 21, issue 6, pp. 762-74; http://virtualmuseum.seg.org/bio_joseph_zemanek.html; and http://petrowiki.org/Borehole_imaging

[12] Interview with Robert-Hunt Grubbe; Prensky, Stephen (1999); http://petrowiki.org/Borehole_imaging; and www.glossary.oilfield.slb.com/en/Terms/b/borehole_televiewer.aspx

[13] Allaud, Louis and Maurice Martin (1977), pp. 254-67; Ellis, Darwin and Julian Singer (2008), p. 126; and Oristaglio, Michael and Alexander Dorozynski (2007), p. 261

[14] Interview with Stefan Luthi; and Bateman, Richard (2012), p. 250

[15] Interview with Robert Hunt-Grubbe; and Hunt-Grubbe, Robert (2007): *From little acorns... the story of a fledgling company* (Sondex), Potterne (Wiltshire): unpublished manuscript

[16] Christie, Phil et al. (1995): "Borehole seismic data sharpen the reservoir image," *Oilfield Review*, winter, pp. 18-31

[17] Anstey, Nigel and Turhan Taner (1975): *Broad Line Seismic Profiling*, US Patent no. 3885225 patented May 20; Bateman, Richard (2012), pp. 250; Brewer, Robert (2000): *VSP Data in Comparison to other Borehole Seismic Data*, online presentation: www.searchanddiscovery.com/documents/geophysical/brewer/images/brewer.pdf; EAGE (ed.)(2008): "Virtues of VSP highlighted at Galperin Readings," special topic, *First Break*, vol. 26, August; Galperin, Evsei Iosifovich (1985): *Vertical Seismic Profiling and its Exploration Potentials*, Dordrecht: Kluwer, p. xi; Sheriff, Robert and Lloyd Geldart

(1995), p. 21; www.bairdpetro.com/simple_seismics.htm; and http://en.wikipedia.org/wiki/Nigel_Anstey

[18] Myers, Gary (2007): "Nuclear Logging," in: Holstein, Edward (ed.): *Petroleum Engineering Handbook*, vol. V (a)(Reservoir engineering and petrophysics), Richardson (Texas): Society of Petroleum Engineers pp. 243-87, here p. 262; and www.spec2000.net/07-activationlog.htm

[19] Interviews with Robert Hunt-Grubbe and Larry Jacobson; Allaud, Louis and Maurice Martin (1977), p. 306; National Petroleum Council (ed.)(1967), p. 70; and www.spec2000.net/07-activationlog.htm

[20] Interviews with Darwin Ellis and Larry Jacobson

[21] Interviews with Larry Jacobson and E.C. Thomas; Baker Hughes (ed.)(2007), p. 40; Payne, Darwin (1979), p. 223; Youmans, Arthur et al. (1964): "The Neutron Lifetime Log," conference paper, *SPWLA* 5th Annual Logging Symposium, 13-15 May, Midland, Texas; Baker Hughes (ed.)(2007), p. 40; and http://en.wikipedia.org/wiki/Baker_Hughes

[22] Herron, Susan and Michael Herron (1996): "Quantitative lithology: an application for open and cased hole spectroscopy," conference paper, *SPWLA* 37th Annual Logging Symposium, June 16-19, New Orleans, Louisiana; Hertzog, Russ (1980): "Laboratory and field evaluation of an inelastic neutron scattering and capture gamma ray spectrometry tool," *Society of Petroleum Engineers Journal*, vol. 20, no. 5, pp. 327-40; Horkowitz, John and Darrel Cannon (1997): "Complex reservoir evaluation in open and cased wells," conference paper, *SPWLA* 38th Annual Logging Symposium, June 15-18, Houston; Radtke, R.J. et al. (2012): "A new capture and inelastic spectroscopy tool takes geochemical logging to the next level," conference paper, *SPWLA* 53rd Annual Logging Symposium, June 16-20, Cartagena, Colombia

[23] Kleinberg, Robert et al. (1995): "Nuclear magnetic resonance imaging: technology for the 21st century," *Oilfield Review*, autumn, pp. 19-33; and www.nobelprize.org/nobel_prizes/physics/laureates/1952/

[24] Brown, Robert (2001): "The earth's field NML development at Chevron," *Concepts in Magnetic Resonance*, vol. 13, no. 6, pp. 344-66; Ellis, Darwin and Julian Singer (2008), pp. 415-6; Kleinberg, Robert and Jasper Jackson (2001): "An introduction to the history of NMR well logging," *Concepts in Magnetic Resonance*, vol. 13, no. 6, pp. 340-2; Varian, Russell (1951): *Method and Means for Correlating Nuclear Properties of Atoms and Magnetic Fields*, US Patent no. 2561490A patented October 21; and http://en.wikipedia.org/wiki/Russell_and_Sigurd_Varian

[25] Interviews with Richard Bateman, George Coates, Robert Freedman and E.C. Thomas; Brown, Robert (2001); Ellis, Darwin and Julian Singer (2008), pp. 415-6; Herrick, R.C., S.H. Couturie and D.L. Best (1979): "An improved nuclear magnetism logging system and its application to formation evaluation," conference paper, 54th Annual Fall Technical Conference and Exhibition of the *Society of Petroleum Engineers of AIME*, 23-26 September, Las Vegas; Kleinberg, Robert (2001): "NMR well logging at Schlumberger," *Concepts in Magnetic Resonance*, vol. 13, no. 6, pp. 396-403; Kleinberg, Robert et al. (1995); Timur, Aytekin (1968): "Effective porosity and permeability of sandstones investigated through nuclear magnetic resonance principles", conference paper, *SPWLA* 9th Annual Logging Symposium, 23-26 June, New Orleans, Louisiana; and http://petrowiki.org/Nuclear_magnetic_resonance_(NMR)_logging

[26] Brown, J., Brown, Lee and Jasper Jackson (1981): "NMR measurements on western gas sands core," *SPE/DOE* Low Permeability Symposium, 27-29 May, Denver, Colorado; Ellis, Darwin and Julian Singer (2008), pp. 10, 415-6; Jackson, Jasper (2001): "Los Alamos well logging project," *Concepts in Magnetic Resonance*, vol. 13, no. 6, pp. 368-78; Kleinberg, Robert and Jasper Jackson (2001); and Lyons, William and Gary Plisga (2004):

Standard Handbook of Petroleum and Natural Gas Engineering, Houston: Professional Publishing, p. 5119

[27] Interviews with Richard Bateman and Robert Freedman; Miller, Melvin (2001): "Numar and Numalog overview," *Concepts in Magnetic Resonance*, vol. 13, no. 6, pp. 379-85; and http://petrowiki.org/Nuclear_magnetic_resonance_(NMR)_logging

[28] Interview with Richard Bateman; Kleinberg, Robert and Jasper Jackson (2001); Miller, Melvin (2001); and Miller, Melvin et al. (1990): "Spin echo magnetic resonance logging: porosity and free fluid index determination," conference paper, *SPE Annual Technical Conference and Exhibition*, 23-26 September, New Orleans, Louisiana

[29] Interview with George Coates; Bateman, Richard (2012), pp. 341, 358; Cannon, D.E., C. Cao Minh and R.L. Kleinberg (1998): "Quantitative NMR interpretation," *SPE Annual Technical Conference and Exhibition*, 27-30 September, New Orleans; Ellis, Darwin and Julian Singer (2008), pp. 415-6, 471; Kenyon, W. et al. (1988): "A three-part study of NMR longitudinal relaxation properties of water-saturated sandstones," *SPE Formation Evaluation*, September, pp. 622-36; Kleinberg, Robert (1998): "Nuclear Magnetic Resonance," in: Wong, Po-zen (ed.): *Methods in the Physics of Porous Media*, San Diego: Academic Press, pp. 337-85; Kleinberg, Robert and Jasper Jackson (2001); Kleinberg, Robert et al. (1995), p. 24; Miller, Melvin (2001); and http://petrowiki.org/Nuclear_magnetic_resonance_(NMR)_logging

[30] Bateman, Richard (2009): "Petrophysical data acquisition, transmission, recording and processing: a brief history of change from dots to digits," conference paper, *SPWLA 50th Annual Logging Symposium*, 21-24 June, The Woodlands, Texas

[31] Burgen, Jack and Hilton Evans (1975): "Direct digital laserlogging," conference paper, 50th annual fall meeting of the *Society of Petroleum Engineers of AIME*, 28 September-1 October, Dallas; and Campbell-Kelly, Martin and William Aspray (2004): *Computer: A history of the information machine*, London: Basic Books, pp. 209-17

[32] Interviews with Richard Bateman and Marvin Gearhart; Bateman, Richard (2009); Moseley, L.M. (1976): "Field evaluation of Direct Digital Well Logging," conference paper, 17th *SPWLA* annual logging symposium, June 9-12, Denver; and http://en.wikipedia.org/wiki/Gearhart

[33] Interviews with Richard Bateman and Marvin Gearhart; Bateman, Richard (2009); Burgen, Jack and Hilton Evans (1975); Eaton, F.M. et al. (1976): "The Cyber Service Unit: an integrated logging system," conference paper, *SPE Annual Fall Technical Conference and Exhibition*, 3-6 October, New Orleans; Moseley, L.M. (1976); and http://en.wikipedia.org/wiki/Gearhart

Chapter 17: Controlling the Well

[1] Isaac Newton cited by Bill Rehm, in: Journal of Petroleum Technology (ed.)(2008): "Bill Rehm: From the Shoulders of Giants," *Journal of Petroleum Technology*, special section "Legends of Drilling", December, p. 44

[2] Interview with Bill Rehm; and O'Brian, T.B. and W.C. Goins (1960): *The Mechanics of Blowouts and How to Control Them*, API Drilling and Production Practices P41

[3] Adams, Neal and Alfred Eustes III (2011): "Drilling Problems," in: Mitchell, Robert and Stefan Miska (eds.)(2011): *Fundamentals of Drilling Engineering*, Richardson (Texas): Society of Petroleum Engineers, pp. 625-76, here p. 635; Grace, Robert (2003): *Blowout and Well Control Handbook*, Amsterdam: Gulf Professional Publishing, p. 25; and Rehm, Bill (2011): *Geological Engineering: how we learned to drill safely in the Gulf of Mexico Miocene shale in the GOM: a story of mud and early drilling problems*, paper given at the Missouri University of Science and Technology, November, p. 2

[4] Interview with Bill Rehm; and Rehm, Bill (2011), p. 2

[5] Schempf, F. Jay (2007), p. 162; Singerman, Philip (1990): *An American Hero: the Red Adair story*, Boston: Little, Brown and Company, pp. 25, 58; Baltimore Sun (ed.)(2003): "Preparing to battle infernos," *Baltimore Sun*, March 17; and Houston Chronicle (ed.)(2010): "'Coots' Matthews, oil well firefighter, dies at 86," *Houston Chronicle*, April 1

[6] Grace, Robert (2003), p. 439

[7] Grace, Robert (2003), p. 309; and Weaver, Bobby (2010), p. 160

[8] Singerman, Philip (1990), pp. 75, 123-5

[9] Schempf, F. Jay (2007), pp. 161-2; and Weaver, Bobby (2010), p. 161

[10] www.freerepublic.com/focus/f-news/2484939/posts

[11] Interviews with Erik Nelson and Mario Zamora

[12] www.fundinguniverse.com/company-histories/Weatherford-International-Inc-Company-History.html; and http://en.wikipedia.org/wiki/Weatherford_International

[13] Interview with Alistair Oag; and Walker, James (1985): *Weld-On Casing Connector*, US patent no. 4509777 patented Apr. 9

[14] Interview with Erik Nelson

[15] Lyons, William and Gary Plisga (2004), p. 4448; Millikan, Charles (1961), p. 487; Nelson, Erik, Michel Michaux and Bruno Drochon (2006): "Cementing Additives and Mechanisms of Action," in: Nelson, Erik and Dominique Guillot (eds.)(2006): *Well Cementing*, Sugar Land: Schlumberger, pp. 49-91, here pp. 58, 62; and Sweatman, Ron (2011), p. 147

[16] Millikan, Charles (1961), pp. 487-8; and Millikan, Charles (1971): "Cementing," in: Brantly, John (1971): *History of Oil Well Drilling*, Houston: Gulf Publishing, pp. 1306-41, here p. 1336

[17] Halliburton annual report 1957; Millikan, Charles (1961), p. 488; and Rodengen, Jeffrey (1996), pp. 57-8, 64

[18] Interview with Erik Nelson; Stiles, David (2006): "Annular Formation Fluid Migration," in: Nelson, Erik and Dominique Guillot (eds.)(2006): *Well Cementing*, Sugar Land: Schlumberger, pp. 289-317, here p. 307; Tinsley, John (1980): "Study of factors causing annular gas flow following primary cementing," *Journal of Petroleum Technology*, vol. 32, no. 8, pp. 1427-37; www.halliburton.com/ps/default.aspx?pageid=1143&navid=2183; and https://www.trademarkia.com/gaschek-73229169.html

[19] Interview with Alistair Oag; Parcevaux, Philippe, Bernard Piot and Claude Vercaemer (1985): *Cement Compositions for Cementing Wells, Allowing Pressure Gas-Channeling in the Cemented Annulus to be Controlled*, US patent no. 4537918 patented Aug. 27; and Stiles, David (2006), p. 308

[20] Interview with Leon Robinson; Bommer, Paul (2008), p. 133; and Robinson, Leon and Joe Heilhecker (1973): *Method and Apparatus for Treating a Drilling Fluid*, US patent no. 3766997 patented Oct. 23

[21] Interviews with Keith Millheim and Mario Zamora; and Darley, Henry, George Gray and Ryen Caenn (2011), pp. 58-59, 170, 370-1

[22] Interview with Ted Bourgoyne; interview of Martin Chenevert by Mario Zamora (American Association of Drilling Engineers); interview of Tommy Mondshine by Mario Zamora (American Association of Drilling Engineers; and Chenevert, Martin (1970): "Shale control with balanced-activity oil-continuous muds," *Journal of Petroleum Technology*, October 1970, pp. 1309-16

[23] Darley, Henry, George Gray and Ryen Caenn (2011), p. 71

[24] Interview with Alistair Oag; and Darley, Henry, George Gray and Ryen Caenn (2011), p. 68

[25] www.epa.gov/oecaerth/civil/cwa/index.html

[26] Bleier, Roger, Arthur Leuterman and Cheryl Stark (1993): "Drilling Fluids: making peace with the environment," *Journal of Petroleum Technology* 45, 1 (January 1993), pp. 6-10; Bleier, Roger, Arvind Patel, Raymond McGlothlin and H.N. Brinkley (1993): *Oil Based Synthetic Hydrocarbon Drilling Fluid*, US Patent no. 5189012 patented Feb. 23; and Zamora, Mario (2001), slide no. 3

[27] Bleier, Roger, Arthur Leuterman and Cheryl Stark (1993); Duncan, William, Jr. (1996): *Organic Liquid Base Drilling Fluid with Terpene*, US Patent no. 5559085 patented Sep. 24; Lyons, William and Gary Plisga (2004), p. 4117; Mercer, James and Laura Nesbit (1992): *Oil-Base Drilling Fluid Comprising Branched Chain Paraffins such as the Dimer of 1-Decene*, US Patent no. 5096883 patented Mar. 17; and Rudnick, Leslie (ed.)(2005): *Synthetics, Mineral Oils, and Bio-Based Lubricants: chemistry and technology*, Boca Raton (Florida): CRC Press, p. 71

[28] Duncan, William, Jr. (1996); and Lyons, William and Gary Plisga (2004), p. 4117

[29] Interview with Mario Zamora; Bleier, Roger, Arthur Leuterman and Cheryl Stark (1993); Candler, John et al. (1995): "Seafloor monitoring for synthetic-based mud discharged in the western Gulf of Mexico," conference paper, *SPE/EPA Exploration and Production Environmental Conference*, March 27-29, 1995, Houston; James, R.W. and Bjørn Helland (1992): "The Greater Ekofisk area: addressing drilling fluid challenges with environmental justifications, conference paper," *European Petroleum Conference*, November 16-18, 1992, Cannes, France; Lyons, and William and Gary Plisga (2004), p. 4117

[30] Interviews with Keith Millheim and Mario Zamora; interview of Ronald K. Clark by Mario Zamora (American Association of Drilling Engineers); and Darley, Henry, George Gray and Ryen Caenn (2011), pp. 58-9, 170, 370-1

[31] M-I Swaco (ed.)(2011): *Offshore TCC Hammermill System*, Houston: M-I Swaco; Williamson, Bob et al. (2004): "Offshore hammermill process meets OSPAR discharge limit," *Oil & Gas Journal*, October 5; and www.thermtech.no/home/the_tcc/how-does-a-tcc-reg_-work

Chapter 18: Going Horizontal

[1] Interview with Ted Bourgoyne; Morgan, James (1992): "Horizontal drilling applications of petroleum technologies for environmental purposes," *Ground Water Monitoring & Remediation*, summer 1992, pp. 98-102, here p. 102; Raymond, Martin and William Leffler (2006), p. 16; The Texas Ranger Dispatch (ed.)(2002): "Drilling on the slant," *The Texas Ranger Dispatch*, no. 7 (summer), pp. 33-4; and www.tshaonline.org/handbook/online/articles/doe01

[2] Devereux, Steve (1999): *Drilling Technology in Nontechnical Language*, Tulsa (Oklahoma): PennWell Books, p. 175

[3] Eastman, H. John (1971): "Directional Drilling, 1960s," in: Brantly, John (1971): *History of Oil Well Drilling*, Houston: Gulf Publishing, pp. 1182-1209, here p. 1209; Larson, Henrietta and Kenneth Porter (1959), pp. 403-4; and Miska, Stefan (2011): "Directional Drilling," in: Mitchell, Robert and Stefan Miska (eds.)(2011): *Fundamentals of Drilling Engineering*, Richardson (Texas): Society of Petroleum Engineers, pp. 449-583, here p. 449

[4] Interview of Alexander Grigoryan by Mark Schmidt (Offshore Energy Center); and Gelfgat, Yakov, Mikhail Gelfgat and Yuri Lopatin (2003), vol. 1, p. 10

[5] Interview with Alistair Oag; Baker, Ron (1998), p. 38; Dorozynski, Alexander (2011), p. 132; Gelfgat, Yakov, Mikhail Gelfgat and Yuri Lopatin (2003), vol. 1, p. 14; Maurer, William et al. (1977): *Downhole Drilling Motors: technical review, final report*, Houston: Maurer Engineering, pp. 19-20, 28; and Miska, Stefan (2011), p. 501

[6] Dorozynski, Alexander (2011), p. 23; and Gelfgat, Yakov, Mikhail Gelfgat and Yuri Lopatin (2003), vol. 1, p. 53

[7] Interview with Walt Aldred; and Congress of the United States Office of Technology Assessment (ed.)(1985): *Technology & Soviet Energy Availability*, Washington (DC): US Government, p. 39

[8] Beaton, Tim, Dan Calnan and Rocky Seale (2007): "Identifying applications for turbodrilling and evaluating historical performances in North America," *Journal of Canadian Petroleum Technology*, vol. 46, no. 6 (June); Dorozynski, Alexander (2011), p. 23; Gelfgat, Yakov, Mikhail Gelfgat and Yuri Lopatin (2003), vol. 1, p. 55; and Maurer, William et al. (1977), pp. i, 2-3

[9] Congress of the United States Office of Technology Assessment (ed.)(1985), p. 40; Gelfgat, Yakov, Mikhail Gelfgat and Yuri Lopatin (2003), vol. 1, p. 57; and Lyons, William and Gary Plisga (2004), p. 4276

[10] Gelfgat, Yakov, Mikhail Gelfgat and Yuri Lopatin (2003), vol. 1, pp. 16, 18, 21; and Lyons, William and Gary Plisga (2004), p. 4276

[11] Gelfgat, Yakov, Mikhail Gelfgat and Yuri Lopatin (2003), vol. 1, pp. 24-25, 88-89; Lyons, William and Gary Plisga (2004), p. 4276; and Maurer, William et al. (1977), pp. 12-13, 19

[12] Bommer, Paul (2008), p. 118; and Gelfgat, Yakov, Mikhail Gelfgat and Yuri Lopatin (2003), vol. 1, pp. 21, 154-5

[13] Interview with Hobie Smith; Lyons, William and Gary Plisga (2004); Maurer, William et al. (1977), p. I; and www.dyna-drill.com/aboutus

[14] Interview with Hobie Smith; and Maurer, William et al. (1977), p. i

[15] Interview with Hobie Smith; Grace, Robert (1974b): "Rotary Drilling Bits," in: Moore, Preston (1974): *Drilling Practices Manual*, Tulsa (Oklahoma): PennWell Books, pp. 354-80, here p. 354; and Ozbayoglu, Evren (2011): "Rotary Drilling Bits," in: Mitchell, Robert and Stefan Miska (eds.)(2011): *Fundamentals of Drilling Engineering*, Richardson (Texas): Society of Petroleum Engineers, pp. 311-84, here p. 311

[16] Interview with Hobie Smith; Grace, Robert (1974b), p. 355; www.itia.info/history.html; and en.wikipedia.org/wiki/Tungsten_carbide

[17] Baker Hughes (ed.)(2007), pp. 25, 41; Baker, Ron (1998), p. 32; Cannon, George and Thomas Pennington (1971): "Rotary Drag-Type Drilling Bits," in: Brantly, John (1971): *History of Oil Well Drilling*, Houston: Gulf Publishing, pp. 1060-70, here p. 1064; Gerding, Mildred (1986), p. 124; Moore, W.W. (1981), pp. 6-7; and Scott, Floyd (1971), p. 1090

[18] Interview with Keith Millheim; Baker Hughes (ed.)(2007), p. 45; Grace, Robert (1974b), p. 362; and Scott, Floyd (1971), p. 1092

[19] Interviews with David Curry and Hobie Smith; Baker Hughes (ed.)(2007), p. 59; and Grace, Robert (1974b), p. 366

[20] Gelfgat, Yakov, Mikhail Gelfgat and Yuri Lopatin (2003), vol. 1, p. 114; Léonardon, Eugène (1961), p. 505; Oil & Gas Journal (ed.)(2006): "Roller Cones vs. Diamonds: a reversal of roles," *Oil & Gas Journal*, February 20; www.deseretnews.com/article/884049/Obituary-Frank-Langton-Christensen.html; www.referenceforbusiness.com/history2/99/Christensen-Boyles-Corporation.html; and en.wikipedia.org/wiki/Hughes_Christensen

[21] Interviews with Keith Millheim and Dan Scott; Baker Hughes (ed.)(2007), p. 54; Oil & Gas Journal (2006); and Scott, Dan (2006): "The history and impact of synthetic diamond cutters and diamond enhanced inserts on the oil and gas industry," *Industrial Diamond Review*, vol. 66, no. 1

[22] Oil & Gas Journal (ed.)(1992): "Antiwhirl PDC bit designs reduce vibrations," *Oil & Gas Journal*, November 30; Scott, Dan (2006), p. 5; and Warren, Tommy et al. (1993): "Field testing of low-friction gauge PDC bits," *SPE Drilling Engineering*, vol. 8, no. 1 (March), pp. 21-27

[23] Interviews with Marvin Gearhart, Leon Robinson and Dan Scott; Oil & Gas Journal (ed.)(1995): "News: industry pushes use of PDC bits to speed drilling, cut costs," *Oil & Gas Journal*, August 14; and Oil & Gas Journal (2006)

[24] Interview with George Boyadjieff; and en.wikipedia.org/wiki/National_Oilwell_Varco

[25] Interview with George Boyadjieff; Dyke, Kate van (1997); and Hyne, Norman (2001), p. 255

[26] Interviews with Jacques Bosio and George Boyadjieff; and www.nov.com/aboutnov.aspx?id=5239&linkidentifier=id&itemid=5239

[27] Interview with George Boyadjieff; Burleson, Clyde (1999), p. 127; www.glossary.oilfield.slb.com/Display.cfm?Term=kelly; and www.glossary.oilfield.slb.com/Display.cfm?Term=swivel

[28] Interview with George Boyadjieff; Frank Williford's written answer to Mark Mau's interview questions; and www.nov.com/aboutnov.aspx?id=5239&linkidentifier=id&itemid=5239

[29] Interview with George Boyadjieff; Aldred, Walt et al. (2005): "Changing the way we drill," *Oilfield Review*, spring, pp. 42-9, here p. 43; and Ford, John (2000), pp. 107-8

[30] Interviews with Walt Aldred, George Boyadjieff, Keith Millheim and David Reid

[31] Bosworth, Steve et al. (1998): "Key issues in multilateral technology," *Oilfield Review*, winter, pp. 14-28, here p. 16; Gaddy, Dean (1998): "Pioneering work, economic factors provide insights into Russian drilling technology," *Oil & Gas Journal*, vol. 96, no. 27 (July 6), pp. 67-9; Gelfgat, Yakov, Mikhail Gelfgat and Yuri Lopatin (2003), vol. 2, p. 69; and www.aapg.org/explorer/2000/09sep/horiz_drill.cfm

[32] Interview with Robert Mitchell; Gaddy, Dean (1998); Gadelle, Claude and Gerard Renard (1999): "*Increasing oil production through horizontal and multilateral wells*," paper presented at the workshop on Enhanced Production of Old Oil Fields, Surgut, Russia, March 17-18, 1999; Gelfgat, Yakov, Mikhail Gelfgat and Yuri Lopatin (2003), vol. 2, p. 69

[33] Bosio, Jacques (2011): "A life led horizontally," *Offshore Engineer*, January; Gaddy, Dean (1998); and Gadelle, Claude and Gerard Renard (1999)

[34] Interview with Jacques Bosio; Bosio, Jacques (2011); and Jahn, Frank, Mark Cook and Mark Graham (2008), p. 68

[35] Bosworth, Steve (1998), pp. 16-17; Dorozynski, Alexander (2011), pp. 132-3; Gelfgat, Yakov, Mikhail Gelfgat and Yuri Lopatin (2003), vol. 2, pp. 64, 68; Hill, A.D., Ding Zhu and Michael Economides (2008): *Multilateral Wells*, Richardson (Texas): Society of Petroleum Engineers, p. 3; and Schempf, F. Jay (2007), p. 202

[36] Bosworth, Steve (1998), p. 14; and Hill, A.D., Ding Zhu and Michael Economides (2008), p. 3

[37] Salamy, Salam et al. (2008): "Maximum reservoir contact wells performance update: Shaybah Field, Saudi Arabia," *SPE Production & Operation*, vol. 23, no. 4, November, pp. 439-43; Saleri, Nansen (2000): *Re-engineering reservoir management for the new Millennium*, speech given to the Dhahran Geological Society, Dhahran, February; Saleri, Nansen, Salam Salamy and S.S. Al-Otaibi (2003): "The extending role of the drill bit in shaping the subsurface," *Journal of Petroleum Technology*, December 2003, pp. 53-8; and Saleri, Nansen et al. (2004): "Shaybah-220: a maximum-reservoir-contact (MRC) well and its implications for developing tight-facies reservoirs," *SPE Reservoir Evaluation & Engineering*, August, pp. 316-21

Chapter 19: Intelligent Drilling

[1] Interviews with Keith Millheim, Alistair Oag and John Thorogood; and Eastman, H. John (1971), p. 1202

[2] Carden, Richard and Robert Grace (2007): *Horizontal and Directional Drilling*, seminar catalogue, Tulsa (Oklahoma): Petroskills, p. 11; Eastman, H. John (1971), p. 1202; Grace, Robert (1974b), p. 366; Miska, Stefan (2011), p. 449; and Moore, W.W. (1981), pp. 179

[3] Interviews with John Cook and Geoff Downton; Baker Hughes INTEQ (ed.)(1992): *Advanced Wireline & MWD Procedures Manual*, Houston: Baker Hughes Technical Publications Group, pp. 21-22; Bommer, Paul (2008), p. 180; Fontenot, John (1986): "Measurement while drilling: a new tool," *Journal of Petroleum Technology*, vol. 38, no. 2, pp. 128-30, here p. 128; Spotkaeff, Matthew (2007): *Logging While Drilling*, workshop presentation, SPE Queensland, p. 4; Suman, John (1961), p. 79

[4] Baker Hughes (1992), pp. 21-22; Fontenot, John (1986), p. 128; and http://aimehq.org/programs/award/bio/jan-j-arps

[5] Interviews with Walt Aldred, David Curry, Keith Millheim, Alistair Oag, Leon Robinson and John Thorogood; Baker Hughes (1992), p. 22; Baker Hughes INTEQ (1997): *Baker Hughes INTEQ's Guide to Measurement While Drilling*, Houston: Baker Hughes INTEQ Technical Communications Group, p. 15; Roberts, Andrew, Robert Newton and Frederick Stone (1982): "MWD field use and results in the Gulf of Mexico," conference paper, *SPE Annual Technical Conference and Exhibition*, September 26-29, New Orleans, Louisiana, p. 1; Spinnler, R.F., F.A. Stone and C. Ray Williams (1978): *Mud Pulse Logging while Drilling Telemetry System: design, development, and demonstrations*, Bartlesville (Oklahoma): Teleco Oilfield Services, Inc.; and http://articles.courant.com/1991-11-30/business/0000208844_1_oil-wells-oil-industry-oil-exploration

[6] Interviews with Marvin Gearhart and Keith Millheim; and Lyons, William and Gary Plisga (2004), p. 4300

[7] Interviews with John Cook, Andrew Gould and Leon Robinson; Angehrn, J. and S. Sie (1987): "A high data rate fiber optic well logging cable," *The Log Analyst*, vol. 28, no. 2; Bateman, Richard (2009), p. 4; Holcomb, David, Robert Hardy and David Glowka (1997): *Disposable Fiber Optics Telemetry for Measuring While Drillings*, Albuquerque (New Mexico): Sandia National Laboratories; Wharton, Russell (1983): *Well Logging Fiber Optic Communication System*, US patent no 4,389,645 patented June 21; www.nov.com; www.sandia.gov/media/NewsRel/NR2001/fiber.htm; and www.worldoil.com/June-2014-Whats-new-in-well-logging-and-formation-evaluation.html

[8] Interview with Roland Chemali; and Allen, David et al. (1987): "Logging while drilling," *Oilfield Review*, vol. 1, no. 1, pp. 4-17, here p. 4

[9] Arps, J.J. (1963): "Continuous logging while drilling," conference paper, *SPE Annual Fall Meeting*, October 6-9, New Orleans; Arps, J.J. and J.L. Arps (1964): "The Subsurface Telemetry Problem: a practical solution," *Journal of Petroleum Technology*, vol. 16, no. 5, pp. 487-93; Karcher, J.C. (1933): *Method and Apparatus for Exploring Bore Holes*, US Patent no. 1927664 patented Sept. 19; Lyons, William and Gary Plisga (2004), p. 4300; and Meador, Richard (2009): "Logging-while-drilling: a story of dreams, accomplishments, and bright futures," conference paper, *SPWLA* 50th Annual Logging Symposium, 21-24 June, The Woodlands, Texas

[10] http://petrowiki.org/Electromagnetic_logging_while_drilling; http://aimehq.org/programs/award/bio/jan-j-arps; and www.fundinguniverse.com/company-histories/nl-industries-inc-history/

[11] Interviews with Roland Chemali and Tom Zimmerman; Antonov, Yu. N. and D.S. Daev (1965): "Dielectric logging equipment," *Geophysical Equipment*, Nm Nedram Rel 26; Bateman, Richard (2012), p. 270; Daev, D.S., and S.B. Denisou (1970): "About high frequency induction logging," *Geophysical Equipment*, M. Rel 42; Lyon, William and Gary Plisga (2004), p. 4328; Meador, Richard (2009); and Meador, Richard and P.T. Cox (1975): "Dielectric constant logging: a salinity independent estimation of formation water volume," conference paper, fall meeting of the *Society of Petroleum Engineers of AIME*, 28 September-1 October, Dallas

[12] Interview with Brian Clark and Billy Hendricks; and Meador, Richard (2009)

[13] Interviews with Roland Chemali and Brian Clark

[14] Interview with Brian Clark

[15] Interview with Brian Clark; and Clark, Brian (1983): *Well Logging Apparatus and Method using Transverse Magnetic Mode*, European Patent Application, no. 83401907.7, September 29

[16] Interviews with Brian Clark and Billy Hendricks; and Prensky, Stephen (1999), p. 18

[17] Interviews with John Cook, Andrew Gould, Charles Ingold, Keith Millheim, Alistair Oag, Volker Reichert and Peter Sharpe; Baker Hughes (ed.)(2007), p. 67; Feenstra, Robijn and Anthony Kamp (1984): *Downhole Motor and Method for Directional Drilling of Boreholes*, US Patent no. 4485879 patented Dec. 4; Lesso, William (2009): "Geosteering," in: Aadnoy, Bernt et al. (eds.): *Advanced Drilling and Well Technology*, Richardson (Texas): Society of Petroleum Engineers, pp. 458-73, here pp. 458, 460; Miska, Stefan (2011), pp. 505-507; and Warren, Tommy (2006): "Steerable motors hold out against rotary steerables," paper prepared for the 2006 *SPE Annual Technical Conference and Exhibition* held in San Antonio (Texas), September 24-27

[18] Interview with Walt Aldred; and Haugen, Jonny (1998): "Rotary steerable system replaces slide mode for directional drilling applications," *Oil & Gas Journal*, February 3

[19] Interviews with John Cook and David Curry; Baker Hughes (ed.)(2007), p. 5; Bram, Kurt at al. (1995): "The KTB Borehole: Germany's superdeep telescope into the Earth's crust," *Oilfield Review*, January, pp. 4-22; Emmermann, Rolf and Jörn Lauterjung (1997): "The German Continental Deep Drilling Program KTB: overview and major results," *Journal of Geophysical Research*, 102,B8, August 10, pp. 18179-201; Haugen, Jonny (1998); and Warren, Tommy (2006)

[20] Allen, Frank et al. (1997): "Extended-reach drilling: breaking the 10-km barrier," *Oilfield Review*, winter, pp. 32-47, here p. 38; Haugen, Jonny (1998); Miska (2011), p. 449; and Williams, Mike (2004): "Better turns for rotary steerable drilling," *Oilfield Review*, spring, pp. 4-9, here p. 5

[21] Interview with David Llewelyn; Alfsen, T.E. (1995): "Pushing the limits for extended reach drilling: new world record from platform Statfjord C, Well C2," *SPE Drilling & Completion*, vol. 10, no. 2, pp. 71-6; and Jahn, Frank, Mark Cook and Mark Graham (2008), p. 69

[22] Interviews with David Curry and Geoff Downton; Allen, Frank et al. (1997), pp. 33, 42; Allen, Frank, Tony Meader and Graham Riley (2000): "The secret of world-class extended-reach drilling," *Journal of Petroleum Technology*, June 2000, pp. 41-3, here p. 41; Journal of Petroleum Technology (ed.): "People: John Barr," *Journal of Petroleum Technology*, April 2007, p. 108; and Scott, Dan (2006)

[23] Interview with David Curry; Downton, Geoff et al. (2000): "New directions in rotary steerable drilling," *Oilfield Review*, spring, pp. 18-29; and www.slb.com/about/history/2000s.aspx

[24] Edmundson, Henry (1986b): "Basics of failure mechanics," *The Technical Review*, vol. 34, no. 3 (October), pp. 10-19

[25] Interviews with Sid Green and Jean-Claude Roegiers

[26] www.slb.com/about/rd/technology/bgc.aspx; and www.slb.com/about/history/2000s.aspx

Chapter 20: Modern Completion and Production

[1] Bellarby, Jonathan (2009): *Well Completion Design*, Amsterdam: Elsevier, pp. 198-201; Penberthy, W.L. Jr. and C.M. Shaughnessy (1992): *Sand Control*, Richardson (Texas): Society of Petroleum Engineers, pp. 1, 3; and http://petrowiki.org/Sand_control

[2] Beecher, C.E. and H.C. Fowler (1961), p. 762; and Penberthy, W.L. Jr. and C.M. Shaughnessy (1992), pp. 3-4

[3] Interview with Harry O'Neal McLeod; Penberthy, W.L. Jr. and C.M. Shaughnessy (1992), pp. 5, 11; Vance, Harold (1961a), pp. 606-7; http://petrowiki.org/History_of_gravel_packs; and http://petrowiki.org/Sand_control_techniques

[4] Interview with Harry O'Neal McLeod; Penberthy, W.L. Jr. and C.M. Shaughnessy (1992), p. 5; and Vance, Harold (1961a), p. 608

[5] Underdown, D., K. Das and H. Nguyen (1985): "Gravel packing highly deviated wells with a crosslinked polymer system," *Journal of Petroleum Technology*, vol. 37, no. 12, pp. 2197-202; Uren, Lester (1945): *Apparatus for Placing Gravel in Wells*, US Patent no. 2372461 patented March 27; and http://petrowiki.org/History_of_gravel_packs

[6] Interview with George King; Penberthy, W. and E. Echols (1993): "Gravel placement in wells," *Journal of Petroleum Technology*, vol. 45, no. 7, pp. 612-74; and http://petrowiki.org/History_of_gravel_packs

[7] Interviews with George King and Martin Rylance; and http://petrowiki.org/History_of_gravel_packs

[8] Bellarby, Jonathan (2009), pp. 201-9; Cirigliano, A.J. and R.E Leibach (1967): "Gravel packing in Venezuela," conference paper, 7th *World Petroleum Congress*, 2-9 April, Mexico City; Ellis, Richard (1998): "An overview of frac packs: a technical revolution (evolution) process," *Journal of Petroleum Technology*, vol. 50, no. 1, pp. 66-8; and Roodhart, L.P. et al. (1993): "Frack and pack stimulation: application, design, and field experience from the Gulf of Mexico to Borneo," conference paper, *SPE Annual Technical Conference and Exhibition*, 3-6 October, Houston

[9] Middle East & Asia Reservoir Review (ed.)(2007): "Frac packing: fracturing for sand control," *Middle East & Asia Reservoir Review*, no. 8, pp. 36-49, here p. 41; and Syed, Ali et al. (2002): "Combined stimulation and sand control", *Oilfield Review*, summer, pp. 30-47, here p. 31

[10] Interview with Larry Harrington; Bellarby, Jonathan (2009), p. 89; Montgomery, Carl and Michael Smith (2010), p. 11; Middle East & Asia Reservoir Review (ed.)(2007), p. 41; Syed, Ali et al. (2002), pp. 31-2; and http://en.wikipedia.org/wiki/Western_Company_of_North_America

[11] Edmundson, Henry (1986a): "Horizontal fractures: debunking a myth", *The Technical Review*, vol. 34, no. 3 October), pp. 4-8, here p. 7; and Montgomery, Carl and Michael Smith (2010), p. 4

[12] Interview with Thomas Perkins; Edmundson, Henry (1986a), pp. 7-8; Hubbert, Marion King and David Willis (1957): "Mechanics of hydraulic fracturing," *AIME Petroleum Transactions*, vol. 210, pp. 153-68; and Montgomery, Carl and Michael Smith (2010), p. 4

[13] Interviews with Larry Harrington and Thomas Perkins; Barenblatt, G.I., S.A. Christianovich, Y.P. Zheltov and G.K. Maximovich (1959): "Theoretical principles of hydraulic fracturing of oil strata," conference paper, 5th *World Petroleum Congress*, 30 May-5 June, New York; Perkins, Thomas and Loyd Kern (1961): "Widths of hydraulic

fractures," *Journal of Petroleum Technology*, vol. 13, no. 9, pp. 937-49; and http://fracking-controversy.weebly.com/

[14] Interview with Mike Prats; Prats, Mike (1961): "Effect of vertical fractures on reservoir behavior: incompressible fluid case," *Society of Petroleum Engineers Journal*, vol. 1, no. 2, pp. 105-18; and http://petrowiki.org/Post-fracture_well_behavior

[15] Interviews with Larry Bell, Larry Harrington, Stephen Holditch and Thomas Perkins; and www.spe.org/industry/history/timeline.php

[16] Interviews with Claude Cooke and Larry Harrington; Cooke, Claude (1977): "Fracturing with a high-strength proppant," *Journal of Petroleum Technology*, vol. 29, no. 10, pp. 1222-6; and http://geology.com/minerals/bauxite.shtml

[17] Raymond, Martin and William Leffler (2006), pp. 209, 211

[18] Interview with George King; www.combinedops.com/pluto.htm; and http://petrowiki.org/History_of_coiled_tubing_technology

[19] Baker Hughes (ed.)(2007), p. 26; Calhoun, George and Herbert Allen (1951): "*Equipment for Inserting Small Flexible Tubing into High-Pressure Wells*, US Patent no. 2567009, patented September 4; Offshore Energy Center (2009): *The Star*, third quarter, p. 2; www.oceanstaroec.com/fame/2009/rike.htm; and http://petrowiki.org/History_of_coiled_tubing_technology

[20] Interview with George King; Afghoul, Ali Chareuf et al. (2004): "Coiled tubing: the next generation," *Oilfield Review*, spring, pp. 38-57, here p. 42; Wright, Russell and Alexander Sas-Jaworsky (1998): *Coiled Tubing Handbook*, Houston: Gulf Publishing Company, p. 7; www.icota.com/historyct.htm; and http://petrowiki.org/History_of_coiled_tubing_technology

[21] Afghoul, Ali Chareuf et al. (2004); Clegg, Joe Dunn and Erich Klementich (2007): "Tubing Selection, Design, and Installation," in: Clegg, Joe Dunn (ed.): *Petroleum Engineering Handbook*, vol. IV (Production operations engineering), Richardson (Texas): Society of Petroleum Engineers, pp. 105-48, here p. 141; Dyke, Kate van (1997), p. 192; Retalic, Ian, Andy Laird and Angus McLeod (2009): "Coiled-Tubing Drilling," in: Aadnoy, Bernt et al. (eds.): *Advanced Drilling and Well Technology*, Richardson (Texas): Society of Petroleum Engineers, pp. 764-85, here p. 765; Varhaug, Matt (2014): "Big reels at the wellsite," *Oilfield Review*, summer, pp. 63-4, here p. 63; www.oceanstaroec.com/fame/2008/productiontech.html; www.scribd.com/doc/45169970/BJ-Coiled-Tubing-Equipment-Manual-Version-1; and www.icota.com/historyct.htm

[22] Interview with Larry Behrmann; Bakker, Eelco et al. (2003): "The new dynamics of underbalanced perforating", *Oilfield Review*, winter, pp. 54-67, here p. 54; Bellarby, Jonathan (2009), p. 75; Handren, P.J., T.B. Jupp and J.M. Dees (1993): "Overbalance perforating and stimulation method for wells," conference paper, *SPE Annual Technical Conference and Exhibition*, 3-6 October, Houston; Huber, T.A. and G.H. Tausch (1953): "Permanent-type well completion", *AIME Petroleum Transactions*, vol. 198, pp. 11-16, here p. 16; and Lebourg, M.P. and G.R. Hodgson (1952): "A method of perforating casing below tubing," *Petroleum Transactions, AIME*, vol. 195, pp. 303-10

[23] Interview with Larry Behrmann; Behrmann, Larry (2007): *After 58 years of Perforating, does it still have a future?* presentation slides, internal Schlumberger presentation, Houston, April 24; Behrmann, Larry et al. (1996): "Quo Vadis, Extreme Overbalance?" *Oilfield Review*, autumn, pp. 18-33, here p. 33; Behrmann, Larry et al. (2000): "Perforating practices that optimize productivity," *Oilfield Review*, spring, pp. 52-74, here p. 53; www.glossary.oilfield.slb.com/en/Terms/t/TCP; www.vannpumping.com/inventor.html; and www.vannpumping.com/img/roy-vann/paper-news.jpg

[24] www.vannpumping.com/inventor.html; www.vannpumping.com/img/roy-vann/paper-news.jpg; www.halliburton.com/public/lp/contents/Books_and_Catalogs/web/TCPCatalog/2005 TCPcatalog/PerforatingSolutions_catalog.pdf; and www.slb.com/services/completions/perforating/tubing_conveyed_perforating.aspx

[25] Baker Hughes (ed.)(2007), p. 39; and Bearden, John, Earl Brookbank and Brown Wilson (2009)

[26] Interview with Roger Hoestenbach; and Bearden, John, Earl Brookbank and Brown Wilson (2009)

[27] Inteview with Roger Hoestenbach; Bearden, John, Earl Brookbank and Brown Wilson (2009); and Bremner, Chad et al. (2006): "Evolving technologies: electrical submersible pumps," *Oilfield Review*, winter, pp. 30-43, here p. 35

[28] Bearden, John, Earl Brookbank and Brown Wilson (2009); http://petrowiki.org/Electrical_submersible_pumps; and http://newsok.com/trw-selling-reda-pump-division/article/2239820

[29] http://petrowiki.org/Electrical_submersible_pumps; and http://petrowiki.org/Artificial_lift

[30] Cholet, Henri (1997): *Progressing Cavity Pumps*, Paris: Editions Technip, pp. 5-6; Lea, James (2007): "Artificial Lift Selection," in: Clegg, Joe Dunn (ed.): *Petroleum Engineering Handbook*, vol. IV (Production operations engineering), Richardson (Texas): Society of Petroleum Engineers, pp. 411-56, here p. 418; Matthews, Cam et al. (2007): "Progressing Cavity Pumping Systems," in: Clegg, Joe Dunn (ed.): *Petroleum Engineering Handbook*, vol. IV (Production operations engineering), Richardson (Texas): Society of Petroleum Engineers, pp. 757-837, here p. 757; Lyons, William and Gary Plisga (2004), p. 6201; http://petrowiki.org/Progressing_cavity_pump_(PCP)_systems; and http://en.wikipedia.org/wiki/Progressive_cavity_pump

[31] Cholet, Henri (1997), pp. 5-6; Dawe, Richard (ed.)(2000): *Modern Petroleum Technology*, vol. 1 (Upstream), New York: John Wiley, p. 295; Matthews, Cam et al. (2007), p. 757; www.ogj.com/articles/print/volume-91/issue-32/in-this-issue/pipeline/progressive-cavity-pumps-prove-more-efficient-in-mature-waterflood-test.html; http://gb.pcm.eu/en/about-us/history-rene-moineau.html; and http://petrowiki.org/Progressing_cavity_pump_(PCP)_systems

[32] Interview with Joe Dunn Clegg; Brill, James and Hemanta Mukherjee (1999): *Multiphase Flow in Wells*, Richardson (Texas): Society of Petroleum Engineers, p. 80; Brown, Kermit and James Lea (1985): "Nodal systems analysis of oil and gas wells," *Journal of Petroleum Technology*, vol. 37, no. 10, pp. 1751-63; and Gilbert, W.E. (1954): "Flowing and gas-lift well performance," conference paper, *API*, Spring Meeting of the Pacific Coast District, Division of Prouction, May 6-7, Los Angeles

[33] Brown, Kermit and James Lea (1985); Gustafson, Thane (2012): *Wheels of Fortune: the battle for oil and power in Russia*, Cambridge (Masachusetts): Harvard University Press, pp. 548-9; and Pappas, James (ed.)(2010): "Legends of Production and Operations," *Journal of Petroleum Technology*, vol. 61, no. 12, pp. 33-47, here p. 42

[34] Interview with Harry O'Neal McLeod; Bellarby, Jonathan (2009), p. 288; Brill, James and Hemanta Mukherjee (1999), p. 78; Gustafson, Thane (2012), pp. 202-3, 212, 548-9; and Mach, Joe, Eduardo Proano and Kermit Brown (1979): "A nodal approach for applying systems analysis to the flowing and artificial lift oil or gas well," *Society of Petroleum Engineers* paper no. 8025

[35] Gelfgat, Yakov, Mikhail Gelfgat and Yuri Lopatin (2003), vol. 1, p. 19; and Gustafson, Thane (2012), pp. 187, 201

[36] Gustafson, Thane (2012), pp. 195, 204-6, 210-1

[37] Hasan, A.R. and C.S. Kabir (2002): *Fluid Flow and Heat Transfer in Wellbores*, Richardson (Texas): Society of Petroleum Engineers, p. 140

[38] Beecher, C.E. and H.C. Fowler (1961), p. 798; Edmundson, Henry (1981): "Production logging," *The Technical Review*, vol. 29, no. 2; Prensky, Stephen (1992), p. 314; Rumble, R.C. (1955): "A subsurface flowmeter," technical note, *Society of Petroleum Engineers*; Schlumberger, Marcel, Henri-Georges Doll and A. Perebinossoff (1937): "Temperature measurements in oil wells," *Journal of the Institute of Petroleum Technologists*, vol. 23, no. 159; and Vincent, R.P., R.M. Leibrock and C.W. Ziemer (1948): "Well flowmeter for logging producing ability of gas sands," *Transactions of the AIME*, vo. 174, no. 1, pp. 305-14;

[39] Edmundson, Henry (1981)

[40] Anderson, R.A. et al. (1980): "A production logging tool with simultaneous measurements," *Journal of Petroleum Technology*, vol. 32, no. 2, pp. 191-8; Hasan, A.R. and C.S. Kabir (2002), p. 140; Tixier, M.P., D. Meunier and J.L. Bonnet (1971): "The production combination tool: a new system for production monitoring," *Journal of Petroleum Technology*, vol. 23, no. 5, pp. 603-13; and Wade, R.T. et al. (1965): "Production logging: the key to optimum well performance," *Journal of Petroleum Technology*, vol. 17, no. 2, pp. 137-44

[41] Bamforth, Steve et al. (1996): "Revitalizing production logging," *Oilfield Review*, vol. 8, no. 4, pp. 44-60; Lenn, Chris, Steve Bamforth, and Hitesh Jariwala (1996): "Flow diagnosis in an extended reach well at the Wytch Farm Oilfield using a new toolstring combination incorporating novel production technology," conference paper, *SPE Annual Technical Conference and Exhibition*, 6-9 October, Denver, Colorado; and http://petrowiki.org/Acquiring_bottomhole_pressure_and_temperature_data

[42] Billingham, Matthew et al. (2011): "Conveyance: down an out in the oil field," *Oilfield Review*, vol. 23, no. 2, pp. 18-31; Hallundbæk, Jørgen et al. (1997): "Wireline well tractor: case histories," conference paper, *Offshore Technology Conference*, May 5, Houston; and Worthington, Paul, K. Boyle and X.D. Jing (2000): "Petrophysics," in: Dawe, Richard (ed.)(2000): *Modern Petroleum Technology*, vol. 1 (Upstream), New York: John Wiley, pp. 131-206, here p. 176

[43] Interview with Robert Hunt-Grubbe

[44] Al-Amer, Abdulhadi et al. (2005): "Tractoring: a new era in horizontal logging for Ghawar Field, Saudi Arabia," conference paper, *SPE Middle East Oil and Gas Show and Conference*, 12-15 March, Kingdom of Bahrain; Hallundbæk, Jørgen (1995): "Reduction of cost with new well intervention technology, well tractors," conference paper, *Offshore Europe*, September 5-8, Aberdeen; www.offshore-publication.com/index.php/companyprofilesmenu/profiles/173-welltec-a-s; and www.welltec.com

[45] Alkhelaiwi, F. and D. Davies (2007): "Inflow control devices: application and value quantification of a developing technology," conference paper, *International Oil Conference and Exhibition*, 27-30 June, Veracruz, Mexico; and Ellis, Tor et al. (2009): "Inflow control devices: raising profiles," *Oilfield Review*, vol. 21, no. 4, pp. 30-37

[46] Alkhelaiwi, F. and D. Davies (2007); Brekke, Kristian and S. Lien (1994): "New and simple completion methods for horizontal wells improve the production performance in high-permeability, thin oil zones," *SPE Drilling & Completion*, vol. 9, no. 3, pp. 205-9; and http://petrowiki.org/PEH%3ACompletion_Systems#Packer-to-Tubing_Seal_Stacks

[47] Interview with Nansen Saleri

Chapter 21: A Seismic Revolution

[1] Interviews with Guus Berkhout and Bob Peebler; Cartwright, Joe and Mads Huuse (2006): "3D seismic technology: the geological 'Hubble'," *Basin Research*, vol. 17, pp. 1-20; Dyke, Kate van (1997), p. 33; Gray, Samuel and Michael O'Brian (1996): "Can we image beneath salt?" *The Leading Edge of Exploration*, vol. 15, no. 1, pp. 17-22; and Schoenberger, Michael (2000), p. 86

[2] Interviews with Milo Backus and Bert Bally; Brown, Alistair (2011): *Interpretation of Three-Dimensional Data*, Tulsa (Oklahoma): The American Association of Petroleum Geologists and the Society of Exploration Geophysicists, p. 2; Lawyer, Lee, Charles Bates and Robert Rice (2001), p. 180, p. 256; Proubasta, Dolores (1997): "Hugh W. Hardy," *The Leading Edge of Exploration*, vol. 16, pp. 481-6; Schneider, William (1998): "3D Seismic: a historical note," *The Leading Edge of Exploration*, vol. 17, pp. 375-80, here p. 380; Walton, G.G. (1972): "Three-dimensional seismic method," *Geophysics*, vol. 37, pp. 417-30; http://en.wikipedia.org/wiki/Hugh_W._Hardy; and www.cgg.com

[3] Schneider, William (1998), p. 375; and Pirtle, Caleb (2005), pp. 18-9

[4] Interviews with Milo Backus and Alistair Brown; Brown, Alistair (2011), pp. 2-3; and www.oceanstaroec.com/fame/2005/marine3d.html

[5] Interview with Jim Hornabrook; Moreton, Richard (ed.)(1995), p. 47; and www.oceanstaroec.com/fame/2005/marine3d.html

[6] Interview with Robin Walker

[7] Interview with Bob Peebler; Brown, Alistair (2011), p. 3; Priest, Tyler (2007), p. 246; Schoenberger, Michael (2000), pp. 86-7, 89; and Sheriff, Robert and Lloyd Geldart (1995), p. 451

[8] Interview with Alistair Brown; Brown, Alistair (2011), pp. 22-3; Schoenberger, Michael (1996): "The growing importance of 3-D seismic technology," conference paper, *Offshore Technology Conference*, May 6-9, Houston; and Sheriff, Robert and Lloyd Geldart (1995), p. 459

[9] Ceruzzi, Paul (2003), pp. 281-2; Lawyer, Lee, Charles Bates and Robert Rice (2001), pp. 202-8; and http://en.wikipedia.org/wiki/Andy_Bechtolsheim

[10] Interviews with Bob Peebler and Walter Ziegler; Burleson, Clyde (1999), p. 182; Cartwright, Joe and Mads Huuse (2006); Lawyer, Lee, Charles Bates and Robert Rice (2001), pp. 202-8; Priest, Tyler (2007), p. 246; Selley, Richard (1998), p. 6; and Sheriff, Robert and Lloyd Geldart (1995), p. 451

[11] Interview with David Jenkins; Ostrander, William (1984): "Plane-wave reflection coefficients for gas sands at nonnormal angles of incidence," *Geophysics*, vol. 49, no. 10 (October), pp. 1637-48; Ostrander, William (2006): "Memoirs of a successful geophysicist," *CSEG Recorder*, June 2006, pp. 38-41; and www.aapg.org/explorer/2001/10oct/destindome.cfm

[12] Ostrander, William (2006)

[13] Ostrander, William (1984); Ostrander, William (2006); and Sheriff, Robert and Lloyd Geldart (1995), p. 73

[14] Interview with Mike Forrest; Rutherford, Steven and Robert Williams (1989): "Amplitude-versus-offset variations in gas sands," *Geophysics*, vol. 54, no. 6 (June), pp. 680-8; and http://en.wikipedia.org/wiki/Tenneco

[15] Interviews with Chris Chapman and Leon Thomsen; Meissner, Rolf (1986), p. 148; Sheriff, Robert and Lloyd Geldart (1995), p. 55; Telford, W. M. et al. (1976), pp. 227-8; www.glossary.oilfield.slb.com/en/Terms/a/anisotropy.aspx; and http://en.wikipedia.org/wiki/Anisotropy

[16] Interview with Leon Thomsen; Helbig, Klaus and Leon Thomsen (2005): "75-plus years of anisotropy in exploration and reservoir seismics: a historical review of concepts and

methods," *SEG 75th Anniversary issue*, pp. 9ND-23ND; www.debretts.com/people/biographies/browse/c/17260/Stuart+CRAMPIN.aspx; and www.geosc.uh.edu/people/faculty/leon-thomsen/index.php

[17] Interview with Chris Chapman; and Thomsen, Leon (1997): "Seismic anisotropy: from constipation to exploration effectiveness," *CSEG Recorder*, December, pp. 4-5

[18] Barclay, Frazer et al. (2008): "Seismic inversion: reading between the lines," *Oilfield Review*, spring, pp. 42-63, here p. 43; and Latimer, Rebecca (2011): "Inversion and Interpretation of Impedance Data," in: Brown, Alistair: *Interpretation of Three-Dimensional Seismic Data*, Tulsa (Oklahoma): The American Association of Petroleum Geologists and the Society of Exploration Geophysicists pp. 309-49, here p. 309

[19] Interview with Roy Lindseth; http://ae.linkedin.com/in/johnlrees; and http://virtualmuseum.seg.org/bio_roy_o__lindseth.html

[20] Interview with Chris Chapman; Krebs, Jerome et al. (2009): "Fast full-wavefield seismic inversion using encoded sources," *Geophysics*, vol. 74, no. 6, pp. 177-88; Margrave, Gary, Robert Ferguson and Chad Hogan (2010): *Full Waveform Inversion with Wave Equation Migration and Well Control*, CREWES research report, vol. 22, Calgary: University of Calgary, p. 1; Margrave, Gary, Robert Ferguson and Chad Hogan (2011): *Full Waveform Inversion using One-Way Migration and Well Calibration*, CREWES convention paper, Calgary: University of Calgary, p. 1; and http://cseg.ca/education/view/what-else-can-the-seismic-waveform-tell-us

[21] Interview with Chris Chapman

[22] Interview with Robin Walker; International Association of Geophysical Contractors (ed.)(2002): *Marine Seismic Operations: an overview*, Houston: International Association of Geophysical Contractors (IAGC), p. 19; Leffler, William, Richard Pattarozzi and Gordon Sterling (2003), p. 48; Schempf, F. Jay (2007), p. 79; Schoenberger, Michael (2000), p. 66; Sheriff, Robert and Lloyd Geldart (1995), p. 18; and www.pgs.com/en/Pressroom/Press_Releases/PGS_Announces_Breakthroughs_i/

[23] Interviews with Guus Berkhout and Robin Walker; Gruffeille, Jean-Paul et al. (2010): "Exploring Oligocene targets using multi-azimuth acquisitions: application of non-linear slope tomography," conference paper, 72nd *EAGE* Conference & Exhibition, Barcelona, June 14-17; Schoenberger, Michael (2000), p. 66; www.pgs.com/en/Pressroom/News/PGS_Awarded_Innovative_Multi-/; www.seg.org/education/misc/continuing-courses/full-curriculum/bios/jack; http://fb.eage.org/publication/content?id=26413; www.glossary.oilfield.slb.com/en/Terms/n/narrow-azimuth_seismic_data.aspx; www.glossary.oilfield.slb.com/en/Terms/s/single-azimuth_towed-streamer_acquisition.aspx; and www.glossary.oilfield.slb.com/en/Terms/m/multiazimuth_towed-streamer_acquisition.aspx

[24] Interviews with Guus Berkhout and Leon Thomsen; Baldock, Simon et al. (2012): "Orthogonal wide azimuth surveys: acquisition and imaging," *First Break*, vol. 30, September, pp. 35-41; Fromyr, Eivind (2010): "The role of wide azimuth in subsalt imaging," conference paper, *Offshore Technology Conference*, Houston, May, 3-6; and www.glossary.oilfield.slb.com/en/Terms/w/wide-azimuth_towed-streamer_acquisition.aspx

[25] http://en.wikipedia.org/wiki/Geophysical_Service; http://de.wikipedia.org/wiki/Prakla-Seismos; and http://en.wikipedia.org/wiki/WesternGeco

[26] Nigel Anstey's email to Mark Mau, February 11, 2014; Buia, Michele et al. (2008): "Shooting seismic surveys in circles," *Oilfield Review*, autumn, pp. 27-8; Proffitt, J.M. (1991); and www.glossary.oilfield.slb.com/en/Terms/f/full-azimuth_towed-streamer_acquisition.aspx

[27] www.statoil.com/en/NewsAndMedia/News/1999/Pages/MedalForSeismicMethod.aspx; and www.geoexpro.com/articles/2008/04/seismic-imaging-technology-part-iii

[28] Amundsen, Lasse et al. (1999): "Multicomponent seabed seismic data: a tool for improved imaging and lithology fluid prediction," conference paper, *Offshore Technology Conference*, 5 March, Houston; Ikelle, Luc and Lasse Amundsen (2005): *Introduction to Petroleum Seismology*, Tulsa (Oklahoma): Society of Exploration Geophysicists, p. 277; Johansen, B., O. Holberg and K. Ovreba (1995): "Sub-sea seismic: impact on exploration and production," conference paper, *Offshore Technology Conference*, 1 May, Houston; and www.eage.org/index.php?evp=3525&epb=9167&oldevp=2477&Opendivs=&ActiveMenu=

[29] Interview with Leon Thomsen; www.geoexpro.com/articles/2008/04/seismic-imaging-technology-part-iii; and www.eage.org/index.php?evp=3525&epb=9167&oldevp=2477&Opendivs=&ActiveMenu=

[30] Morton, Andrew et al. (2003): "Evaluation and impact of sparse-grid, wide-azimuth 4C-3D node data from the North Sea," conference paper, *SEG Annual Meeting*, 26-31 October, Dallas; Pettenati-Auzière, C., C. Debouvry and E. Berg (1997): "Node-based sea-bottom seismic: a new way to reservoir management," conference paper, 1997 *SEG Annual Meeting*, 2-7 November, Dallas; www.statoil.com/en/NewsAndMedia/News/1999/Pages/MedalForSeismicMethod.aspx; www.geoforskning.no/geoprofilen/50-god-pa-bunnen; and www.pnronline.com.au/article.aspx?id=139&p=1

Chapter 22: Engineering the Oceans

[1] Pratt, Joseph, Tyler Priest and Christopher Castaneda (1997): *Offshore Pioneers: Brown & Root and the history of offshore oil and gas*, Houston: Gulf Publishing Company, p. 26; Schempf, F. Jay (2007), p. 32

[2] Leffler, William, Richard Pattarozzi and Gordon Sterling (2011): *Deepwater Petroleum Exploration & Production*, Tulsa (Oklahoma): PennWell, p. 7; Pratt, Joseph, Tyler Priest and Christopher Castaneda (1997), pp. 27-9; Schempf, F. Jay (2007), pp. 82-6; Veldman, Hans and George Lagers (1997): *50 Years Offshore*, Delft: Foundation for Offshore Studies, p. 68; www.oceanstaroec.com/fame/1998/platformspiled.html; and www.mcdermott.com/AboutUs/Pages/History.aspx

[3] Interview with Ken Arnold; Leffler, William, Richard Pattarozzi and Gordon Sterling (2011), p. 152; Priest, Tyler (2007), pp. 237-40; Zanden, Jan Luiten van et al. (2007), vol. 3, p. 186; www.oceanstaroec.com/fame/1998/platformspiled.html; and http://en.wikipedia.org/wiki/Bullwinkle_(oil_platform)

[4] Interview with Ken Arnold; Finn, L.D. (1976): "A new deepwater offshore platform: the guyed tower," conference paper, *Offshore Technology Conference*, 3-6 May, Houston; Leffler, William, Richard Pattarozzi and Gordon Sterling (2011), pp. 155-6; http://petrowiki.org/Offshore_and_subsea_facilities; http://en.wikipedia.org/wiki/Petronius_(oil_platform); and http://en.wikipedia.org/wiki/Burj_Khalifa

[5] Pratt, Joseph, Tyler Priest and Christopher Castaneda (1997), p. 29; and www.oceanstaroec.com/fame/1999/crane.html

[6] Pratt, Joseph, Tyler Priest and Christopher Castaneda (1997), pp. 1-14, 28-9, 42; Schempf, F. Jay (2007), pp. 82-3; and www.oceanstaroec.com/fame/1999/crane.html

[7] Pothoven, Boudewijn and Matthijs Dicker (2012): *Our Own Course: 50 years of Heerema Marine Contractors*, Rotterdam: Uitgeverij De Tijdgeest Publishers, pp. 81-2; and Veldman, Hans and George Lagers (1997), pp. 137-42

[8] Interview with Jan Meek; and Pothoven, Boudewijn and Matthijs Dicker (2012), pp 62-3, 81-2

[9] Interview with Jan Meek; Heerema, Pieter, Alexandre Horowitz and Henricus Willemsen (1980): *Stabilizing System on a Semi-Submersible Crane Vessel*, US Patent no. 4231313A patented Nov. 4; Leffler, William, Richard Pattarozzi and Gordon Sterling (2011), pp. 24, 280-1; Pothoven, Boudewijn and Matthijs Dicker (2012), pp. 82, 86-90, 112, 182; Veldman, Hans and George Lagers (1997), pp. 141-2; http://hmc.heerema.com/content/activities/remarkable-projects/records/; http://gcaptain.com/heerema-looks-build-worlds-largest-crane-vessel/; and http://en.wikipedia.org/wiki/Alexandre_Horowitz

[10] Interview with Gordon Sterling; Baker, Ron (1998), pp. 74-5; and Leffler, William, Richard Pattarozzi and Gordon Sterling (2011), p. 153

[11] Steen, Øyvind (1993): *På Dypt Vann: Norwegian Contractors 1973-1993*, Oslo: Aker, pp. 8-13; Veldman, Hans and George Lagers (1997), pp. 129, 170; and http://petrowiki.org/Offshore_and_subsea_facilities

[12] Interview with Gordon Sterling; Leffler, William, Richard Pattarozzi and Gordon Sterling (2011), p. 153; Pratt, Joseph, Tyler Priest and Christopher Castaneda (1997), pp. 270-1; Steen, Øyvind (1993), pp. 15-9; http://en.wikipedia.org/wiki/Condeep; www.oceanstaroec.com/fame/2004/concretegravitystructure.html; and www.oceanstaroec.com/fame/2007/concrete.html

[13] Veldman, Hans and George Lagers (1997), p. 172; and http://petrowiki.org/Offshore_and_subsea_facilities

[14] Steen, Øyvind (1993), pp. 135-9; https://www.ima.umn.edu/~arnold/disasters/sleipner.html; and http://en.wikipedia.org/wiki/Sleipner_A

[15] Maynard, Lara (ed.)(1997): *Hibernia: promise of rock and sea*, St. John (Newfoundland): Breakwater Books, p. 167; Sédillot, François (1998): "The Hibernia gravity base structure," conference paper, The Eighth *International Offshore and Polar Engineering Conference*, 24-29 May, Montréal; and www.offshore-technology.com/projects/exxon_hebron/

[16] www.oceanstaroec.com/fame/2011/mopu.html

[17] Veldman, Hans and George Lagers (1997), pp. 111, 127, 148, 176; http://petrowiki.org/Offshore_and_subsea_facilities; www.oceanstaroec.com/fame/2011/mopu.html; and www.offshore-mag.com/articles/print/volume-60/issue-2/news/production/mopus-evolving-to-meet-greater-depth-flexibility-challenges.html

[18] Veldman, Hans and George Lagers (1997), p. 176; www.ogj.com/articles/print/volume-112/issue-2/drilling-production/deepwater-gulf-decommissioning-mdash-1-aging-platforms-ownership-changes-pose-special-risks.html; and www.worldrecordacademy.com/technology/deepest_production_platform_Independence_Hub_set_world_record_70661.htm

[19] Interviews with Ken Arnold and Ed Horton; Brewer, John (1975): "The Tension Leg Platform Concept," conference paper, *API*, annual meeting papers, division of production, 7-9 April, Dallas; and www.oceanstaroec.com/fame/2005/tensionlegplatform.html

[20] Baker, Ron (1998), p. 76; Veldman, Hans and George Lagers (1997), pp. 176-7; Zanden, Jan Luiten van et al. (2007), vol. 3, p. 187; http://petrowiki.org/Offshore_and_subsea_facilities; and http://en.wikipedia.org/wiki/Tension-leg_platform

[21] Interviews with Ken Arnold and Ed Horton; and http://petrowiki.org/Offshore_and_subsea_facilities

[22] Interviews with Ed Horton, Brian Skeels and Gordon Sterling; Pothoven, Boudewijn and Matthijs Dicker (2012), p. 146; www.mcdermott-investors.com/phoenix.zhtml?c=96360&p=irol-newsArticle_Print&ID=493183&highlight= www.oceanstaroec.com/fame/2005/sparplatform.htm; http://en.wikipedia.org/wiki/Brent_Spar; and www.shell.com/global/aboutshell/major-projects-2/perdido/overview.html#iframe-L1dlYkFwcHMvRGVlcF9XYXRlci9pbmRleC5odG1s

[23] Veldman, Hans and George Lagers (1997), pp. 174-5; http://petrowiki.org/Offshore_and_subsea_facilities; www.oceanstaroec.com/fame/2011/fpso.html; and www.offshore-technology.com/features/feature40937/

[24] Teekay Petrojarl (ed.)(2011): *Petrojarl I*, Trondheim: Teekay Petrojarl; Veldman, Hans and George Lagers (1997), p. 175; and http://petrowiki.org/Offshore_and_subsea_facilities

[25] Veldman, Hans and George Lagers (1997), p. 175; www.oceanstaroec.com/fame/2011/fpso.html; and www.offshore-technology.com/features/feature40937/

[26] Carlson, Burt (1979): "First OCS Subsea Completion," *Petroleum Engineer*, August, pp. 98-9; Hillegeist, Paul (2012): *Subsea Acceleration: fathoming new technologies*, Hong Kong: CLSA, p. 17; Schempf, Jay (2007), pp. 110-4; Veldman, Hans and George Lagers (1997), p. 82; www.europeanoilandgas.co.uk/article-page.php?contentid=14808&issueid=436; and http://petrowiki.org/Offshore_and_subsea_facilities

[27] Marine Technology Society (ed.) (2013): *Advances in Marine Technology*, Washington D.C.: Marine Technology Society, pp. 9-13, 30-9; and Pratt, Joseph, Tyler Priest and Christopher Castaneda (1997), pp. 137-47

[28] Leffler, William, Richard Pattarozzi and Gordon Sterling (2011), p. 21; http://aoghs.org/offshore-history/offshore-robot/; and www.academia.edu/672914/Subsea_Robots

[29] Interview with Brian Skeels; and Burkhardt, J.A. and T.W. Michie (1979): "Submerged Production System," conference paper, *Offshore Technology Conference*, April 30, Houston

[30] Interviews with drew Michel and Brian Skeels; www.seatrepid.com/files/UT_PAPER_FINAL-WERNLI.pdf; http://magazines.marinelink.com/Magazines/MaritimeReporter/197709/content/operating-seafloor-production-210671; and http://m.myiwf.com/otc/DisplayingDoc.aspx?Annual=1

[31] Interviews with Brian Skeels and Ken Arnold; Priest, Tyler (2007), p. 187; Veldman, Hans and George Lagers (1997), p. 148; www.petrobras.com/en/about-us/our-history/; www.fmctechnologies.com/SubseaSystems/GlobalProjects/South%20America/Brazil/PetrobrasAlbacora.aspx; and www.ogj.com/articles/print/volume-94/issue-46/in-this-issue/exploration/petrobras-may-have-record-water-depth-find.html

[32] Interviews with Drew Michel; and Schempf, F. Jay (2007), p. 181

[33] Interviews with Drew Michel and Brian Skeels

[34] Interviews with Murray Burns; Leffler, William, Richard Pattarozzi and Gordon Sterling (2011), p. 307; Veldman, Hans and George Lagers (1997), p. 125; www.fundinguniverse.com/company-histories/coflexip-s-a-history/; www.technip.com/en/entities/draps/profile; and www.oceanstaroec.com/fame/2006/flexpipe.htm

[35] Bai, Yong and Qiang Bai (2012): *Subsea Engineering Handbook*, Amsterdam: Gulf Professional Publishing, p. 35; Lund, G.G. et al. (2009): "Advanced flow assurance system for the Ormen Lange subsea gas development," conference paper, *Offshore*

Technology Conference, 4-7 May, Houston; http://petrowiki.org/Offshore_and_subsea_facilities; www.fmctechnologies.com/SubseaSystems/GlobalProjects/Europe/Norway/StatoilOrmenLange.aspx; and www.epmag.com/Production-Field-Development/Shell-Sets-World-Record-Deepest-Subsea-Producing-Well_91835

[36] Jacobs, Trent (2014): "Pioneering subsea gas compression offshore Norway," *Journal of Petroleum Technology*, vol. 66, no 2, pp. 58-65; Ramberg, Rune, Simon Davies and Hege Rognoe (2013): "Steps to the Subsea Factory," conference paper, *Offshore Technology Conference*, 29-31 October, Rio de Janeiro; www.maximizerecovery.com/Subsea-Processing-Projects/Statoil-Tordis; and www.statoil.com/en/technologyinnovation/fielddevelopment/aboutsubsea/Pages/Lengre%20dypere%20kaldere.aspx

Chapter 23: Reservoir Engineering Comes of Age

[1] Hassler, Gerald (1944): *"Method and Apparatus for Permeability Measurements*, US Patent no. 2345935, patented April 4; Hyne, Norman (2001), p. 157; Lyons, William and Gary Plisga (2004), p. 555; Worthington, Paul, K. Boyle and X.D. Jing (2000), p. 145; and www.glossary.oilfield.slb.com/en/Terms/p/permeameter.aspx

[2] Interview with Koen Weber; Weber, Koenraad and R. Eijpe (1971): "Geological note: mini-permeameters for consolidated rock and unconsolidated sand," *AAPG Bulletin*, vol. 55, no. 2 (February), pp. 307-9; Worthington, Paul, K. Boyle and X.D. Jing (2000), p. 146; and www.glossary.oilfield.slb.com/en/Terms/p/permeameter.aspx

[3] Ertekin, Turgay, Jamal Abou-Kassem and Gregory King (2001): *Basic Applied Reservoir Simulation*, Richardson (Texas): Society of Petroleum Engineers, pp. 320-1; Leverett, M.C. (1941): "Capillary behavior in porous solids," *Transactions of the AIME*, vol. 142, pp. 152-69; Purcell, W.R. (1949): "Capillary Pressures: their measurement using mercury and the calculation of permeability therefrom," *Journal of Petroleum Technology*, vol. 1, no. 2, pp. 39-48; www.scaweb.org/about_awards.shtml; and http://petrowiki.org/Capillary_pressure_models

[4] Interviews with Pierre Adler and Scott Tinker; Weinbrandt, R.M. and Irving Fatt (1969): "Scanning electron microscope study of the pore structure of sandstone," conference paper, 11th *U.S. Symposium on Rock Mechanics*, 16-19 June, Berkeley, California; and www.iceht.forth.gr/staff/payatakes.html

[5] Sen, P.N., C. Scala, and M.H. Cohen (1981): "A self-similar model for sedimentary rocks with application to the dielectric constant of fused glass beads," *Geophysics*, vol. 46, no. 5, pp. 781-95

[6] Interview with Pierre Adler; Adler, Pierre (2013): *A Lévy Flight through Porous Media*, presentation slides, University of Wyoming, Laramie, Wyoming, September 19-20; Adler, Pierre and Howard Brenner (1992): *Porous Media: geometry and transports*, Stoneham (Massachusetts): Butterworth-Heinemann, pp. xv-xvi; and Adler, Pierre, Christian Jacquin and Jean Quiblier (1990): "Flow in simulated porous media," *International Journal of Multiphase Flow*, vol. 16, no. 4, pp. 691-712

[7] Spanne, P. et al. (1994): "Synchrotron computed microtomography of porous media," *Physical Review Letters*, vol. 73, no. 14, pp. 2001-4

[8] Interviews with Pierre Adler and Richard Bateman; www.fei.com; www.lithicon.com; and www.ingrainrocks.com/history/

[9] Interview with Pierre Adler; Lemmens, Herman, Alan Butcher and P.W. Botha (2010): "FIB/SEM and automated mineralogy for core and cuttings analysis," conference paper, *SPE* Russian Oil and Gas Conference and Exhibition, 26-28 October, Moscow; Prensky, Stephen (2012): "What's new in well logging and formation evaluation, part 2,"

World Oil, July, pp. 107-12; www.microscopy-analysis.com/blog/blog-articles/focused-ion-beam-%E2%80%93-all-grown; http://whatis.techtarget.com/definition/atomic-force-microscopy-AFM; www.microscopemaster.com/atomic-force-microscope.html; http://machinemakers.typepad.com/machine-makers/2011/05/advantages-and-disadvantages-of-atomic-force-microscopy.html; http://thebreakthrough.org/archive/interview_with_dan_steward_for; and www.mydigitalpublication.com/article/George_P._Mitchell_And_The_Barnett_Shale/1535436/179598/article.html

[10] Interview with Alain Gringarten; Gringarten, Alain (2008): "From straight lines to deconvolution: the evolution of the state of the art in well test analysis," *SPE Reservoir Evaluation & Engineering*, vol. 11, no. 1, pp. 41-62; Horner, D.R. (1951): "Pressure build-up in wells," conference paper, 3rd *World Petroleum Congress*, 28 May-6 June, The Hague; Kuchuk, Fikri, Mustafa Onur and Florian Hollaender (2010): *Pressure Transient Formation and Well Testing: convolution, deconvolution and nonlinear estimation*, Amsterdam: Elsevier, pp. xv-xx; Lee, John (1982), p. 1; Lyons, William and Gary Plisga (2004), p. 5152; and Miller, C.C., A.B. Dyes and C.A. Hutchinson (1950): "The estimation of permeability and reservoir pressure from bottom hole pressure build-up characteristics," *Journal of Petroleum Technology*, vol. 2, no. 4, pp. 91-104

[11] Gringarten, Alain (2008); Conrad, K.M. (1962): "Application of the wireline formation tester," conference paper, *SPE Drilling and Production Practices Conference*, 5-6 April, Beaumont, Texas; Lebourg, M., R.Q. Fields and C.A. Doh (1957): "A method of formation testing on logging cable," *Society of Petroleum Engineers* paper; and www.slb.com/about/history/1950s.aspx

[12] Dake, Laurence (2001): *The Practice of Reservoir Engineering*, Amsterdam: Elsevier, pp. 34, 58, 138, 151; Gringarten, Alain (2008); http://en.wikipedia.org/wiki/Pressure_measurement; Schultz, A.L., W.T. Bell and H.J. Urbanosky (1975): "Advancements in uncased-hole, wireline formation-tester techniques," *Journal of Petroleum Technology*, vol. 27, no. 11, pp. 1331-6; and www.sensorland.com/HowPage059.html

[13] Interview with Alain Gringarten; Ramey, Hank (1992): "Advances in practical well-test analysis," *Journal of Petroleum Technology*, vol. 44, no. 6, pp. 650-9; http://news.stanford.edu/pr/93/931122Arc3036.html; and www.slb.com/about/history/1970s.aspx

[14] Interviews with Ram Agarwal, Alain Gringarten, Roland Horne, Leif Larsen and Rudolf Shagiev; interview of Fikri Kuchuk by Amy Esdorn (Society of Petroleum Engineers); and Bourdet, Dominique et al. (1983): "A new set of type curves simplifies well test analysis," *World Oil*, May, pp. 95-106

[15] Interview with Leif Larsen; and interview of Fikri Kuchuk by Amy Esdorn (Society of Petroleum Engineers)

[16] Interview with Tom Zimmerman; Badry, Rob et al. (1994): "Downhole optical analysis of formation fluids," *Oilfield Review*, January, pp. 21-8; Dake, Laurence (2001), p. 34; Smits, A.R. et al. (1993): "In-situ optical fluid analysis as an aid to wireline formation sampling," *SPE Formation Evaluation*, vol. 10, no. 2, pp. 91-8; and www.slb.com/services/characterization/reservoir/wireline/modular_formation_dynamics_tester.aspx

[17] Gringarten, Alain (2008); and Schroeter, Thomas von, Florian Hollaender and Alain Gringarten (2001): "Deconvolution of well test data as a nonlinear total least squares problem," conference paper, *SPE Annual Technical Conference and Exhibition*, 30 September-3 October, New Orleans

[18] Interviews with Pierre Adler, Alain Gringarten and Fred Stalkup

[19] Interviews with Khalid Aziz and Donald Peacenan; Mills, R. (1920): "Experimental studies of subsurface relationships in oil and gas fields," *Economic Geology*, vol. 15, no. 5, pp. 398-421; Peaceman, Donald (1990): "A Personal Retrospection of Reservoir Simulation," in: Nash, Stephen (ed.): *A History of Scientific Computing*, New York: ACM Press, pp. 106-29; and Reistle, C.E. (1961), pp. 815-6

[20] Peaceman, Donald (1957): "Application of large computers to reservoir engineering problems," *Journal of Petroleum Technology*, vol. 9, no. 10, pp. 14-18; Peaceman, Donald (1990); Peaceman, Donald and Henry Rachford (1955): "The numerical solution of parabolic and elliptic differential equations," *Journal of the Society for Industrial and Applied Mathematics*, vol. 2, no. 1, pp. 28-41; and http://en.wikipedia.org/wiki/Alternating_direction_implicit_method

[21] Interviews with Pierre Adler, Alain Gringarten and Leif Larsen; Barenblatt, G. I., Yu. P. Zheltov and I.N. Kochina (1960): Basis concepts in the theory of seepage of homogeneous liquids in fissured rocks," *Journal of Applied Mathematics* (Soviet), vol. 24, no. 5, pp. 1286-1303; Kazemi, H. et al. (1976): "Numerical simulation of water-oil flow in naturally fractured reservoirs," *Society of Petroleum Engineers Journal*, vol. 16, no. 6 (December), pp. 317-26; Warren, J.E. and P.J. Root (1963): "The behavior of naturally fractured reservoirs," *Society of Petroleum Engineers Journal*, vol. 3, no. 3, pp. 245-55; and http://math.berkeley.edu/~gibar/

[22] Interviews with Khalid Aziz, George Hirasaki, Tommy Miller, Donald Peaceman, Fred Stalkup and Paul Willhite; Coats, Keith (1982): "Reservoir simulation: state of the art," *Journal of Petroleum Technology*, vol. 34, no. 8, pp. 1633-42; www.aimehq.org/programs/award/bio/keith-h-coats; www.aimehq.org/programs/award/bio/hossein-kazemi; www.cmgl.ca; and http://en.wikipedia.org/wiki/Reservoir_simulation

[23] Interviews with Jonathan Holmes and Tommy Miller; Blair, P.M. and C.F. Weinaug (1969): "Solution of two-phase flow problems using implicit difference equations," *Society of Petroleum Engineers Journal*, vol. 9, no. 4, pp. 417-24; Cheshire, Ian et al. (1980): "An efficient fully implicit simulator," conference paper, *European Offshore Technology Conference and Exhibition*, 21-24 October, London; Sonier, F. and P. Chaumet (1974): "A fully implicit three-dimensional model in curvilinear coordinates," *Society of Petroleum Engineers Journal*, vol. 14, no. 4, pp. 361-70; and http://en.wikipedia.org/wiki/Atomic_Energy_Research_Establishment

[24] Interviews with Jonathan Holmes and Tommy Miller; and www.linkedin.com/pub/ted-daniels/16/340/795

[25] Interviews with Alain Gringarten, Jonathan Holmes and Tommy Miller; Edwards, David et al. (2011): "Reservoir simulation: keeping pace with oilfield complexity", *Oilfield Review*, vol. 23, no. 4 (winter), pp. 4-15; and Gilman, James and Chet Ozgen (2013): *Reservoir Simulation: history matching and forecasting*, Richardson (Texas): Society of Petroleum Engineers, pp. 5-6

[26] Interviews with Nick Koutsabeloulis and Robin Walker; and Teufel, Lawrence, Douglas Rhett and Helen Farrell (1991): "Effect of reservoir depletion and pore pressure drawdown on in situ stress and deformation in the Ekofisk Field, North Sea," conference paper, 32nd US Symposium on Rock Mechanics, *American Rock Mechanics Association*, 10-12 July, Norman, Oklahoma

[27] Interviews with Roland Horne and Rudolf Shagiev; Baker, Alan et al. (1995): "Permanent monitoring: looking at lifetime reservoir dynamics," *Oilfield Review*, winter, pp. 32-46; Horne, Roland (2007): "Listening to the reservoir: interpreting data from permanent downhole gauges," *Journal of Petroleum Technology*, vol. 59, no. 12, pp. 78-86; Schlumberger (ed.)(2013): *Schlumberger: 40 years of permanent downhole monitoring*, internal paper; and http://petrowiki.org/Single_well_chemical_tracer_test

[28] Nyhavn, Fridtjof and Anne Dalager Dyrli (2010): "Permanent tracers embedded in downhole polymers prove their monitoring capabilities in a hot offshore well," conference paper, *SPE Annual Technical Conference and Exhibition*, 19-22 September, Florence, Italy

[29] Interviews with Amos Nur and Sven Treitel; Nur, Amos (1982): "Seismic Imaging in enhanced recovery," conference paper, *SPE Enhanced Oil Recovery Symposium*, 4-7 April, Tulsa, Oklahoma; Sheriff, Robert and Lloyd Geldart (1995), pp. 120-1, 499-500, 515; and http://csegrecorder.com/articles/view/a-personal-perspective-on-the-past-present-and-future-of-time-lapse-seismic

[30] http://csegrecorder.com/articles/view/a-personal-perspective-on-the-past-present-and-future-of-time-lapse-seismic

[31] Pullin, Norm, Larry Matthews and Keith Hirsche (1987): "Techniques applied to obtain very high resolution 3-D seismic imaging at an Athabasca tar sands thermal pilot," *The Leading Edge*, vol. 6, no. 12, pp. 10-15; http://csegrecorder.com/articles/view/a-personal-perspective-on-the-past-present-and-future-of-time-lapse-seismic; and http://csegrecorder.com/interviews/view/interview-with-larry-matthews

[32] Interviews with Al Breitenbach and Robin Walker; Breitenbach, E.A., G.A. King and K.N.B. Dunlop (1989): "The range of application of reservoir monitoring," conference paper, *SPE Annual Technical Conference and Exhibition*, 8-11 October, San Antonio, Texas; Breitenbach, E.A. et al. (1991): "Monitoring oil/water fronts by direct measurement," *Journal of Petroleum Technology*, vol. 43, no. 5, pp. 596-602; Wang, Zhijing, Amos Nur and Michael Batzle (1988): "Effect of different pore fluids on velocities of rocks," conference paper, *SEG Annual Meeting*, 30 October-3 November, Anaheim, California; http://csegrecorder.com/articles/view/a-personal-perspective-on-the-past-present-and-future-of-time-lapse-seismic; and http://csegrecorder.com/interviews/view/interview-with-mike-batzle

[33] Rigmor, M. Elde et al. (2000): "Troll West: reservoir monitoring by 4D seismic," conference paper, *SPE European Petroleum Conference*, 24-25 October, Paris; Sonneland, L. et al. (1997): "4D seismic on Gullfaks," conference paper, *Offshore Technology Conference*, 5 May, Houston; http://petrowiki.org/Seismic_time-lapse_reservoir_monitoring; and http://csegrecorder.com/articles/view/a-personal-perspective-on-the-past-present-and-future-of-time-lapse-seismic

[34] Thedy, E.A. et al. (2013): "Jubarte permanent reservoir monitoring: installation and first results," conference paper, 13th International Congress of the *Brazilian Geophysical Society*, August 26-29, Rio de Janeiro

[35] Albright, James et al. (1988): *The Crosswell Acoustic Surveying Project*, Los Alamos (New Mexico): Los Alamos National Laboratory, p. v; Pettitt, Roland (1979): "Completion of Hot Dry Rock Geothermal Well Systems," conference paper, *SPE Annual Technical Conference and Exhibition*, 23-26 September, Las Vegas; and http://petrowiki.org/Seismic_profiling

[36] Crampin, Stuart and David Taylor (1994): "The potential for monitoring the progress of production fronts across hydrocarbon reservoirs," conference paper, *Rock Mechanics in Petroleum Engineering*, 29-31 August, Delft, Netherlands; Marion, Bruce (2014): "Crosswell imaging offers higher resolution," *The American Oil & Gas Reporter*, January; Northrop, David and Karl-Heinz Frohne (1990): "The Multiwell Experiment: a field laboratory in tight gas sandstone," *Journal of Petroleum Technology*, vol. 42, no. 6, pp. 772-9; and Sandia Corporation (ed.)(1988): *Multi-Well Experiment MWX-3: as-built report*, Albuquerque (New Mexico): U.S. Department of Energy, p. 24

[37] Levesque, Cyrille (2006): "Crosswell electromagnetic resistivity imaging: illuminating the reservoir," *Middle East & Asia Reservoir* Review, no. 7, pp. 24-33; Science & Technology Review (ed.)(1996): "Exploring oil fields with crosshole electromagnetic induction,"

Science & Technology Review, August 1996, pp. 20-3; Wilt, Michael et al. (1995): "Crosshole electromagnetic tomography: a new technology for oil field characterization," *The Leading Edge*, March, pp. 173-7; https://e-reports-ext.llnl.gov/pdf/236063.pdf; and www.slb.com/services/characterization/reservoir/wireline/other/deeplook_em.aspx

[38] Shen, Jinsong et al. (2008): "Application of 2.5D cross-hole electromagnetic inversion in Gudao Oil Field, East China," *Applied Geophysics*, vol. 5, no. 3, pp. 159-69

[39] Al-Ali, Zaki et al. (2009): "Looking deep into the reservoir," *Oilfield Review*, vol. 21, no. 2, pp. 38-47; Marion, Bruce (2014); Nalonnil, Ajay and Bruce Marion (2010): "High resolution reservoir monitoring using crosswell seismic," conference paper, SPE Asia Pacific Oil and Gas Conference and Exhibition, 18-20 October, Brisbane, Queensland, Australia; and www.slb.com/services/characterization/geophysics/wireline/deeplook_cs.aspx

[40] Interview with Bob Peebler; Lawyer, Lee, Charles Bates and Robert Rice (2001), pp. 202-8; and http://en.wikipedia.org/wiki/Petrel_(reservoir_software)

Chapter 24: The Boom in Unconventionals

[1] Alvarez, Jose, Raul Moreno and Ronald Sawatzky (2014): "Can SAGD be exported? Potential challenges," conference paper, SPE Heavy and Extra Heavy Oil Conference: Latin America, 24-26 September, Medellín, Colombia; www.canadianpetroleumhalloffame.ca/roger-butler.html; www.suncor.com/en/about/252.aspx; www.cnrl.com/operations/north-america/north-american-crude-oil-and-ngls/thermal-insitu-oilsands; and http://en.wikipedia.org/wiki/Steam-assisted_gravity_drainage

[2] Bokserman, A.A., V.P. Filippov and V.Yu. Filanovskii (1998), pp. 145-50; and www.ogj.com/articles/print/volume-102/issue-21/special-report/sagd-drilling-parameters-evolve-for-oil-sands.html

[3] Interview with Paul Willhite; Alvarez, Jose, Raul Moreno and Ronald Sawatzky (2014); Butler, Roger (1994): *Horizontal Wells for the Recovery of Oil, Gas and Bitumen*, Houston: Gulf Publishing Company, p. 18; Butler, R.M. and D.J. Stephens (1981): "The gravity drainage of steam-heated heavy oil to parallel horizontal wells," *Journal of Canadian Petroleum Technology*, vol. 20, no. 2, pp. 90-6; Butler, R.M., G.S. McNab and H.Y. Lo (1981): "Theoretical studies on the gravity drainage of heavy oil during in-situ steam heating," *The Canadian Journal of Chemical Engineering*, vol. 59, no. 4, pp. 455-60; www.albertaoilmagazine.com/2006/04/an-interview-with-roger-butler/2/; and https://www.spe.org/twa/print/archives/2010/2010v6n2/05_Tech_101.pdf

[4] Deutsch, C.V. and J.A. McLennan (2005): *Guide to SAGD (Steam Assisted Gravity Drainage) Reservoir Characterization Using Geostatistics*, Edmonton: University of Alberta; Smalley, Craig (2000): "Heavy Oil and Viscous Oil," in: Dawe, Richard (ed.)(2001): *Modern Petroleum Technology*, vol. 1: Upstream, Chichester: Wiley, pp. 409-35, here p. 427; www.canadianpetroleumhalloffame.ca/roger-butler.html; www.ogj.com/articles/print/volume-102/issue-21/special-report/sagd-drilling-parameters-evolve-for-oil-sands.html; and http://en.wikipedia.org/wiki/Steam-assisted_gravity_drainage

[5] Interview with Richard Bateman; Al-Asimi, Mohammad et al. (2002): "Advances in well and reservoir surveillance," *Oilfield Review*, winter, pp. 14-35; Brown, George (2004): "Permanent reservoir monitoring using fiber optic distributed temperature measurements," SPE Distinguished Lecturer presentation, 2004-2005; Carnahan, B.D. et al. (1999): "Fiber optic temperature monitoring technology," conference paper, SPE Western Regional Meeting, 26-27 May, Anchorage, Alaska; Kersey, Alan, James Dunphy and Arthur Hay (1998): "Optical reservoir instrumentation system," conference

paper, *Offshore Technology Conference*, 5 April, Houston; and http://en.wikipedia.org/wiki/Distributed_temperature_sensing

[6] Interview with Paul Willhite; Alvarez, Jose, Raul Moreno and Ronald Sawatzky (2014); Holly, Christopher, Martin Mader and Jesse Toor (2012): *Alberta Department of Energy: oil sands production profile, 2002-2010*, Edmonton (Alberta): Government of Alberta; Smalley, Craig (2000), p. 427; Speight, James (2012): *Oil Sand Production Processes*, Oxford: Gulf Professional Publishing, pp. 122-9, pp. 122-9; http://albertainnovates.ca/media/20420/sagd_technologies_ogm_lightbown.pdf; www.albertaoilmagazine.com/2006/04/an-interview-with-roger-butler/2/; http://imperialoil.ca/Canada-English/about_who_stories_rogerbutler.aspx; and http://en.wikipedia.org/wiki/Steam-assisted_gravity_drainage

[7] Speight, James (2013): *Shale Gas Production Processes*, Oxford: Gulf Professional Publishing, pp. 11-13

[8] Interview with Robert Freedman; Kutchin, Joseph (2001): *How Mitchell Energy & Development Corp. got its start and how it grew*, Boca Raton: Universal Publishers, pp. 15-6; www.tshaonline.org/handbook/online/articles/dom05; and www.mydigitalpublication.com/article/George_P._Mitchell_And_The_Barnett_Shale/1535436/179598/article.html

[9] Henry, James (1982): "Stratigraphy of the Barnett Shale (Mississippian) and associated reefs in the northern Fort Worth Basin," *Petroleum Geology of the Fort Worth Basin and Bend Arch Area*, Dallas Geological Society, pp. 157-77; Zuckerman, Gregory (2013): *The Frackers: the outrageous inside story of the new energy revolution*, London: Portfolio Penguin, p. 33; and www.tshaonline.org/handbook/online/articles/dom05

[10] Interviews with Claude Cooke, George King and Martin Rylance; Zuckerman, Gregory (2013), p. 35; http://thebreakthrough.org/archive/interview_with_dan_steward_for; http://thebreakthrough.org/images/main_image/Where_the_Shale_Gas_Revolution_Came_From2.pdf; and http://www.rff.org/RFF/documents/RFF-DP-13-12.pdf

[11] Interviews with Claude Cooke and Nansen Saleri; Bowker, Kent (2003): "Recent developments of the Barnett Shale play, Fort Worth Basin," *West Texas Geological Society Bulletin*, vol. 42, no. 6, pp. 4-11; Zuckerman, Gregory (2013), pp. 35, 38; http://thebreakthrough.org/archive/interview_with_dan_steward_for; www.panamop.com/brnetovrvw.htm; and www.mydigitalpublication.com/article/George_P._Mitchell_And_The_Barnett_Shale/1535436/179598/article.html

[12] Interviews with Claude Cooke and Martin Rylance; http://thebreakthrough.org/archive/interview_with_dan_steward_for; http://thebreakthrough.org/images/main_image/Where_the_Shale_Gas_Revolution_Came_From2.pdf; and www.mydigitalpublication.com/article/George_P._Mitchell_And_The_Barnett_Shale/1535436/179598/article.html

[13] Interview with Claude Cooke; Goins, W.C., K.R. Webster and S.C. Berry (1965): "A Continuous Multistage Tracing Technique," *Journal of Petroleum Technology*, vol. 17, no. 6, pp. 619-25; Yergin, Daniel (2011), p. 327; Zuckerman, Gregory (2013), pp. 75-6, 196, 252; http://thebreakthrough.org/blog/Where_the_Shale_Gas_Revolution_Came_From.pdf; http://theenergycollective.com/jimpierobon/257691/george-p-mitchell-founder-shale-gas-here-s-how-he-and-his-team-did-it; and www.beg.utexas.edu/pttc/archive/barnettshalesym/notsosimple.pdf

[14] King, George (2010): "Thirty years of gas shale fracturing: what have we learned?" conference paper, *SPE Annual Technical Conference and Exhibition*, 19-22 September, Florence, Italy; Zuckerman, Gregory (2013), pp. 77-81, 91, 93-4; http://thebreakthrough.org/archive/interview_with_dan_steward_for;

www.ogj.com/articles/2000/07/mitchell-succeeds-with-light-sand-fracing-in-barnett-shale.html; and www.theatlantic.com/business/archive/2013/11/breakthrough-the-accidental-discovery-that-revolutionized-american-energy/281193/

[15] Andersen, Svend Aage et al. (1990): "Exploiting reservoirs with horizontal wells: the Maersk experience," *Oilfield Review*, vol. 2, no. 3, pp. 11-21; Yergin, Daniel (2011), p. 328; Zuckerman, Gregory (2013), pp. 67, 196, 199-202; http://theenergycollective.com/jimpierobon/257691/george-p-mitchell-founder-shale-gas-here-s-how-he-and-his-team-did-it

[16] Maxwell, Shaun et al. (1998): "Microseismic logging of the Ekofisk reservoir," conference paper, *SPE/ISRM Rock Mechanics in Petroleum Engineering*, 8-10 July, Trondheim, Norway; Maxwell, Shaun et al. (2002): "Microseismic imaging of hydraulic fracture complexity in the Barnett Shale," conference paper, *SPE Annual Technical Conference and Exhibition*, 29 September-2 October, San Antonio, Texas; http://csegrecorder.com/interviews/view/interview-with-shawn-maxwell; and www.seismics.co.uk/MicroseismicSurveyNorthSea.pdf

[17] Prud'Homme, Alex (2014): *Hydrofracking: what everyone needs to know*, Oxford: Oxford University Press, p. 20; www.eia.gov/todayinenergy/detail.cfm?id=13491; and www.bloomberg.com/news/2014-07-04/u-s-seen-as-biggest-oil-producer-after-overtaking-saudi.html

[18] Interview with Larry Harrington; Yergin, Daniel (1991), pp. 677, 698; and www.wellservicingmagazine.com/featured-articles/2013/03/when-the-u-s-government-believed-synthetic-fuels-could-save-the-world/

[19] Meissner, Fred (1978): "Petroleum geology of the Bakken Formation Williston Basin, North Dakota and Montana," in: Estelle, D. and R. Miller (eds.): *The Economic Geology of the Williston Basin*, Billings (Montana): Montana Geological Society, pp. 207-30; Zuckerman, Gregory (2013), pp. 154-63; http://en.wikipedia.org/wiki/Continental_Resources; and www.petroleumnewsbakken.com/pntruncate/612342886.shtml

[20] Yergin, Daniel (2011), p. 261; Zuckerman, Gregory (2013), p. 167-70; www.petroleumnewsbakken.com/pntruncate/612342886.shtml; and http://en.wikipedia.org/wiki/Bakken_formation

[21] Zuckerman, Gregory (2013), pp. 203-8, 229-32, 250-4, 272-4, 310, 312, 329; and www.petroleumnewsbakken.com/pntruncate/612342886.shtml

[22] Zuckerman, Gregory (2013), pp. 318, 367; www.bp.com/statisticalreview www.slb.com/~/media/Files/resources/oilfield_review/ors13/sum13/03_liquid_rich.pdf ; and www.forbes.com/sites/christopherhelman/2014/04/16/harold-hamm-billionaire-fueling-americas-recovery/

Chapter 25: Looking Ahead

[1] www.neftex.com

[2] www.aapg.org/explorer/2013/05may/ace_welte0513.cfm

[3] International Energy Agency (2012): *World Energy Outlook 2012: executive summary*, Paris: International Energy Agency

[4] Interviews with Phil Christie and Robin Walker; and Carstens, Halfdan (2006): "Quantum leap for seismic," *GeoExpro*, May 2006, pp. 22-4

[5] Interview with Phil Christie; Harries, Steve (2012): "Reading between the lines: point-receiver data, isometrically sampled in both crossline and inline directions, fully captures the three-dimensional seismic wavefield for the first time," *GeoExpro*, October 2012, pp. 50-2; and www.theogm.com/2013/04/03/the-first-true-3d-marine-seismic-system/

[6] www.oilandgastechnology.net/upstream-news/pgs-launches-world%E2%80%99s-widest-seismic-vessel

[7] Interview with Guus Berkhout

[8] Beasley, Craig et al. (2012): "Simultaneous sources: the inaugural full-field, marine seismic case history," conference paper, *SEG* Annual Meeting, Las Vegas, November 4-9; Berkhout, Guus (2008): "Changing the mindset in seismic data acquisition," *The Leading Edge*, vol. 27, July 2008, pp. 924-38; Treitel, Sven (2005), p. 8; Womack, J. E. (1988): "Simultaneous Vibroseis encoding techniques," conference paper, *SEG Annual Meeting*, Anaheim, California, October 30-November 3; and http://virtualmuseum.seg.org/bio_daniel_silverman.html

[9] Interview with Guus Berkhout; and Berkhout, Guus (2013): "Decentralized blended acquisition: are networks the next big step in seismic data collection?" *EAGE* London meeting, June, talk summary, in: www.earthdoc.org/publication/publicationdetails/?publication=68347

[10] Interview with Guus Berkhout; and www.seg.org/education/lectures-courses/distinguished-lecturers/spring2013/schusterinterview

[11] Interview with Alistair Brown; www.aapg.org/explorer/2010/02feb/gpc0210.cfm; and http://en.wikipedia.org/wiki/Bright_spot

[12] Constable, Steven and Leonard Srnka (2007): "An introduction to marine controlled-source electromagnetic methods for hydrocarbon exploration," *Geophysics*, vol. 72, no. 2, pp. WA3-WA12; and Constable, Steven (2010): *Seafloor Electromagnetic Methods Consortium: a research proposal*, La Jolla (California); and Srnka, Leonard (1986): *Method and Apparatus for Offshore Electromagnetic Sounding Utilizing Wavelength Effects to Determine Optimum Source and Detector Positions*, US Patent no. 4,617,518 patented Oct. 14

[13] Constable, Steven and Leonard Srnka (2007); Constable, Steven (2010); Tisot, Jean-Paul (ed.)(2008): *Understanding the Future: geosciences serving society*, Strasbourg: Editions Hirlé, pp. 45-6; and http://marineemlab.ucsd.edu/semc.html

[14] Constable, Steven and Leonard Srnka (2007); Jahn, Frank, Mark Cook and Mark Graham (2008), p. 24; and Sengupta, Souvik (2011): "Target detectability of marine controlled source electromagnetic method: insights from 1D modeling," conference presentation, *GEO India*, New Delhi, January 12-14

[15] Zanden, Jan Luiten van et al. (2007), vol. 3, pp. 433-4

[16] Interview with Peter Sharpe; and Filippov, Andrei et al. (1999): "Expandable tubular solutions," conference paper, *SPE Annual Technical Conference and Exhibition* held in Houston, Texas, October 3-6

[17] Interview with Mike Jellison; Offshore Magazine (ed.)(2000): "Expandable casing program helps operator hit TD with larger tubulars," *Offshore Magazine*, no. 1; www.enventuregt.com/about-us/history-milestone; and en.wikipedia.org/wiki/Expandable_tubular_technology

[18] Coolidge, Robert et al. (2007): "Special report: BP, Baker run first expandable monobore liner extension system," *Oil & Gas Journal*, December 2

[19] Gala, Deepak and Steve Nas (2009): "Underbalanced Drilling Operations," in: Aadnoy, Bernt et al. (eds.): *Advanced Drilling and Well Technology*, Richardson (Texas): Society of Petroleum Engineers, pp. 678-702, here p. 678; and Shell (ed.)(2009): "Underbalanced drilling offers more," *EP Technology*, no. 2, pp. 14-16

[20] Gelfgat, Yakov, Mikhail Gelfgat and Yuri Lopatin (2003), vol. 2, p. 200

[21] Interviews with Robert Mitchell, John Mogford and Bill Rehm; Eaton, Ben (1969): "Fracture gradient prediction and its application in oilfield operations," *Journal of Petroleum Technology*, vol. 21, no. 10 (October), pp. 1353-60

[22] Devereux, Steve (1999), pp. 259-60; Gelfgat, Yakov, Mikhail Gelfgat and Yuri Lopatin (2003), vol. 2, pp. 188, 200, 212; Hannegan, Don (2009): "Managed-Pressure Drilling," in: Aadnoy, Bernt et al. (eds.): *Advanced Drilling and Well Technology*, Richardson (Texas): Society of Petroleum Engineers, pp. 750-64, here p. 751; Rehm, Bill and Jim Hughes (2008): "Equipment Common to MPD Operations," in: Rehm, Bill et al. (eds.): *Managed Pressure Drilling*, Houston: Gulf Publishing, pp. 227-59, here p. 228; and Shell (ed.)(2009)

[23] Interview with Ted Bourgoyne; Gala, Deepak and Steve Nas (2009), p. 678; Gelfgat, Yakov, Mikhail Gelfgat and Yuri Lopatin (2003), vol. 2, p. 188; Shell (ed.)(2009); and Vidrine, D.J. and E.J. Benit (1968): "Field verification of the effect of differential pressure on drilling rate," *Journal of Petroleum Technology*, vol. 20, no. 7, pp. 675-82

[24] Interviews with Geoff Downton and Mike Dyson; Gelfgat, Yakov, Mikhail Gelfgat and Yuri Lopatin (2003), vol. 2, p. 212; Frink, Philip (2006): "Managed pressure drilling: what's in a name?" *Drilling Contractor*, March/April, pp. 37-9; Hannegan, Don and K. Fisher (2005): "Managed pressure drilling in marine environments," conference paper, *International Petroleum Technology Conference*, November 21-23, Doha, Qatar; and Shell (2009)

[25] Interview with Bill Rehm; Hannegan, Don (2009), p. 751; Hannegan, Don and K. Fisher (2005); Malloy, Kenneth (2007): "Managed pressure drilling: what is it anyway?" *World Oil*, March 2007, pp. 27-34; Rehm, Bill et al. (2008): "The Why and Basic Principles of Managed Pressure Drilling," in: Rehm, Bill et al. (eds.): *Managed Pressure Drilling*, Houston: Gulf Publishing, pp. 1-38, here pp. 1, 2, 20, 23; and www.drillingcontractor.org/perspectives-mpd-pioneer-don-hannegan-credits-work-to-early-industry-adopters-iadc-committee-23765

[26] Almeida, Rob (2011): "Dual gradient technology: a game-changer for offshore drilling," *gCaptain*, December 20; Gault, Allen (1996): "Riserless drilling: circumventing the size/cost cycle in deepwater," *Offshore Magazine*, May, pp. 49-54; Hannegan (2009), p. 756; Oil & Gas Journal (ed.)(2012): "Chevron's dual gradient drillship arrives in Gulf of Mexico," *Oil & Gas Journal*, May 7; Redden, Jim (2010): "Dual-gradient drilling promises to change the face of deepwater," *Offshore Magazine*, vol. 70, issue no. 50; and Schubert, Jerome and Brandee Elieff (2009): "Deepwater Dual-Gradient Drilling," in: Aadnoy, Bernt et al. (eds.): *Advanced Drilling and Well Technology*, Richardson (Texas): Society of Petroleum Engineers, pp. 621-34, here pp. 623-24

[27] Oil & Gas Journal (ed.)(2012); Schubert, Jerome and Brandee Elieff (2009), pp. 623-4; Smith, Ken (2009): "Dual gradient drilling: has its time finally come?" presentation paper at the *AADE* Emerging Technologies Forum, April 22; Smith, Ken et al. (1999): "SubSea MudLift Drilling JIP: achieving dual-gradient technology," *World Oil*, August; and www.api.org/environment-health-and-safety/environmental-performance/public-private-partnerships/environmental-partnerships/technology-development/subsea-mudlift-drilling-system.aspx

[28] Interview with Greg Myers; Consortium for Ocean Leadership (ed.)(2007): *Ocean Drilling Program: final technical report 1983-2007*, Washington (D.C.): Consortium for Ocean Leadership; Hannegan (2009), p. 759; Hannegan, Don and Roger Stave (2006): "The time has come to develop Riserless Mud Recovery technology's deepwater capabilities," *Drilling Contractor*, September/October, pp. 50-4; Myers, Greg (2008): "Ultra-deepwater riserless mud circulation with dual gradient drilling," *Scientific Drilling*, no. 6 (July), pp. 48-51, here p. 48; Schubert, Jerome and Brandee Elieff (2009), p. 624; and www.iodp.org/about

[29] www.deepwater.com/fw/main/Firsts-and-Records-4.html; and www.deepwater.com/fw/main/Dhirubhai-Deepwater-KG2-458C16.html?LayoutID=17

[30] Interviews with Gregers Kudsk and David Reid; and Offshore Engineer (ed.)(2003): "Automating the drill floor," *Offshore Engineer*, June 10

[31] Interview with David Reid; National Oilwell Varco (ed.)(2010): *Drilling Evolution 1985-2010*, Houston: National Oilwell Varco; Offshore Engineer (2003); and Submarine Telecoms Forum (ed.)(2008): "Why we don't automate drilling," *Submarine Telecoms Forum*, issue 40 (Offshore oil & gas issue), pp. 11-14, here p. 13

[32] Interviews with Walt Aldred and Keith Millheim; Aldred, Walt et al. (2005); Greenberg, Jerry (2012): "Holistic approach to drilling means going beyond rig equipment: an interview with David Reid," *Drilling Contractor*, May 14; and Reid, David (2007): "Drilling automation: are we there yet?" *E&P Magazine*, January 16

[33] www.drilltec.de/eng/hydrocarbon/home.html

[34] www.aogr.com/magazine/cover-story/advances-in-nanotechnology-hold-huge-potential-promise-in-upstream-applicat

[35] Al-Shehri, Abdullah et al. (2013): "Illuminating the reservoir: magnetic NanoMappers, conference paper, *SPE* Middle East Oil and Gas Show and Conference, 10-13 March, Manama, Bahrain; Prensky, Stephen (2013): "What's new in well logging and formation evaluation, part 2," *World Oil*, July, pp. 71-78; www.spe.org/events/13akyo/pages/about/; and www.aogr.com/magazine/cover-story/advances-in-nanotechnology-hold-huge-potential-promise-in-upstream-applicat

[36] Interview with Scott Tinker; Cookson, Colter (2014): "Nanotech sensors to reveal reservoir," *The American Oil & Gas Reporter*, July; Denney, Dennis (2014): "Nanotechnology applications for challenges in Egypt," *Journal of Petroleum Technology*, vol. 66, no. 2, pp. 123-6; and www.geoexpro.com/articles/2011/11/mighty-thoughts-of-small-nanotechnology-for-upstream-applications

[37] Interview with Scott Tinker; Cookson, Colter (2014); and www.beg.utexas.edu/aec/

[38] http://petrowiki.org/Nanotechnology_in_hydrogen_sulfide_detection; and http://news.rice.edu/2014/04/18/nanoreporters-tell-sour-oil-from-sweet-2/

[39] Cookson, Colter (2014)

[40] www.beg.utexas.edu/aec/projects.php

[41] Gelfgat, Yakov, Mikhail Gelfgat and Yuri Lopatin (2003), vol. 2, p. 286; and www.icdp-online.org/front_content.php?idcat=695

[42] Foster, Brandon, Doug Hupp and Hal Martens (2011): *ENI Nikaitchuq: on the edge of the envelope at the end of the earth, an ERD case history*, symposium paper, Schlumberger Drilling Symposium, figure 2 (Industry ERD envelope); and www.sakhalin-1.com/Sakhalin/Russia-English/Upstream/default.aspx

[43] www.shell.com/global/aboutshell/major-projects-2/stones.html; www.offshore-technology.com/projects/julia-field-gulf-mexico/; and http://en.wikipedia.org/wiki/Lula_oil_field

Epilogue

[1] Bret-Rouzaut, Nadine and Michel Thom (2005): "Technology strategy in the upstream petroleum supply chain," *Institut Français du Pétrole*, Les cahiers de l'économie, no. 57 (March); and Neal, W. Howard et al. (2007): *Oil and Gas Technology Development*, Global Oil & Gas Study, US National Petroleum Council, July 18, published in: www.npc.org/study_topic_papers/26-ttg-ogtechdevelopment.pdf

[2] Mitchell, John, Valerie Marcel, and Beth Mitchell (2012): *What Next for the Oil and Gas Industry*, Chatham House, October, published in: www.chathamhouse.org/publications/papers/view/186327

[3] Bates, T. and G. Coyle (2005): "Accelerating technology acceptance: nucleating and funding E&P technology," conference paper, *SPE Annual Technical Conference and Exhibition*, October 9-12, Dallas

Index

A.P. Møller Group, 127
Abegg and Reinhold Company, 199
Abercrombie, James, 66
Adair, Red, 181, 182
Adams, Leason Heberling, 72
Adams, Wade, 72, 319
ADCO, 204
Adler, Pierre, 277, 284
Advanced Energy Consortium, 319, 321
AEG, 131
Aero Service Corporation, 95
Agip, 214
AGR Drilling Services, 316
Aiken, Charles, 108
Aker Solutions, 260, 274
Alberta Energy Company Ltd., 297
Alberta Oil Sands Technology and Research Authority, 296
Alberti, Friedrich August von, 15
Alexander, Clyde, 80
Alexeev, Major, 7, 321
Allen, Herbert, 227
All-Union Petroleum Scientific-Research Geological Prospecting Institute, Moscow, 154
Alta Vista Hydraulic Company, 81
Amerada Geophysical Research Corporation, 208
Amerada Petroleum Corporation, 30, 36, 37, 72, 73, 113, 208
American Association of Petroleum Geologists (AAPG), 39, 41, 85, 86, 100, 107, 153, 307
American Institute of Mining Engineers (AIME), 26
American Petroleum Institute (API), 80, 96, 233
Amoco, 119, 136, 138, 140, 152, 162, 169, 176, 198, 204, 222, 227, 233, 237, 241, 247, 290, 309, 323, 324
Anchor Drilling Fluids, 189
Andrews, Ebenezer Baldwin, 16
Anstey, Nigel, 145, 172
Antonov, Yuriy, 209
Apache Corporation, 227, 309
Appleyard, John, 286, 287
Applin, Esther, 42
Arabian Oil Company, 204
Archbold, John Dustin, 17

Archie, Gustavus Erdman, 106, 107, 165, 323
Armstrong, Ramsey, 118
Arnold, Ken, 256
Arps, J.J., 205
Arutunoff, Armais, 78, 79, 80, 81, 231
Ashton Valve Company, 83
Asiatic Petroleum Corporation, 233
Aspdin, John, 62
Atlantic Refining Company, 77, 78, 161, 162, 225, 279
Atlantic Richfield Company (ARCO), 158, 162, 169, 204, 222, 225, 282, 290, 324
Atomic Energy Research Establishment, 112, 286
Australian National University, 278
Azerbaijan Petroleum Institute, 105, 192
Aziz, Khalid, 285, 286
Azneft Trust, 193
Backus, Milo, 134, 135, 136, 241, 242
Badische Anilin- und Soda-Fabrik (BASF), 131
Baker Casing Shoe Company, 61
Baker Hughes, 4, 198, 214, 215, 222, 232, 240, 251, 312
Baker International, 198, 232
Baker Oil Tools, 61, 183
Baker, M.C. and C.E., 20
Baker, Reuben "Carl", 60
Baku Company of Russian Oil, 193
Baku Oil Trust, 105
Baldenko, Dmitri, 195
Bally, Bert, 93
Banjavich, Mark, 270
Barenblatt, Grigory, 224, 284
Barnes, Kenneth, 45
Barnickel, William, 84
Barnsdall Oil Company, 56, 109
Baroid Corporation, 188, 189, 208
Barr, John, 215
Bart Manufacturing Company, 80
Bateman, Richard, 176
Bechtolsheim, Andy, 244
Beeby-Thompson, Arthur, 77, 412
Beecher, C.E., 71
Behrmann, Larry, 229
Bell Laboratories, 133
Bell, Larry, 225
Benit, E.J., 314

Berg, Eivind, 252, 254
Berkhout, Guus, 241, 309
BHP Company, 287
Big Lake Oil Company, 191
Biot, Maurice Anthony, 116
Birdwell, 167
Birrell, N.D. "Scotty", 267
Black, Walter G., 21
Bloch, Felix, 174
Blue Water Drilling Corporation, 125
Böckh, Hugo de, 28
Bolt Technology Inc., 143
Bosio, Jacques, 4, 203, 324
Botset, Holbrook, 41, 106
Bouma, Arnold, 148
Bourdet, Dominique, 280
Bourgoyne, Ted, 129
Boyadjieff, George, 4, 199, 200, 201, 316
BP, 4, 93, 126, 128, 136, 140, 142, 147, 149, 151, 169, 170, 183, 189, 214, 215, 222, 223, 237, 242, 249, 250, 254, 262, 288, 292, 299, 306, 312, 316, 319, 324
Bradbury, Robert, 130
Breitenbach, Al, 285, 290
Brenner, Howard, 277
British Insulated Callenders, 270
British-American Oil Producing Company, 205
Brongniart, Alexandre, 15
Brons, Folkert, 110
Brookhaven National Laboratory, 278
Broussard, Martha Lou, 99
Brown & Root, 257
Brown Oil Tools, 200, 227, 228
Brown, Alistair, 244, 310
Brown, Charley, 80
Brown, Cicero, 200
Brown, Hermann, 257
Brown, Kermit, 234, 235
Brown, Robert, 175
Brundred Petroleum Inc., 159
Brunton, David W., 92
Bullock, Milan, 43
Bureau of Geophysical Prospecting, Beijing, 156
Burg, John, 136
Burg, Ken, 143
Burgen, Jack, 178
Burke, Jack, 167
Burkhardt, Joe, 271
Burma Western, 262
Burns, Bob, 117
Butler, Roger, 295

Byron Jackson Company, 119
Calhoun, George, 227
California Oil Company, 227, 262
California Research Corporation, 175
California-Arabian Standard Oil Company, 87, 88, 89
Camco Drilling Group Ltd., 215, 232
Camco International, 215, 232
Cameron Iron Works, 66, 128, 269, 271
Cameron, Harry S., 66
Carborundum Abrasives, 226
Carle, Joe, 231
Carll, John Franklin, 44, 85, 157
Carter Oil Company, 113, 147
Carter, US President Jimmy, 113, 147, 175, 303
Casey, Dave, 306
Cavendish, Henry, 27
Champlin Oil Company, 246
Chaney, Reverend John C., 22
Chapman, Chris, 248, 249
Chelminski, Stephen "Bolt", 143
Chemali, Roland, 210
Chenevert, Martin, 187, 188
Chesapeake Energy, 302
Cheshire, Ian, 286
Chevron, 87, 137, 138, 140, 162, 175, 207, 223, 241, 246, 262, 287, 312, 315, 316, 324, 325
Chiles, Eddie, 223
China National Petroleum Company (CNPC), 163
Christensen Diamond Products, 197, 198, 213
Christensen, Frank, 196
Christensen, George, 197
Christie, H. Merlyn, 298
Christie, Phil, 150
Christman, Stan, 313
Cidra Corporation, 297
Cities Service Oil Company, 71
Claerbout, Jon, 4, 138, 139
Clapp, Frederick Gardner, 85
Clark, Brian, 211
Clark, Ronald K., 189
Clavier, Christian, 166, 167
Coal Generation Institute at the Academy of Sciences of the USSR, 155
Coates, George, 166, 167, 176
Coats, Keith, 285
Coflexip, 273
Collipp, Bruce, 125
Columbia University, 102, 116, 143, 150

Compagnie Générale de Géophysique (CGG), 37, 136, 140, 254
Compañía Shell de Venezuela, 160
Computer Modelling Group, 284
Conoco, 129, 176, 221, 243, 251, 253, 259, 267, 315, 324
ConocoPhillips, 140
Constable, Steven, 312
Continental Oil Company, 124, 131, 144, 145, 221, 313
Continental Resources, 304, 305
Cooke, Claude, 226, 299
Core Laboratories Inc., 275
Coulomb, Charles-Augustin de, 27, 216
Cox, Charles, 310
Crampin, Stuart, 247
Crawford, John, 145
Creole Petroleum Corporation, 222
Cross, John Ross, 10
CSIRO, 239
Cunningham, George, 89, 152
Cunningham-Craig, Edward, 152
Curie, Pierre and Marie, 110
Curtis, L. B. "Buck", 267
CUSS consortium, 129
Daev, Dmitri, 209
Dahm, Cornelius, 244
Dallas Geological Society, 298
Daniels, Ted, 286, 287
Darcy, Henry, 40
Davies, Roger, 306
Decker, Harry R., 66
Deep Oil Technology Company, 265
DeGolyer, Everette Lee, 29, 30, 31, 36, 37, 39, 90, 133, 323
Delong Corporation, 56
Delong, Colonel Leon B., 56
Dempsey, J.C., 167
Deutsche Erdöl AG, 30
Devon Energy, 301
Dewan, John, 112
Diamond Offshore Drilling, 262
Dickinson, William, 150
Digicon, 136
Digitalcore, 278
Dinsmoor, James, 160, 161
Doll, Henri-Georges, 47, 50, 71, 73, 75, 108, 109, 110, 165, 169, 323
Dominion Energy plc, 267
Dotson, George, 318
Doty, Bill, 145
Dow Chemical, 82, 162, 186
Dow, Wallace, 152
Dowell Chemical Company, 217, 221
Dowell Schlumberger, 186

Dragon Technology, 248
Drake, Colonel Edwin, 8, 10, 44, 60, 76, 86, 122, 323
Drake, Eldon, 231
Dresser Atlas, 210, 291
Dresser Company, 11
Dresser, Solomon Robert, 10
Drilling & Service Company, 197, 215
Drilling Well Control, 180, 314
Dumanoir, Jean, 166
Dunham, Robert J., 97
DuPont Corporation, 119
Dyes, A.B., 279
Dyke, J.W. van, 82
Dyna-Drill Technologies, 195
Eastman Industries, 194
Eastman, H. John, 191
Eaton, Ben, 313
Ecole des Mines, 31
Edison, Thomas, 35
Eindhoven University of Technology, 258
Ekstrom, Mike, 169
El Paso Natural Gas Company, 188
ElectroMagnetics Instruments Inc., 292
Electro-Mechanical Research, 108
Elf Aquitaine, 202, 203, 205, 206, 273, 282, 323, 324
Elliott, John "Brick", 43
Ellisor, Alva, 43
Engineering Consultants Limited (ECL), 286, 287
Engineering Data Processors, 136
Engineering Laboratories Inc., 109
Engle, Wendel, 291
Eni S.p.A., 214, 252, 268
Environmental Protection Agency, 188
EOG Resources, 304
Eötvös, Baron Roland von, 28, 29
Esso, 151, 205, 241, 289
Esso Production Research, 241, 289
European Petroleum Company, 77
Evans, Sam, 137
Everdingen, Antonius van, 282
Ewing, Maurice, 102
Exploration Consultants Limited, 286
Exxon, 98, 101, 103, 104, 146, 147, 151, 154, 155, 164, 187, 200, 206, 221, 222, 226, 246, 247, 254, 256, 267, 271, 299, 302, 303, 311, 313, 321, 322, 324
ExxonMobil, 262
Fahmy, E.H., 164
Fairchild Aviation Corporation, 88
Fairfield Industries, 254

Fairhurst, Charles, 217
Fancher, George, 45
Farris, Floyd, 120
Fast, Robert, 120
Feenstra, Robijn, 213
Fermi, Enrico, 110
Fessenden, Reginald Aubrey, 35, 36, 37, 102
Fettke, Charles, 44
Fielder, Coy, 198
Flopetrol, 280
FMC Technologies, 271, 273
Folinsbee, Professor Bob, 103
Forero, Manuel, 242
Forrest, Mike, 141, 142
Fox, Robert Were, 50
Franklin Electric Company, 231
Frasch, Hermann, 81, 82
Freedman, Robert, 166, 298
Gage, Arthur, 81
Galloway, Professor J.J., 43
Galperin, Evsei Iosifovich, 171, 172
Gansser-Biaggi, Augusto, 93
Gearhart, Marvin, 178, 179, 206
Gearhart-Owen Industries, 178, 179, 206
Geco-Prakla, 251
General Electric, 197
Geo Vann, Inc., 230
Geophysical Analysis Group, 134
Geophysical Company of Norway (Geco), 243, 245, 251
Geophysical Research Corporation (GRC), 36
Geophysical Service Incorporated (GSI), 37, 60, 89, 131, 134, 135, 136, 137, 140, 143, 241, 242, 243, 244, 251
GeoQuest Systems, 245, 293
Geosciences Incorporated, 136
Gianzero, Stan, 169
Gilbert, W.E., 233
Giles, Ben, 143
Giliasso, Captain Louis, 54, 55
Glenn, John, 181
Glennie, Ken, 94, 101
Global Marine Exploration Company, 125, 200
Goins, W.C., 180, 181
Gola Nor Offshore A/S, 269
Goodman, Clark, 111
Gould, Andrew, 2, 207
Goulds Pumps Corporation, 231
Graebner, Bob, 135, 241, 244
Gray, Sam, 138
Green, Cecil, 89, 134

Green, Sid, 217
Green, William G., 37, 109
Grigoryan, Alexander, 192, 202, 203
Grimnes, Jan, 293
Gringarten, Alain, 280, 281, 282
Grzybowski, J zef, 42
Gulf Marine Fabricators, 256
Gulf Oil Corporation, 34, 72, 82, 93, 167, 180, 234, 282, 301
Gulf Oil Research and Development Corporation, 165, 180, 282
Gulf Refining Company, 54
Gunning, Thomas, 77
Gusman, Mikhail, 193, 194, 195
Gypsy Oil Company, 72, 82
Hager, Dorsey, 17, 412
Hall Sr., Jesse E., 183
Halliburton, 4, 63, 64, 65, 70, 71, 78, 120, 121, 177, 179, 183, 185, 186, 211, 221, 225, 230, 240, 251, 293, 306, 312, 323
Halliburton, Erle Palmer, 4, 63, 64, 69, 70, 71, 120, 220, 323
Hallundbæk, Jørgen, 239
Hamill, Curt and Al, 21, 22, 67
Hamilton Brothers Oil and Gas Ltd, 263
Hamm, Harold, 6, 303, 304, 305
Hammett, Dillard, 125
Hannah, Bob, 223
Hansen, Asger "Boots", 182
Hardison & Stewart Oil Company, 62
Hardison, Wallace, 62
Hardy, Hugh, 241
Harrington, Larry, 226
Harth, Phillip, 67
Haseman, William Peter, 36
Hassler, Gerald, 275
Hatchett, Charles, 13
Hawthorn, D.G., 208
Hayward, John T., 56
Headington Oil and Lyco Energy Corporation, 304
Heerema Marine Contractor, 258
Heerema, Pieter, 258, 259, 260
Heezen, Bruce, 102
Helmerich & Payne Inc., 318
Hempstead, Norman, 242
Hendricks, Billy, 210
Heng, Zhang, 32
Hennion, John, 143
Henry, Krug, 88
Herron, Michael and Susan, 174
Hertzog, Russ, 173
Hess, Harry, 102, 104
Hickey, J.R., 167

Higgins, Patillo, 17, 18
Hildebrand, Andy, 245
Hill, John R., 61
Hill, Ross, 140
Hingle, Tom, 115
Hirasaki, George, 162, 164, 285
Hitec Products A/S, 317
Hoestenbach, Roger, 232
Höfer, Hans, 152
Holmes, Arthur, 97
Holmes, Jonathan, 286, 287
Hoover, J.W. "Soak", 88
Hopper, Richard, 106
Hornabrook, Jim, 136, 142, 242
Horner, D.R., 279
Horowitz, Alexandre, 258
Horton, Ed, 265, 267
Høyer-Ellefsen A/S, 260
Hoyle, Bill, 167
Hubbert, Marion King, 3, 99, 100, 224, 225, 323
Hughes Aircraft Company, 270
Hughes Jr., Howard, 25, 271
Hughes Sr., Howard, 4, 24, 77, 323
Hughes Tool Company, 25, 60, 183, 196, 197, 198, 232
Humble Oil and Refining Company, 24, 42, 67, 72, 85, 106, 108, 113, 165, 187, 227, 229, 236, 276, 282, 283, 323
Hunt, Thomas Sterry, 16
Hunt-Grubbe, Robert, 168, 169, 239
Hurst, William, 282
Hutchinson, C.A., 279
Hydril Corporation, 128, 269
Hydro Products Inc., 271
IG Farben, 162
Illing, Vincent C., 98, 99, 100
Imperial College London, 249, 277, 281
Imperial Oil, 101, 295, 296
Ingeniør F. Selmer A/S, 260
Ingeniør Thor Furuholmen A/S, 260
Ingrain Inc., 278
Institut de Physique du Globe de Paris, 249
Institut Français du Pétrole (IFP), 127, 153, 202, 233, 249, 273, 277
Institute for Petroleum and Organic Geochemistry, Germany, 153
Institute of Oceanographic Sciences, UK, 151
Institute of Petroleum at the Academy of Sciences, Moscow, 284
Institute of Physics of the Earth at the Academy of Sciences, Moscow, 171

Intera Information Technologies Corporation, 287
Intercomp Resources Development and Engineering Inc., 285
International Energy Agency (IEA), 307
Ioannesyan, Rolen, 193, 194
J. Ray McDermott Inc., 255
Jabiol, Marcel, 48, 105
Jack, Ian, 250
Jackson, Byron, 119
Jackson, Jasper, 175
Jacquin, Christian, 277
Jakosky, John, 106
Jenkins, David, 93, 140
Jervey, Mac, 147
Johnson, Pete, 124
Johnston, John, 72
Johnston, Mordica, 69
Jordan & Taylor Inc., 78
Jost, Roger, 47
Joy, Bill, 244
Kapelyushnikov, Matvei Alkuvich, 193
Karcher, John Clarence, 36, 37, 138, 144, 208
Kay, Marshall, 150
Kazemi, Hossein, 284
Keggin, Jim, 250
Kelvin, Lord, 71
Kern, Loyd, 225
Kerr, Richard, 55, 56, 58, 88, 267
Kerr-McGee Oil Industries Inc., 55, 56, 58, 267
Kewanee Oil Company, 231
Khordokhovsky, Mikhail, 235
Khosla, Vinod, 244
King Abdullah University of Science and Technology, 310
King, George, 119, 120, 227, 228
Kingston, Dave, 91, 94, 95, 147, 150, 151, 249
Kinley, Myron M., 182
Kirchhoff, Gustav, 140
Knox, Granville Sloan, 128
Kochic, Jim, 304
Kochina, Irina, 284
Koebel Diamond Tools, 197
Kogan, Vladimir I., 105, 107
Kokesh, Frank, 113
Kosloff, Dan, 140
Koster, Jan, 43
Koutsabeloulis, Nick, 288
Krüger, Volker, 213
KTB project, 213
Kuchuk, Fikri, 281
Kudsk, Gregers, 127

Kuenen, Philip, 148
Laborde, Alden J. "Doc", 56
Lago Petroleum, 55
Lailly, Patrick, 249
Lamont-Doherty Geological Observatory, 143
Lanagan, Mike A., 65, 66
Landmark Graphics, 243, 245, 285, 293
Lane, Bill, 74, 75
Lane-Wells Company, 75, 110, 173, 210
Larsen, Leif, 280
Lawrence Livermore Laboratory, 169, 292
Layne-Atlantic Company, 220
Lebourg, Maurice, 279
Lee, Robert E., 191
Lengfield, Lew, 30
Leningrad Mining Institute, 156
Léonardon, Eugène, 75
Leschot, Rodolphe, 43
Leverett, Miles, 106, 276
Levorsen, Arville Irving, 86
Lewis, James, 45
Lewis, Samuel, 56
Limbaugh, Bob, 245
Lindseth, Roy, 136, 137, 248
Link-Belt Company, 68
Llewelyn, David, 128, 130
Lockett, Andrew, 77
Logan, William Edmond, 15, 16
Lohman, Ralph, 108
Lopatin, Nikolai, 155, 412
Los Alamos National Laboratory, 175, 291
Lovell, Bryan, 147, 150
Lucas, Captain Anthony Francis, 17, 18, 20, 24, 122
Lukoil, 314
Luthi, Stefan, 169
Lyell, Charles, 14
M.M. Kinley Company, 182
Mach, Joe, 234, 235
Maersk Drilling, 127
Maersk Oil, 127, 201, 204, 316, 318
Magcobar, 180
Magnolia Petroleum Company, 113, 115, 131, 159, 255
Manning, Frank, 260
Marathon Oil Company, 163
Marco Polo, 7, 412
Marine Instruments, 59
Marion, Bruce, 3, 99, 207, 224, 293
Maritime Hydraulics, 317
Marland Oil Company, 81

Massachusetts Institute of Technology (MIT), 44, 111, 125, 133, 136, 138, 145, 277
Matson, George, 72
Matthews, Drummond, 102
Matthews, Ed "Coots", 182
Matthews, Larry, 290
Mayerhofer, Mike, 301
Mayne, W. Harry, 132
Maze, Colonel Delamare, 73, 74
McCardell, William, 165, 166
McCollum, Burton, 144
McDermott, Ralph T., 255, 256, 257, 258, 259
McEvoy, Joseph, 77
McKenzie, Dan, 150
McKetta, Steven, 301
McLemore, Robert, 118
McLeod, Harry O'Neal, 221, 235
McNealy, Scott, 244
Meador, Richard, 209, 210
Meek, Jan, 258, 259
Meissner, Fred, 303
Melcher, Arles, 44
Merlat, Henry A., 54
Merten, Eugen, 58
Mexican Eagle Oil Company, 29
M-I Drilling Fluids, 188, 189
Michel, Drew, 272, 273
Miller, Bert, 88
Miller, Charles, 279
Miller, George L., 67
Miller, Melvin, 176
Miller, Tommy, 286
Millikan, Charles, 72
Mills, Robert van A., 282
Milne, John, 32
Mims, Sidney, 73, 74
Mintrop, Ludger, 33, 34, 251
Mitchell Energy and Development Corporation, 298, 299, 300, 301, 302
Mitchell, George, 6, 298, 299, 301, 304
Mitsui Corporation, 258
Mo, Olav, 260
Mobil, 58, 109, 113, 133, 135, 142, 168, 176, 189, 205, 206, 241, 260, 262, 288, 292, 324
Mohaupt, Henry, 117, 119
Mohorovi i , Andrija, 100
Mohr, Christian Otto, 216
Moineau, René, 232, 233
Moore, Lee C., 23
Morley, Lawrence Whitaker, 103
Moscow University, 155
Moulton, John, 245

Munn, Malcolm J., 98
Munroe, Charles Edward, 117
Murchison, Sir Roderick, 15, 259
Murphy Jr., Charles, 56
Murphy Oil, 56, 262
Muskat, Morris, 41, 282
Mutti, Emiliano, 148
National Aeronautics and Space Administration (NASA), 95, 96
National Lead Company, 67, 208
National Oilwell Varco, 207
National Supply Company, 83
Natural Gas Pipeline Company of America, 300
Neftex, 306
Neill, T.A., 23
Nelson, Erik, 183
Nelson, Royce, 245
Newell, Frederick Haynes, 44
Newpark Resources, 189
Neyrpic SA, 194
Nippon Kokan Koji Corporation, 269
Nixon, Jeddy, 78
NL Industries, 208, 209
Nobel, Robert and Ludvig, 9, 10
Noble, Orange, 65
Noik, Simon, 237
Nolen & Associates, 285
Nolen, John, 285
Norsk Hydro, 239, 269, 290
Northern Caucasian Mining Institute, 155
Northwestern University, 146
Norton Christensen Inc., 213
Norwegian Contractors, 260
Numar, 176, 212
Numerical Rocks, 278
Nutting, Perley Gilman, 40, 41
Oag, Alistair, 183
Ocean Drilling and Exploration Company (ODECO), 56, 127, 262
Oceanit Laboratories Inc., 321
Oceanographic Engineering Corporation, 168
Offshore Company, 130
Ohio Oil Company, 82
Oil Base Drilling Company, 67
Oil Center Tool, 123
Oil Drilling Inc., 298
Oil Dynamics Inc., 231
Oil Well Supply Company, 83, 160
Oldham, Richard Dixon, 100
Oldham, Thomas, 15
Øren, Pål-Eric, 278
Orstrand, C.E. van, 72

Oshman, Jake, 298
Ostrander, Bill, 246
Owen, Harold, 178
Owen, J.E., 208
Owsley, Bill, 121
Pan-Geo Atlas Corporation, 172, 175, 210
Parkhurst, Ivan, 71
Parry, George, 158
Paslay, Roy, 58
Patnode, Homer, 165
Pauling, Randy, 265, 267
Payatakes, Alkiviades, 277
Peaceman, Donald, 283
Pechelbronn Oil Company, 48
Peebler, Bob, 243
Peerless Pump Company, 231
Penberthy, Wally, 222
Penick, Arthur and Kirby, 123
Pennsylvania State University, 45
Perkins Oil Well Cementing Company, 62
Perkins, Almond A., 62
Perkins, Thomas, 225
Peterman, Charles, 315
Petrobras, 264, 268, 271, 290, 322, 325
Petroleum Geo-Services (PGS), 250, 290, 307, 308
Petrolite, 84
Petty Geophysical Engineering, 34, 132, 136
Pfleumer, Fritz, 131
Phillips Petroleum Company, 80, 81, 189, 221, 260, 302
Phillips, Frank, 80
Phillips, John, 15
Placid Oil, 264
Pohlé, Julius, 77
Poiseuille, Jean Léonard Marie, 41
Pompes Compresseurs Mécanique SA, 233
Pontecorvo, Bruno, 110, 112
Pope, Gary, 164
Posamentier, Henry, 147
Poupon, Andre, 167
Prakla-Seismos AG, 251
Prats, Mike, 159, 225
Pratt, Gerhard, 249
Pratt, Wallace, 85
Prikarpatburneft Drilling Company, 197
Proano, Eduardo, 235
Pullin, Norm, 290
Purcell, Bob, 276
Purcell, Edward Mills, 174
Pye, David, 162

Quality Tubing Inc., 228
Rabson, W.R., 172
Rachford, Henry, 283
Ramey, Hank, 280
Raymond Precision Industries, 206
Reading, Harold, 149
Red Adair Company, 182
REDA Pump Company, 78, 79, 81, 231, 232
Reed Hycalog, 215
Reed, Donald, 41
Rees, John, 248
Rehm, Bill, 118, 180
Reichert, Volker, 281
Reinhold, Baldwin, 199
Reinhold, Ben, 199
Resman A/S, 289
Rey-Grange, André, 127
Rice University, 93, 99, 162, 164, 319, 320
Richard E. Smalley Institute, 319
Rieber, Frank, 131
Rike, Jim, 227
Rio Bravo Oil Company, 42
Roberts, Count Hayden, 64
Roberts, David, 151
Robinson, Enders, 134, 136, 145
Robinson, Leon, 187
Rockwell, Donald, 241
Roegiers, Jean-Claude, 217
Romanovsky, 61
Root, Dan, 257
Root, Paul, 284
Ross, Rex, 245
ROV Technologies, 272, 273
Roxana Petroleum, 46
Ruffner, David and Joseph, 10
Russell, William, 44
Rutherford, Ernest, 97
Rutherford, Steven, 247
Rylance, Martin, 222, 299
Saleri, Nansen, 204, 240
Salvatori, Henry, 38, 58
Salzgitter AG, 194
Sandia National Laboratories, 207
Sandiford, Burton, 162
Santos Ltd, 287
Saudi Aramco, 87, 204, 209, 240, 283, 293, 319, 325
Sauvage, Raymond, 105
Savage, George, 130
Savit, Carl, 141
Savre, Wayland, 167
Scharpenburg, C.C., 193
Scheele, Carl Wilhelm, 195
Scheibli, Charles, 47
Scherbatskoy, Serge, 109, 110
Schilling Robotics, 272, 273
Schilling, Tyler, 272, 273
Schlumberger, 31–32, 46–52, 60, 71–73, 75, 105–16, 165–70, 173–79, 183–86, 215–17, 229, 232–40, 243–45, 248–54, 277, 279–81, 288, 292–97, 306–7, 319
Schlumberger Macco, 234
Schlumberger, Conrad, 31, 32, 33, 37, 46, 47, 48, 50, 113, 169, 323
Schlumberger, Marcel, 3, 31, 32, 37, 46, 48, 50, 51, 71, 73, 75
Schlumberger-Doll Research, 109, 111, 166, 169, 173, 211, 277
Schmidt, Arthur, 167
Schneider, William, 140, 241
Schuster, Gerard, 309
Scientific Software-Intercomp, 285, 290
Sclater, John, 150
Scott, Dan, 196
Scripps Institute of Oceanography, 96, 267, 310, 311
SeaBed Geophysical, 254
Segré, Emilio, 110
Seiscom Delta, 172
Seismograph Service Corporation, 37, 109, 114, 145, 251
Seismos AG, 34, 35, 251
Sensa, 297
Shaffer Tool Works, 66, 128, 313
Shaffer, William D., 66
Sharland, Peter, 306
Sharp, Walter, 25, 77
Sharp-Hughes Tool Company, 25
Shell, 21, 30, 43, 46, 54–58, 67, 72, 73, 77, 82, 93–100, 106–7, 110, 115–16, 124–25, 131, 136, 141, 142, 149, 152, 159, 160, 162–69, 181, 188–90, 204, 213, 221, 223–25, 231, 232–37, 246, 253–57, 260, 261, 267–76, 279, 282, 285, 312–15, 319–22
Shelly Dean Oil Company, 303
Shepard, Francis Parker, 96
Sheriff, Bob, 137
Sherwood, John, 138
Shockley, William, 133
Silverman, Daniel, 208, 309
Simmons, John, 69, 70
Sinclair Oil and Gas Company, 72, 159
Single Buoy Moorings Inc., 268
Sinha, Martin, 311
Sinopec, 292
Skeels, Brian, 271, 272, 273

Sloss, Lawrence, 146, 147, 150
Smith Tool Company, 195, 196
Smith, John R., 291
Smith, Mark, 143
Smith, Samuel, 9
Smith, William A., 8, 9
Smits, Lambert, 166
Société de Prospection Electrique, 32
Société Géophysique de Recherches Minières, 37
Society of Exploration Geophysicists (SEG), 39, 137, 140, 246
Society of Petroleum Engineers (SPE), 26
Socony-Vacuum Oil Company, 109
Soddy, Frederick, 97
Sohio, 222, 223
Solar Oil, 81, 82
Sondex Limited, 170, 239
South Penn Oil Company, 23
Southeast Drilling Company (SEDCO), 125, 128, 201
Southwestern Pipe, 228
Sperry Sun Drilling Services, 51, 210, 211
Sperry Sun Well Surveying Company, 51
Srnka, Leonard, 311
Stalkup, Fred, 158, 162, 282
Standard Development Company, 159
Standard Oil Company of California, 21, 72, 87, 89, 175, 193, 265
Standard Oil Company of Indiana, 82
Stanford Exploration Project, 139
Stanford University, 86, 87, 138, 139, 140, 150, 174, 217, 224, 244, 278, 280, 285, 289
Stanolind Oil and Gas Company, 82, 110, 119, 120, 121, 141, 208, 236
Statoil, 214, 239, 252, 253, 254, 271, 273, 278, 280, 287, 288, 290, 308, 311, 322, 325
Steinbeck, John, 124
Steineke, Max, 87, 89
Steinsberger, Nick, 301
Stensen, Niels, 14
Steward, Dan, 299, 300
Stewart, Lyman, 62
Stoneley, Robert, 115, 116
Strata Bit Corporation, 198
Streicher AG, 318
Stroud, B.K., 67
SubseaCo, 254
Sun Oil Company, 242
Superior Oil, 58, 124, 255

Swan, John, 73
Sydansk, Robert "Bob", 163
Tarantola, Albert, 249
Taylor Diving & Salvage, 270
Taylor, Edward Lee, 270
Technical University Delft, 169, 241, 258
Technical University of Denmark, 239
Technip SA, 267, 273
Technoguide A/S, 293
Teekay Corporation, 269
Teknica Resource Development, 248
Tel Aviv University, 140
Teleco Oilfield Services, 206
Teledyne Geotech Inc., 178
Tenneco Oil Company, 247
Terratek, 217
Terzaghi, Karl von, 217
Texaco, 54, 135, 205, 209, 220, 241, 257, 258, 324
Texas A&M, 319
Texas Company, 54, 55, 220, 221
Texas Instruments, 134, 135, 178, 242
Texas Pacific Oil Company, 244
Thalén and Tiberg, 27
Thomas, E.C., 115, 166
Thompson, Larry, 210
Thomsen, Erik, 90, 141
Thomsen, Leon, 90, 141, 247
Thorogood, John, 126
Threadgold, Philip, 169
Timken Roller Bearing Company, 119
Timken, Henry, 119
Timur, Aytekin "Turk", 175
Tinker, Scott, 319
Tiraspolsky, Wladimir, 194
Tittman, Jay, 111, 112
Tolson, Eugene, 118
Tomsk Technological Institute, 193
Tono, Henrique, 278
Torrey, Paul, 44
Total, 189, 319, 324
Trauzl AG, 194
Treitel, Sven, 134, 135, 136, 139
Trostle, M.E. 'Shorty', 241
TRW Inc., 231
Turboservice, 194
Tuzo-Wilson, John, 103
Udden, Johan August, 85, 86
UK Department of Energy, 286
Union Oil Company of California (Unocal), 62, 188, 241, 303
Union Pacific Resources, 301
University of Alberta, 103
University of Berkeley, 225

University of Budapest, 28
University of California at Berkeley, 164, 219
University of Cambridge, 102, 150, 224, 311
University of Groningen, 148
University of Houston, 90, 137, 211, 277
University of Krakow, 42
University of Leoben, 152
University of Manchester, 288
University of Michigan, 285
University of Minnesota, 217
University of New South Wales, 278
University of Oklahoma, 29, 36
University of Oxford, 149, 286
University of Princeton, 101, 102, 104
University of Purdue, 108
University of Southern California, 265
University of Texas at Austin, 43, 164, 187, 319
University of Tulsa, 234
University of Turin, 148
Uren, Lester, 164, 219
US Atomic Energy Commission, 195
US Bureau of Mines, 40, 282
US Defense Advanced Research Projects Agency, 310
US Department of Defense, 217
US Department of Energy, 164, 291, 321
US Geological Survey (USGS), 29, 40, 44, 72, 85, 98, 99, 305
US National Bureau of Standards, 144
US Navy, 58, 95, 125, 227, 250, 270, 272, 311
Vail, Peter, 4, 146, 147, 306
Vann Tool Company, 229
Vann, Roy, 229
Varco International, 199, 200, 201, 207, 316, 318
Varian, Russell, 174
Vector International Processing System, 288
Venezuelan Ministry of Mines and Hydrocarbons, 222
Vidrine, D.I., 314
Vine, Frederick, 102, 103, 138
Vining, Bernie, 154, 155
Vorontsov, Count Mikhail Semyonovich, 7, 321

Wahl, John, 112
Walker, Jay, 84
Wang, Zhijing, 290
Warner, William, 158
Warren, Joseph, 284
Warren, Tommy, 198
Waxman, Monroe, 165, 166
Weatherford Inc., 69, 183, 315
Weber, Koen, 275
Wegener, Alfred, 101
Welex, 119, 120
Well Explosives Company Inc., 118, 119
Well Surveys Incorporated, 109, 110, 112, 173
Wells, Walt, 74
Welltec A/S, 239
Welte, Dietrich, 152, 153, 154, 306
Wengchong, Zeng, 292
Wert, Samuel van, 80
Western Company of North America, 223, 226
Western Geophysical, 37, 141, 251, 290
WesternGeco, 250, 251, 307, 308, 309
Westinghouse Electric Corporation, 79
Whipstock Company, 194
White, Israel, 17
Whitmore, Dan, 140
Wiener, Norbert, 136
Williams, Henry Lafayette, 53
Williams, Robert, 247
Williford, Frank, 128
Willis, David, 224
Wilson Supply Company, 78
Wilson, James Lee, 97
Wilt, Michael, 292
Winsauer, Weldon, 165, 166
Wolcott, Edson, 159
Woodside Petroleum Limited, 287
Wyckoff, Ralph, 41, 106
Wyllie, Malcolm, 115, 165
York, Victor, 21
Yukos, 235, 236
Zemanek, Joseph, 168
Zhang, Yu, 140
Zheltov, Yuri, 284
Ziegler, Martin, 94, 97, 146
Ziegler, Walter, 101, 103, 155, 245
Zimmerman, Tom, 210, 281
Z-Seis Corporation, 293